CONSCIOUSNESS

A STANDARD REFERENCE MODEL

On the relation between reality and imagination and
the corresponding diversity of human philosophy.

BRYANT JOHNSON

Grosvenor House
Publishing Limited

This book is published by
Grosvenor House Publishing Ltd
Link House
140 The Broadway, Tolworth, Surrey, KT6 7HT.
www.grosvenorhousepublishing.co.uk

A CIP record for this book
is available from the British Library

ISBN 978-1-83975-873-7

DEDICATION

This book is dedicated to the memory of
my dear Mother and Father

CONTENTS

Contents

PREFACE

From an early age the author has been interested in establishing a model of existence relating reality to imagination, the objective as opposed to the subjective, which explains the relationship between the world as seen by the individual, that seen by others and ultimately that apparent to the human race as a whole. However, such a model was only a work in progress, not even properly formed in the author's mind, yet alone written down.

In addition to the four-dimensional space-time reality familiar to the common experience based on local causal continuous properties, scientific discoveries in the first half of the 20th century relating to quantum mechanics introduced non-local properties which were at variance with the classical physics of Newton and Einstein.

Furthermore, though the mathematical description of the quantum state could be regarded as deterministic prior to experimental observations, when such observations were made, blatantly non-local indeterminate properties were evident, associated with the process of converting the quantum state to the classical world.

In the latter part of the century a range of theories were formulated to explain what has been termed the quantum dilemma. In terms of theoretical physics the existing situation is well illustrated by the range of opinion, which is divided between determinism and indeterminism, locality and non-locality, the reality and nature of the wave function etc.

Though all parties are engaged in complex mathematical thinking, however noteworthy, the majority have not proceeded with due regard to the fact that they have no theory of consciousness on which to base that thinking, nor indeed do many feel they need to.

Nor do they have any satisfactory explanation for the relationship between human and what is referred to as artificial intelligence, nor the absence of manifest intelligence of either sort elsewhere in the visible universe, commonly assuming it is vanishingly rare.

The scientific majority are in general secularists, who by definition have little or no comprehension of religion, despite the evident widespread belief in such in the human population. They support the objective and feel justified by their confidence in humanism as a philosophy in preference to religion and by the vast improvement in our comprehension of the reality of the four-dimensional space-time world including biology, which science has provided, especially in the last 100 years.

In this way science prides itself on maintaining a generally united front (the debate between four-dimensional reality and the quantum state not withstanding). It is rightly nervous of bad science, that which cannot be measured and which it perceives to be outside the spatiotemporal framework referred to as nature.

While allowing for the uncertainty principle these scientists are best described as measurers who leave metaphysics to philosophers.

Science it should be noted, is relatively recent in origin, ignoring the improvement in technology which has taken place over thousands of years, it dates for practical purposes to the last five hundred years and in particular, as already stated, to the last hundred years.

It is in a constant state of change with new discoveries overtaking or improving existing theories, though some withstand the test of time.

Nevertheless this author is convinced that science shows little sign of casting its net wide enough to explain the deficiencies referred to above, so it remains a partial theory.

Meanwhile a minority of scientists proposed alternative theories which did not place restrictions on non-locality, incorporating proposals which included consciousness as a fundamental feature which could be used to relate observed reality to what were termed explicit orders and imagination to what were termed implicit orders, all based on the more general principle of order.

They published books, which had both scientific and metaphysical implications, beyond the limited understanding which then prevailed and still prevails in both secular and non-secular modes of thinking. They were however expressed in a form too complex for the average individual to relate to. In particular they did not engage with the everyday characteristics of human nature nor account for the diversity of human philosophy, politics and economics etc.

Nevertheless they acted as a starting point from which a coherent model taking account of those characteristics and that diversity then crystallised in the author's mind, in the form described in the chapters which follow this preface.

It is appropriate at this point to turn to the proponents of the subjective, the main religions for example, to see if they can do better.

Just as there is good and bad science, there is good and bad religion. Clearly bad religion does not do better as it is a source of destructive thinking on a mind-boggling scale.

The established, organised religions by maintaining divisions one with another, obviously create a problem for each.

However it is perfectly possible to deduce the constructive elements which these belief systems have in common. The relevant holistic principles have changed little since their formation centuries ago. Religion is essentially unchanging and regards the immeasurable as fundamental, the exact opposite of science.

For these reasons it has done little to either encourage science or to take an interest in it. For example sermons from the pulpit concerning scientific thinking as a proportion of the total are almost as rare as hen's teeth. While many great minds of a religious conviction have indeed made major contributions to science (Newton being just one example), they do so as if in a different mode of thinking to religious belief, a bifurcation which needs an explanation.

It is highly significant that organised religion is also unsuccessful in explaining consciousness and intelligence (human or artificial) or their prevalence in the observable universe, assuming they can only be explained by God, who they place beyond human comprehension. They are primarily concerned with human feelings and behaviour rather than metaphysics, which like scientists they leave to philosophers.

So science and religion share the same deficiencies, which is even more significant.

As a result of all this the dichotomy between the two perpetuates with no sign of resolution despite the holistic nature of religion and of the scientific theories of general relativity and the quantum state.

It is the purpose of this book to sweep away this dichotomy, to demonstrate clearly that a comprehensible metaphysical model of existence, combining science and religion can be constructed, which furthermore can be shown to be supported by the human race as a whole, as opposed to that of the individual or groups of individuals, large or small.

Accordingly there are no statements, written, uttered or merely thought of by human individuals that cannot be explained in principle by that model.

Of course the book cannot have such an infinite scope so a wide selection of subject matter is examined to support this principle.

It is useful at this point to draw a suitable analogy. A bell tent is supported on a single central pole, the latter in turn supported by a number of guy ropes leading to pegs hammered in to the ground. The pegs represent the individuals pulling in all directions, while the central pole represents the human race as a whole i.e. the mean of all the individual differences distributed around it.

If it were not for the individual pegs the central pole would collapse rendering the pegs irrelevant. Hence the human condition always arguing with each other, while the whole set of individuals (the human race) acting as the mean has never before been listened to.

In this regard the model may be regarded not as a theory of everything (TOE) since there is no such thing as everything, but it may be regarded as an attempt to provide a theory of nearly everything (TONE), in musical terms a fundamental tone corresponding to that expressed by the human race.

Now is the point at which our increasingly fine-grained knowledge of reality has stimulated our imagination to the point where new ideas and comprehension can relate the objective and subjective, rather than exacerbate the existing division.

This book could not have been written without being based not only on metaphysical ideas in general, but also on recent scientific discoveries made in the last few decades.

In other words it could not have been written prior to this time. That time is now.

The new metaphysical model breaks through the limitations imposed by prevailing scientific and religious tenets and expands exponentially the range of human enquiry.

It matters not who the author is, since he is merely reporting what should be obvious.

It is the opinion and belief of the human race, which by definition supersedes that expressed by individuals or, as has been said above, groups of individuals large or small.

Sources of information for this book are drawn not only from a wide range of previous authors in a variety of subjects, but from freely accessible internet sources to which this author is indebted and without which the writing of this book would have been that much more difficult. This is another reason why it has been possible to do so at this time and not earlier, before the internet so valuable to the dispersion of knowledge became available. Nevertheless, the reader should be at pains to cross check such internet sources against academic sources.

This book is a precursor for others to continue the lines of enquiry it pursues and to improve on the model described herein.

The description of the model obtained is quite short, in effect a modus operandi of existence, as opposed to a complete description which would be infinitely long and therefore beyond human comprehension.

It is this method of operation which can be understood by the human mind, when put to the test, providing the individual concentrates on the problem and is not distracted.

The central thesis is described to the reader in less than 50 pages with a similar number of figures as illustration in parallel with the text; alternatively it can be explained to an audience in a one-hour lecture with the illustrations projected on to a screen.

For this reason the author hopes the book will generate a readership among the curious who do not wish to live in ignorance, and over time more diversely, thereby combating the fragmentary approaches referred to which have applied to date.

1.0   INTRODUCTION

The central purpose of this book is to disperse the smokescreen of variously confused, misguided or dogmatic thinking, which effectively hides the essential truths of religion and to simultaneously expand the partial theory that is modern science, so as to cure it of the persistent tunnel vision with which it sees existence.

To provide a clear, coherent and unified model of the world, not inordinately lengthy or over complicated, yet a rich picture based on combinations of the best thinking from all such sources, to replace the unbridled confusion which presently prevails.

It is not expected that such a model will act as a cure-all either overnight or even in the foreseeable future, the proponents of disorder will see to that, but it should be viewed as a point of reference, amendments being made where future discoveries warrant but on the same basic principles as the original.

This book sets out to explain the relation between what is considered the real spatiotemporal world i.e. that which is inherently explicit, meaning measurable and the non-spatiotemporal world i.e. that which is inherently implicit, meaning that which cannot be measured. The book also puts forward a basic description relating these worlds to those of philosophy, politics, economics, mathematics, art and literature, etc all of which of course have both real and imaginary aspects.

In particular the book describes basic principles of ordering and disordering processes and from these forms a relationship between the explicit and implicit which allows construction of what is termed a Standard Reference Model. It will be shown that nevertheless the Model is not limited in either an objective or subjective sense to any one view point.

The exercise is carried out keeping complex mathematical approaches in mind but nevertheless in the background in the initial sections of the book, which will be a relief to many individuals. Nevertheless it is made clear that some basic comprehension of the mathematics involved is essential if the reader is to acquire a full

1

understanding of what is put forward in this volume. It is for the reader to decide whether he agrees with the author that the Model reduces the existing cloud of confusion rather than increasing it, particularly as it shows the various objective and subjective approaches can be related to form a rich picture, whereas generally they are regarded as mutually exclusive.

In this respect the use in principle of all human statements whether verbal, written or otherwise expressed without exclusion, is paramount in the formation of the Reference Model which is based on the further principle that all such statements are indicative of an ordering sequence fundamental to existence. Statements with greater coherence rise in the ordering sequence and vice versa statements with lesser coherence fall in that sequence.

It is not possible for conscious entities to make statements without them being in some degree related to existence. The sequence does not exclude therefore statements made with the deliberate intention of an individual to cheat, lie, distort, or conspire against the truth, since the truth rises in the ordering sequence while falsehood falls.

To facilitate understanding of the above consider a dictionary of a given language. The dictionary is constructed using symbols i.e. letters or hieroglyphs arranged or ordered in a principal (alphabetical) sequence. One or more of the symbols makes up a word, each word being ordered again from first to last, such that all words starting with the first symbol and then each subsequent symbol are ordered in accordance the principal sequence.

Each word is ascribed a meaning in terms of other words in the dictionary, a meaning which of necessity involves a linkage between the rules of construction of the dictionary and the language itself. The whole set of words and their meanings being referred to as the language.

The makeup of the words, their meanings defined by other words, sentence construction and the construct of language itself is based on the basic principle of sequencing.

The dictionary is a clear example of manifest order. Manifest order apparent to the eye in an explicit form and order generally in its implicit form is therefore to be understood, to at least some degree, as a sequence, and higher degrees of order approaching an optimum as sequences of sequences, all these being various arrangements of differences and similarities. In the opposite sense manifest disorder and disorder generally can be considered to be lower degrees of order which exhibit a lack of sequencing and increasingly so relative to the optimum sequence.

In this context the Reference Model corresponds to the description of the rules governing the construction of the dictionary, statements or sentences correspond to the words in the dictionary, explanations of those statements in terms of the Model, correspond to the meanings of the words and finally the basic principles of ordering and disordering processes correspond to the whole language.

In essence what is being said is that a higher level of dictionary to the conventional version can be constructed, making it possible to relate any one statement to any other statement, as distinct to relating one word to any other word.

This implies any one philosophy can be related to any other philosophy, since each is simply a combination of statements. The word Reference is used as what is being attempted here is the establishment of a theory that houses all philosophies under one roof, as a reference library houses books.

The Reference Model is intended therefore to represent the total range of philosophies so as to act as a standard against which each philosophy can be judged, hence the term Standard Reference Model.

However the Model has to overcome many hurdles to justify such a theory; two are immediately apparent, namely the twin dichotomies of firstly the concept of the objective versus the subjective and secondly the concept of absolute truth versus relative truth.

The Reference Model uses basic cornerstones of mathematical comprehension to demonstrate these concepts, namely the principles of the Argand diagram, the Central Limit theorem and the normal (Gaussian) distribution (see section 2.0).

These principles are however too symmetrical in nature to explain consciousness which is inherently asymmetrical as the book demonstrates.

What is required therefore is a means of incorporating that asymmetry in combinations of the objective and subjective, measurable and immeasurable. The means used is to compare a distribution curve in which difference is unfolded at the expense of similarity (a process of differentiation) and a second distribution curve in which similarity is unfolded at the expense of difference (a process of integration). In the combined case these two distributions reinforce each other to form a single curve unfolding difference on one side and similarity on the other.

The convention used is to regard the left-hand side of the curve as the differentiation process and the right-hand side the integration process, to illustrate the inherent asymmetry of the combined curve (see Section 2.6).

Lines drawn between the peak and base of the distribution curve parallel to the base represent levels or degrees of ordering.

Order and disorder generally is to be associated with the mind (just as these properties in their manifest form is to be associated with the brain), with consciousness in fact, which involves some degree of indeterminacy. However the motion of particles of which we are in an explicit sense supposedly composed, show no comprehension of the carnage of war. The very nature of such motion is not obviously sequential in character and apparently involves a profusion of discrete values.

The question then arises of how to incorporate in the Model a description of the relationship between that which has degrees of ordering and those states variously described as chaotic (primarily deterministic) and random (completely indeterminate).

Also it is not sufficient to consider only our atomistic (particle) concepts. In addition the Model must incorporate in a comprehensive manner waves, wave motion and wave harmonics, including the principle of reinforcement when in phase and cancellation when out of phase, these being central to an understanding of life at all levels whether at cosmological, human or quantum scales. The distribution curves used in the Model should therefore be thought of as probability waves which deal with averages first and foremost.

The book proposes means of dealing with these matters which lead to the concept of the Ordering Centre and explain its development.

The Ordering Centre is related in the Model to the cosmological concept of energy density, the highly ordered state referred to in scientific terms as a big bang. A big bang involves conditions in which the scientific models of General Relativity and Quantum Theory come into conflict. The Model proposes means of resolving the present dilemma between cosmological theories related to the theory of relativity in four-dimensional space-time (including its classical extension in multidimensional phase space) and the non-classical but equally multidimensional concept of Hilbert space associated with the unitary quantum state; all of which can be considered at least causal and determinate in character.

To achieve this, the Model incorporates an additional multidimensional space of an implicit nature, where non-local, non-causal and non-continuous properties apply, involving a degree of indeterminacy.

This implicit space is used to describe the manner in which the non-classical unitary quantum state transforms into the classical state, via the conscious process of observation, as will be discussed in later chapters.

An Ordering Centre is considered to have two aspects, one inanimate represented by a grid structure and the other animate represented by movement across the grid.

The structure of an Ordering Centre is represented in the Model by a reinforced distribution curve, the vertical axis of the distribution (the ordinate) representing the degree of ordering (the complexity of ordered sequencing, s) and the horizontal axis (the abscissa) representing the base variation (the variance, v).

The abscissae under the curve represent the tendency to form a hierarchical or differentiated system of levels while the ordinates represent the opposite tendency to form a holoarchical or integrated system which dissipates those levels. The structure at a given level comprises that unfolding as a differential (the left-hand side of the curve) and that enfolding as an integral (the right-hand side of the curve).

The unfolding differentiated structure represents a condition in which there is a preponderance of difference over similarity. The enfolding integrated structure represents a condition in which there is a preponderance of similarity over difference. The vertical line at the peak of the reinforced distribution curve represents the condition where difference and similarity have equal influence and where the capacity for sequential ordering is greatest.

The distribution curve is then used to establish a relation between the inanimate and the animate; this involves the requirement to associate the inanimate with the unified Ordering Centre being in a state of symmetry or antisymmetry i.e. balanced about a mean condition of greatest available order.

The animate is then considered to be a complementary asymmetric state to the inanimate associated with movements as unfoldments or conversely enfoldments, all such movements involving distortions of the Ordering Centre from the inanimate state. Animate states are associated with extending the spread of the Ordering Centre at any given level (the variance) and with the unfoldment of order or disorder with change of level in the Ordering Centre (the complexity of sequencing). The movement of the animate system is always divergent from that of the inanimate maintained in motion with it.

The Model then uses the concept of wave motion relative to the grid structure to explain the unfolding of consciousness and the further development of intelligence in the form of objective knowledge and subjective belief, all unfolded from a mean level in the Ordering Centre associated with the act of measurement, where the increase of order with decreasing variance matches the decrease of order with increasing variance.

At this mean level consciousness is enfolded in the inanimate state where the mixture of reality and imagination is in a state of balance, where in fact the capacity for holoarchical and hierarchical consciousness is also in balance.

Above this mean level holoarchical consciousness is predominant and is enhanced by increasing implicit order and reducing explicit disorder, where imagination dominates reality.

Below this mean level hierarchical consciousness is predominant but ultimately dissipated by increasing implicit disorder and reducing explicit order where reality dominates imagination.

In particular it is shown that order and disorder are to be associated with consciousness and intelligence. As the degree of intelligence increases such orders are increasingly interwoven. Accordingly the Model also shows that although increasing intelligence can be associated with increasing order (i.e. good), the opposite can also apply. By reversing the situation so that the ordinate represents the degree of disorder, it is apparent that intelligence can be used to increase disorder (i.e. evil).

Disorder in its manifest explicit sense also takes the form of chaotic disorder, i.e. that not necessarily connected to consciousness, but simply the inevitable operation of the second law of thermodynamics.

The marriage of those concepts to the use of a distribution curve is made on the basis that irrespective of the existence of complex wave patterns, on an overall basis the law of averages ultimately applies.

All of this can be viewed as a system unfolding a given property in one direction about a state of symmetry and unfolding a state of antisymmetry in the reverse direction. It is the growth of animacy and consciousness that introduces asymmetry.

Such unfolding and enfolding properties are directly related in the Model to the principle of the hologram. The primary thesis is that order unfolds from disorder and disorder unfolds from order; likewise that objectivity is primarily a process of unfolding explicit order at a mean level while enfolding order at other levels and conversely that subjectivity is a process of enfolding that explicit order while unfolding orders at other levels.

The Model therefore effectively distinguishes between the objective and the subjective and between implicit order/disorder and explicit order/disorder and crucially identifies a relation between absolute and relative truth, the former represented by the peak of the Ordering Centre and the latter by the elements of the Ordering Centre lying to either side of that peak. The absolute is then associated with the property of discreteness (discontinuity) and the relative with indiscreteness (continuity).

The random (completely indeterminate) state is then associated with the absence of both order and disorder in any sense, involving the collapse of the Ordering Centre scenario.

With this comes the corresponding breakdown of ordering processes firstly into chaotic elements and then to a state of totally random fluctuations as dimension itself collapses.

The relationship between the Reference Model and the objective philosophies is examined in detail. In particular how the Model relates to physics, cosmology, biology and other branches of science. The corresponding relationship between the Reference Model and the subjective philosophies is also examined. In particular how the Model relates to mainstream belief, non-belief and suspended belief systems namely theism, atheism and agnosticism. The Model shows that there is a relation between Darwinian theories of evolution and religious belief in intelligent design and resolves the present dilemma by coupling the former to wave motion which accentuates variance and the latter to wave motion which accentuates order, all relative to the mean level of order. The belief in intelligent design should not be confused with so called 'creation science'.

A similar situation applies to the debate between human intelligence and artificial intelligence, which is resolved in the same manner.

The Model goes on to explain why the universe appears silent to the human race, with no explicit evidence of intelligent extra-terrestrial life.

To achieve all this the Model removes the concept of absolute boundaries from existence, which arises from its fractal nature and goes on to apply this principle to both science and religion, which have insisted on absolute reality or absolute God respectively, to support their conclusions. This does not give rise to unbridled relativism, as appropriate boundaries can be applied where local properties obtain.

These relationships are however only apparent when the limitations of Cartesian coordinates are appreciated. The Reference Model replaces them with non-Cartesian concepts involving but not limited to, the replacement of the Cartesian background with a holographic fractal description, having both discrete and indiscrete properties, which ensures that the Model is independent of any absolute description.

While the Model therefore shares the principle of relativity with modern physics, it exposes the limitations of the sciences, which by definition are tied to local realistic theories in the case of General Relativity, or make limited allowance for non-locality in the case of quantum theory, all based on the Cartesian system.

Non-Cartesian concepts require that non-spatiotemporal states apply in addition to those spatiotemporal in character. This in turn implies an indefinite approach to a preferred ordering/disordering axis, as opposed to fixing that axis in an absolute sense.

In short the Model represents what may be termed a holographic fractal composed of differences and similarities arranged in an infinite variety of ways from extreme simplicity to ever greater degrees of complexity of wave motion.

These ideas, difficult though they may be, offer a means of relating the physics of macro systems (relativity) with the physics of micro systems (the quantum world) and both of these with the classical physics associated with everyday human existence: in particular a relationship linking local, causal and continuous systems with those which are non-local, non-causal and non-continuous in character.

The Model establishes just such a relationship between relativity, the determinate multidimensional states and the implicit state associated with the development of consciousness, all governed by the Ordering Centre.

An Ordering Centre corresponds to a given cosmological energy density. Since this density can in principle take a range of values from the negligible upwards our big bang is not unique.

It is therefore not absolute either; the apparent singularity is not absolute, so relativity and the quantum state do not have to be in conflict as the Reference Model explains.

There are a vast set of apparent big bangs in fact, that set being both holographic and fractal in nature. Such Ordering Centres may split into a primarily fractal set with a profusion of discrete peaks of energy density each with different physics or enfold together to form a unified distribution curve which may be said to constitute the Whole Ordering Centre.

Higher peaks of energy density then allow the unfolding of higher degrees of consciousness than that comprehensible to Homo sapiens.

In terms of the Reference Model the Whole Ordering Centre is considered to be in motion about its inanimate mean axes which support the grid structure referred to earlier.

Summarising it is the purpose of this book therefore to construct a Standard Reference Model showing that arising from the arguments introduced above that all objective and subjective philosophies can be related on the basis of one theory, using the basic principles of ordering and disordering processes referred to earlier.

That construction has to be carried out by maintaining a balance between reason and imagination, all without a priori assumptions, allowing statements to have degrees of validity, yet supporting concepts of both relative and absolute truth, all in terms of a process which explains the transformation from disorder into order and vice versa.

The Model has to be constructed by an intelligent (thinking) entity i.e. the Reference Modeler without biasing that theory in favour of his own opinions.

In principle this is achieved in two ways, by housing all known schools of thought under the roof of one theory based on the basic principles of ordering and disordering processes, as previously stated analogous to a reference library and by using a Modeler who devises a method of relegating his own opinions to being merely one of a set of opinions or beliefs, all subject to that same theory.

In other words the Modeler must find a means of relating each belief to the alternatives.

The commonly held opinion, for example that the objective philosophies such as science and technology can be separated from the primarily subjective philosophies including politics and the main religions such that one or the other approach to what is termed the truth will ultimately prevail, is effectively demolished. In other words the truth turns out to be a rich picture, a moveable feast related to a range of philosophies rather than any particular one.

The Reference Model is to be tested by the Modeler using a Reference Library including in principle all records made by Homo sapiens whether unfolded explicit (spatiotemporal) records regarded as measurable or enfolded implicit (non-spatiotemporal) records regarded as non-measurable. Such records may be in an explicit form on paper or other recording materials or thoughts unfolded and communicated via the senses or held enfolded in the mind without outward expression.

They will relate to both the animate and the inanimate; conversations, statements considered to be what are termed facts, opinions, beliefs, feelings of all kinds whether political, economic, scientific, religious or emotional, serious or comical, on each and every subject, together with all forms of literary, musical and artistic expression.

Clearly there are further records not included above whether past or future, including those not associated with Homo sapiens for example, which may also be termed enfolded records.

The Model inherently includes and allows for these also.

Inevitably only a sample of such records can be utilised for reasons of practicality, nevertheless that sample should be randomly selected without deliberate omissions by the Modeler, since he is required by definition to be close to a neutral position.

The choice of records is therefore to be made in a mathematically random manner, at least in principle. The wider unrestricted use of records for input to the Model is not unrelated to the concept of the input to a universal Turing machine.

In practice the author playing the role of the Modeler is forced in the first instance to make his own choice of records for examination. The principle referred to above is then retrieved by the readership who free to make their own choice, ultimately restore random selection the wider that readership becomes.

All philosophies on the other hand are supported by the philosopher deliberately restricting the records to be accepted in different ways, in part to render the deficiencies of his doctrine less apparent. It is of course this deliberate omission or at least disregard of available records which leads directly to conflict between those philosophies.

Nevertheless they are fundamental to human enquiry and reflect the necessity for diversity; furthermore the Reference Model is dependent on them since it represents in effect the Mean of those philosophies, though is not itself one of them.

The Model accepts in principle all records for processing while the individual philosophies each accept only a proportion, as a group they nevertheless distribute the records among themselves. It will be shown that this distribution is in accordance with the concept of a unified Ordering Centre when a full range of philosophies is examined.

The book incorporates figures designed to support the main text and provide means of visual clarification and assistance to comprehension.

In accordance the above a range of Reference subject matter derived from the library records, is then discussed utilising the work of other authors and conclusions derived from the Model, essentially in the same manner as one might solve mathematically a set of simultaneous equations.

2.0 THE BASIC PRINCIPLES OF ORDERING AND DISORDERING PROCESSES.

2.1 General.

In order to understand these basic principles in the first instance and then the manner in which the Reference Model deals with the problem of relating the philosophies both objective and subjective, the following subject matter needs to be examined and when understood, comprehended in effect simultaneously as one integrated whole.

a) Degrees of existence; bounded v unbounded.
b) Order v disorder; difference v similarity; unfolding v enfolding.
c) Continuous v non-continuous; causality v non-causality; local v non-local; particle v wave; symmetry v antisymmetry.
d) Spatiotemporal v non-spatiotemporal; reality v imagination; explicit v implicit; objective v subjective; measurable v immeasurable.
   Frog's eye view of dimension v bird's eye view using the Argand diagram.
   The Implicate Order.
e) The Central limit theorem and the normal (Gaussian) distribution; difference and similarity sequences; symmetry and asymmetry.
f) Wave motion, wave harmonics, holograms, fractals and the human brain.
g) Hierarchical and holarchical (or holoarchical) structures.

These subjects are introduced and discussed in general terms in this chapter while detailed application of the relevant concepts in terms of the Reference Model is dealt with more rigorously in section 4.0.

2.2 Degrees of existence; bounded v unbounded.

Comprehension of existence is only viable in the context of an animate (meaning alive) or an intelligent (meaning purposeful) entity conscious to a greater or lesser extent of that existence and of itself in relation to it. Likewise living existence has consciousness of the entities that compose it. The more an entity matches the whole of existence the better it should be able to describe it, though as it turns out never precisely.

The inanimate on the other hand is defined as having no consciousness of existence including itself, nevertheless it exists and does so ultimately because animate/intelligent entities place it in a neutral state while also diverging from it, as the Reference Model will explain. The Model implies that animate and inanimate properties are complementary and that existence does not in any overall sense have inanimate properties only nor animate/intelligent properties only.

Both scientific and religious philosophies make fundamental errors in their comprehension of the relation between the animate and inanimate.

These errors have arisen in part from confusion about and misapplication of the concept of boundaries and similarly about what constitutes existence as opposed to what is termed non-existence.

The Model equates the term existence with concepts of difference and similarity and the further concept of order made up of sequences of difference and similarity, sequences which can in turn be equated to energy. It can then be used to explain what it means to unfold ever greater degrees of existence (complexity) from ever lesser degrees of existence (simplicity) and vice versa to enfold existence in the opposite sense.

The Reference Model has a lot to say about boundaries, when they can apply and when they do not. In particular it implies it is meaningless to ascribe a boundary to existence which is considered unbounded. The term non-existence is casually applied to the imaginary when the explicit is treated as an absolute reality whereas the theory shows neither reality nor imagination cannot be isolated in this way.

It is accordingly more meaningful to refer to lesser and lesser degrees of existence and to limit the term non-existence to those cases where appropriate boundaries have been applied.

The common experience is that what is regarded as real things have boundaries. Ideas on the other hand have no measurable boundary.

The theory therefore only applies boundaries to measurable objects which for practical purposes can be considered as in their own three-dimensional space but not to the dimensionally enfolded imaginary world, nor indeed can any such limits be placed on the capacity of the real world to dimensionally unfold as will be shown later.

The use of the word whole typifies the problem as a boundary may be relevant in circumstances explicitly defined as in painting the whole room (though even this is an approximation), but not when referring for example to the whole idea.

The incorrect application of boundaries leads directly to the difficulties that arise when trying to describe reality as opposed to imagination and also the pursuit of the absolute as opposed to the relative.

The Reference Model clearly reveals the insistence on either absolute reality or the opposing view that no objective reality exists (generally termed relativism), are distractions which inhibit comprehension of the relation between reality and imagination.

2.3 Order v disorder; difference v similarity; unfolding v enfolding.

These aspects are related in a fundamental way and will therefore be considered together.

Order and disorder defined in terms of entropy applicable to the real world are related in the Reference Model to the concept of degrees of existence referred to in section 2.2.

Comprehending order and disorder and degrees of existence is difficult but crucial to an understanding of the Reference Model and must include the question of how existence imposes ordered/disordered sequences from a spectrum of differences and similarities, and how those differences and similarities arise. That comprehension must also include the relation between wholeness (ordering), unwholeness associated with the concept of evil (disordering), chaotic deterministic states which range from the predictable to those which involve widely diverging outcomes (another form of unwholeness termed chaotic disordering) and finally indeterminate states associated with randomness, termed stochastic.

Manifest order and disorder can be defined as the extent to which order and disorder become explicitly apparent in what we term the real world revealed by our explicit senses, sight, touch etc.

Chaotic states of disorder, ultimately deterministic, are associated with orders of lower degree but characterised by indiscreteness, a degree of distribution in other words.

The Model regards the random state as being the polar opposite of the order/disorder spectrum, a directionless condition characterised by discreteness, where the concepts of both order and disorder have no meaning. So this spectrum and randomness have an inverse but nevertheless intimate relationship since one cannot exist without the other.

While inanimate order has no knowledge or awareness of wholeness, unwholeness or fragmentation, it will be shown there is however an intrinsic relationship between intelligent/animate order and inanimate order, which forms the core of the Reference Model.

Concepts of order and disorder clearly relate to arrangements of different similarities and similar differences, in other words combinations of difference and similarity in various sequences.

In this context sequences composed largely of differences only or of similarities only exhibit minimal ordering. Alternatively sequences which contain a balance of difference and similarity allow increasing degrees of manifold ordering or disordering.

For example a motor car can be built using a wide variety of materials e.g. wood, glass, steel, leather, rubber and so on, but not an endless variety. Nor could the vehicle be built using a very limited range of materials, glass only for example.

More subtle analogies can be used to convey the meaning of order/disorder sequences and the difference/similarity spectrum, namely the dictionary, playing cards and the carpet roll. Firstly consider the dictionary, involving a comprehensive list of words in alphabetical order.

The maximum ordering is the conventional dictionary but if it is distorted in the direction of increasing difference the words start to change language, at first known languages but then to unknown languages, the profusion of alphabets makes the words increasingly incomprehensible, the sequences are no longer alphabetical. The variation becomes so great that the fabric which normally binds a dictionary namely the ability to relate a given word through a chain of definitions to any other word in the volume breaks down. Ultimately we have an endless sea of incomprehensible symbols, in other words a book with no ordering.

If the original dictionary is now distorted in the opposite direction towards increasing similarity, words start to look like each other, letters become indistinguishable one from another, the number of words decreases and ultimately there is only one word but it is impossible to read what it is, again a book with no ordering.

Repeating this process using playing cards we can lay out the maximum ordering as the four suites arranged as say, hearts to clubs and ace to king. Distorting the cards to the difference end of the spectrum reduces the cards to a random variation of suit and card value with the cards ultimately meaninglessly scattered even torn up, while distortion in the opposite direction shows an increasing tendency for the cards to appear face down until ultimately all are face down one on top of the other, the back of the pack having no discernible pattern. Again the ordering is a maximum at the centre of the range with the minimum ordering to either side.

For a third time we repeat the process, this time using the analogy of a carpet roll.

Consider a set of carpets rolled together.

At the point of maximum ordering the roll exhibits one complex colourful pattern all in one spectacular display but as we move to the difference end of the spectrum the carpets first separate, each with its own pattern etc and then gradually break up into threads until no pattern or weave etc is evident. In the opposite direction the carpets roll together, merging and losing individuality, the patterns and colours slowly fading until ultimately again no pattern or colour is apparent.

The carpet analogy is particularly useful, since it is fundamental to the Reference Model that taking the brain/mind concept as having central importance to a conscious entity, the brain is considered the unfolded mind and conversely the mind the enfolded brain.

In other words the brain/mind should be seen in the context of an unfolding/enfolding system. The physical brain even looks like folded material and has limited dimension while the mind is a subjective concept with no explicit dimension. The relation between the Reference Model and the human brain is discussed in more detail in section 2.7.

It will be shown that existence is only comprehensible when unfolding processes are on average balanced by complementary enfolding processes and in particular that what is understood as inanimate is a neutral state related to the point of balance of those processes.

It is the purpose of the Reference Model to explain the relation between the development of the physical universe as seen explicitly in scientific terms and the imaginary (metaphysical) world.

In particular as has been stated in the introduction objectivity is a process of unfolding explicit order at a given level while enfolding order at other levels and conversely that subjectivity is a process of enfolding that explicit order while unfolding orders at other levels.

The ordering process therefore enfolds disorder while the disordering process enfolds order.

Unfolding and enfolding processes will be related to the concept of the hologram in section 2.7.

2.4   Continuous v non-continuous; causal v non-causal; local v non-local; particle v wave; symmetry v antisymmetry.

In the context of the theory these paradoxes can be considered together.

Continuity refers to the seamless transition between any one element of space-time to any adjacent element such that any point in that space-time can be taken as zero and all other points measured relative to it, on the basis that the space-time is an undivided whole.

Causality requires that if a signal is emitted at any point in the space then that signal will be transmitted from that point initially to the next adjacent point and then to the next point in sequence there being a time lapse for the passage of the signal between points a finite distance apart. The initial emission is then considered the cause and the receipt of the signal at any other point the effect, space-time forming an undivided whole.

Locality requires that the signal is transmitted from a point in space-time to the next adjacent or local point and so on in sequence and not to some distant or non-local point without first moving through all the intervening points.

All these properties are basic to both classical physics where three-dimensional space is taken as absolute and to its extension namely relativistic physics where four-dimensional space-time is taken as absolute.

Classical physics also allows for the extension of three-dimensional space to multidimensional space namely 3n configuration space and 6n phase space, allowing for momentum where n is the number of particles under consideration, all determinate within the context of Hamiltonian mechanics.

Then early in the 20th century the quantum state was discovered involving very small energy changes. Although it is commonly thought to relate to very small orders of magnitude in terms of scale the quantum arena includes not only human scales but scales approaching the size of the visible universe, as experiments have shown.

The basic quantum state termed unitary is again determinate and continuous in a multidimensional space referred to as Hilbert space, in which complex numbers play a fundamental role.

In mathematical terms Hilbert space is associated with symmetry and phase space with anti-symmetry (see section 7.3.7) and the term multidimensional should be regarded as equivalent to infinitely dimensional.

When however attempts are made to relate that quantum state to classical physics (local, causal and continuous) this determinism is effectively overturned and replaced by non-continuous, non-causal, non-local concepts based on probability, but again forming an undivided whole.

Amazingly as the well-known double slit experiment shows, the quantum state happily behaves as a wave if not observed but instantly behaves as a particle if it is.

Quantum physics deduces that the wave can be reduced to a particle to the extent allowed for by the Heisenberg uncertainty principle, by localising the wave in terms of both space and momentum to form wave packets.

At the classical level we are familiar with the concept of waves though they are different from those at the quantum level, and also the concept of particles and their aggregations to form molecules, pebbles and galaxies.

It has already been stated that at least real things, whether particles at the molecular level or macroscopic objects, can be considered for practical purposes to have boundaries. Indeed such boundaries become even more emphasised the deeper the world of the very small (the quantum world) is penetrated. Like an endless set of Russian dolls, these boundaries seem ever more tightly drawn prior to reaching scales of the order of the Planck length (10^{-33} cms).

The Reference Model attempts to explain all this by regarding the system as consisting of a real axis and an imaginary axis orthogonal to it in the conventional manner used to portray complex numbers.

The real axis is considered to be multidimensional in the sense in which phase space is understood and the imaginary axis is also multidimensional but in the sense in which Hilbert space is understood. The Model then introduces implicit space also multidimensional and considered to apply at angles to the grid structure formed by the axes.

Again the term multidimensional in respect of implicit space should be regarded as equivalent to infinitely dimensional.

This implicit space is regarded as the source of animacy and consciousness as opposed to the grid structure which is regarded as inanimate.

The animate and or conscious entity is therefore also multidimensional but reduces the multidimensional state to three spatial dimensions termed the explicit order, when it reduces the real and imaginary state to that of reality only, this being a particular level in a vast range of possible ordering sequences. That three-dimensional real state can then be further reduced to two by using the holographic principle which is discussed later in this section and in more detail in Section 7.0. So the Model can be portrayed in two dimensions on a piece of paper as will be demonstrated in this book.

The grid structure is related in the Model to the concept of the particle and motion across that structure to the concept of the wave.

In their efforts to examine the subatomic world by measurement (a separation process) physicists in effect distort the ordering system into the explicit and are then surprised to find that it obstinately insists on behaving as one integrated whole. More will be said on this later.

Separation in to particles and waves nevertheless while useful for practical purposes are best viewed as distractions from the primary ordering concept, thereby avoiding the pitfalls of absolute reality or at the other extreme relativism.

2.5 Spatiotemporal v non-spatiotemporal; reality v belief and imagination; explicit v implicit; objective v subjective; measurable v immeasurable.

Frog's eye view of dimension v bird's eye view using the Argand diagram.

The Implicate Order.

By spatiotemporal is meant explicit reality, in principle objective and measurable. A state associated with boundaries and the bounded condition, with the concept of particles at the molecular level and the aggregation of particles which form macroscopic objects. A state further associated with the unfolding of a limited number of dimensions with the common-sense notions of a space-time continuum with causal and local properties.

Future time may therefore be construed as being time unfolded and past time as time enfolded as thread on to a reel.

Space is in turn unfolded and enfolded in the same manner and accordingly the Model views space-time as one property coupled in accordance with the requirements of General Relativity.

The additional essential feature of space-time that the Reference Model emphasises is that future and past space-time are in a state of balance about a centreing condition, the present time, about which all unfoldments and enfoldments may be said to occur. The observer always appears to be at this central state, just as all points on the surface of an inflating balloon appear to be at the centre of that inflation.

By non-spatiotemporal is meant implicit and immeasurable certainly, but considered real and objective by Platonists who believe in realism, or consigned to non-existence by nominalists, or considered by many in some sense to exist albeit associated with belief, the imaginary and subjective.

This evidently a state in which spatiotemporal dimensions are not merely enfolded but in some sense fluid as opposed to say abstract mathematical objects which seem to exist in a more solid state independent of consciousness, waiting to be discovered. A state associated with waves and the diffusion of macroscopic objects and the particles of which they are composed in to those waves.

It is therefore pointless (no pun intended) to consider that existence has some starting point or indeed end point based on any concept of absolute time or any fixed geometry based on a concept of absolute space.

It will be shown that the relation between reality, belief and imagination, can only be understood when the concept of unfolding and enfolding processes giving rise to many levels of ordering are related to the distribution arising from those same processes operating at a given level of ordering.

So reality is associated with a particular level in the ordering sequence namely the real axis, imagination is associated with the imaginary axis orthogonal to that and belief intermediate between the two, since it is more confined to practical reality while imagination knows no bounds.

Drawn on a two-dimensional sheet of paper the real axis is conventionally shown as horizontal and the imaginary axis as vertical, a convention adhered to throughout this book (Fig 2.5.1).

In mathematics the corresponding diagram is called an Argand diagram.

The frog's eye view of dimension relates to the explicit world, a particular level of order, while the bird's eye view relates to the implicit world involving many levels on the imaginary axis.

These levels of ordering from which the explicit world unfolds are termed implicit orders, which taken collectively are further termed The Implicate Order, as put forward by the physicist David Bohm in his book *'Wholeness and the Implicate Order'* (Ref 2.5.1).

Bohm's verbal description of the Implicate Order was complex from the outset and became even more so as he and his co-workers attempted with indeed some success to provide an equally complex mathematical theory to support it. This complexity made the task of persuading other physicists to follow the same path difficult and

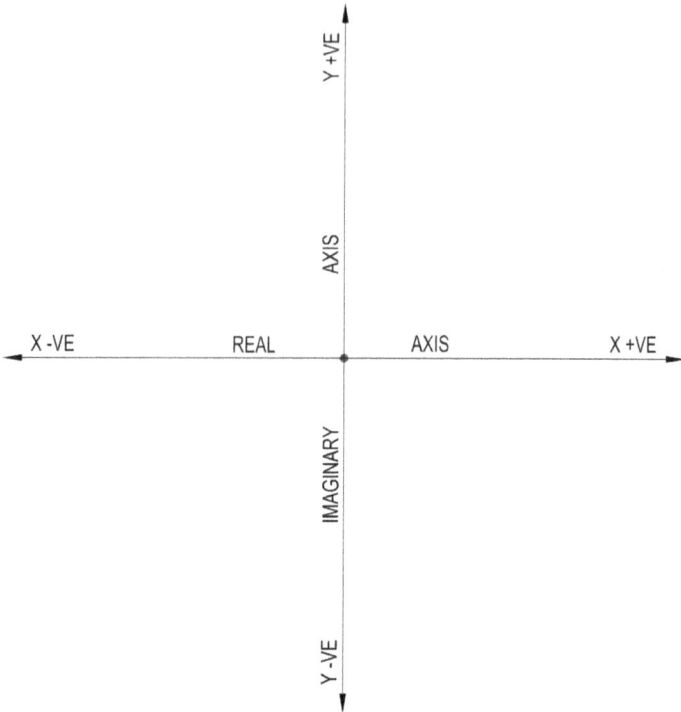

FIG 2.5.1   DIAGRAMMATIC REPRESENTATION OF REALITY (THE OBJECTIVE)
VERSUS IMAGINATION (THE SUBJECTIVE)

although some expressed support (as evidenced by academic literature in reaction to his work), it could not readily be disseminated in a manner comprehensible to the general public.

It is the purpose of the Standard Reference Model to underpin the complex description of the Implicate Order which Bohm originally put forward with a simplified and practical description based on established mathematical axioms, relating the relevant principles to a working model of consciousness made comprehensible to the layman.

While the Implicate Order remains the main support structure for the basic ordering processes which compose the Reference Model, this practical description replaces the complex verbal and mathematical version referred to above. The totality of implicit space in this Model is referred to as Implicate space to conveniently compare it with phase space and Hilbert space again referred to earlier.

2.6   The Central limit theorem and the normal (Gaussian) distribution; difference and similarity sequences; symmetry and asymmetry.

The Reference Model relates the concept of order to the principle of sequencing made up of differences and similarities which can in turn be related to the Central limit theorem and the normal (Gaussian) distribution curves (see Section 7.3.4) or bell curves (Fig 2.6.1).

Such distribution curves are fundamentally symmetric. Alternatively those skewed to the left or right relative to the symmetric case are accordingly asymmetric.

Symmetry and asymmetry (not to be confused with antisymmetry) are related in mathematical terms (see Section 7.3.7).

The familiar bell curve places similarity at the centre with the differences making up the variance (i.e. the range of deviation involved) distributed to either side. Alternatively a similar curve can be formed

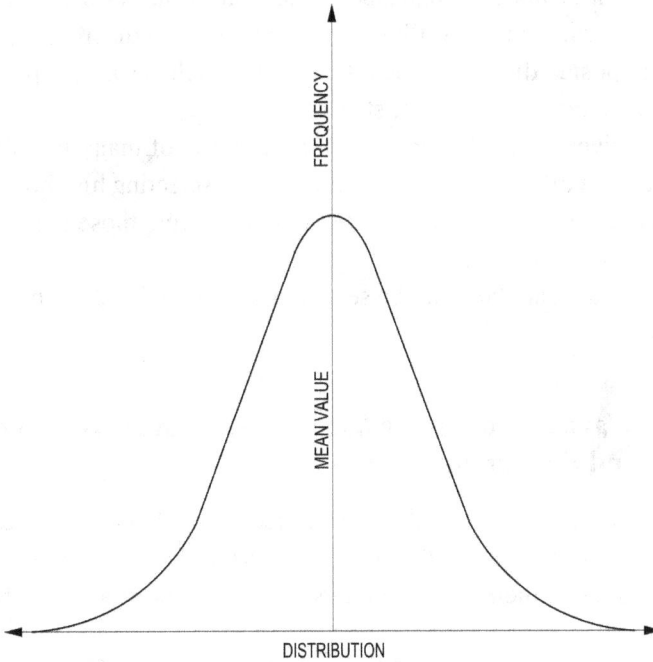

FIG 2.6.1   THE FREQUENCY OF THE MEASURABLE PARAMETER VERSUS THE DISTRIBUTION OF THE PARAMETER

23

placing difference at the centre with similarities to each side. A composite of these curves puts differences to one side and similarities to the other. The sequence of differences and similarities at the central axis then makes up the maximum ordering sequence, as explained in Section 2.3 above.

The differences to one side and similarities to the other represent the variance at any one time and are positioned on the real axis while the sequence referred to lies on the imaginary axis.

The differential side relates to the tendency to emphasise difference between one thing and another while the similarities relate to the tendency to integrate those things.

Although difference and similarity is defined in terms of the real axis they also have meaning along the imaginary axis since it represents those differences and similarities ordered in a particular way.

Order can be seen as an integral and disorder as a differential, since the process of ordering can be equated to endeavouring to select the optimum or integral sequence as the sum of the set of less perfect sequences each seen as a differential. The process of disordering works in the opposite direction such that that which seemed optimal is converted to many less perfect states.

At a given level of order the future consists of many possibilities each seen as a differential the individual entity selecting his choice as to which is best at a given present time thus converting those possibilities into one integrated past.

The future can therefore be seen as a differential and the past as an integral.

2.7   Wave motion, wave harmonics, holograms, fractals and the human brain.

As has been indicated the Reference Model incorporates the principles of wave motion, the hologram and the fractal in a fundamental way and relates them to our understanding of the brain and how it operates.

Many readers will be familiar with the principles of wave motion, reinforcement and cancellation as shown in Fig 2.7.1 and the further principles of periodic waveforms.

The Basic Principles of Ordering and Disordering Processes

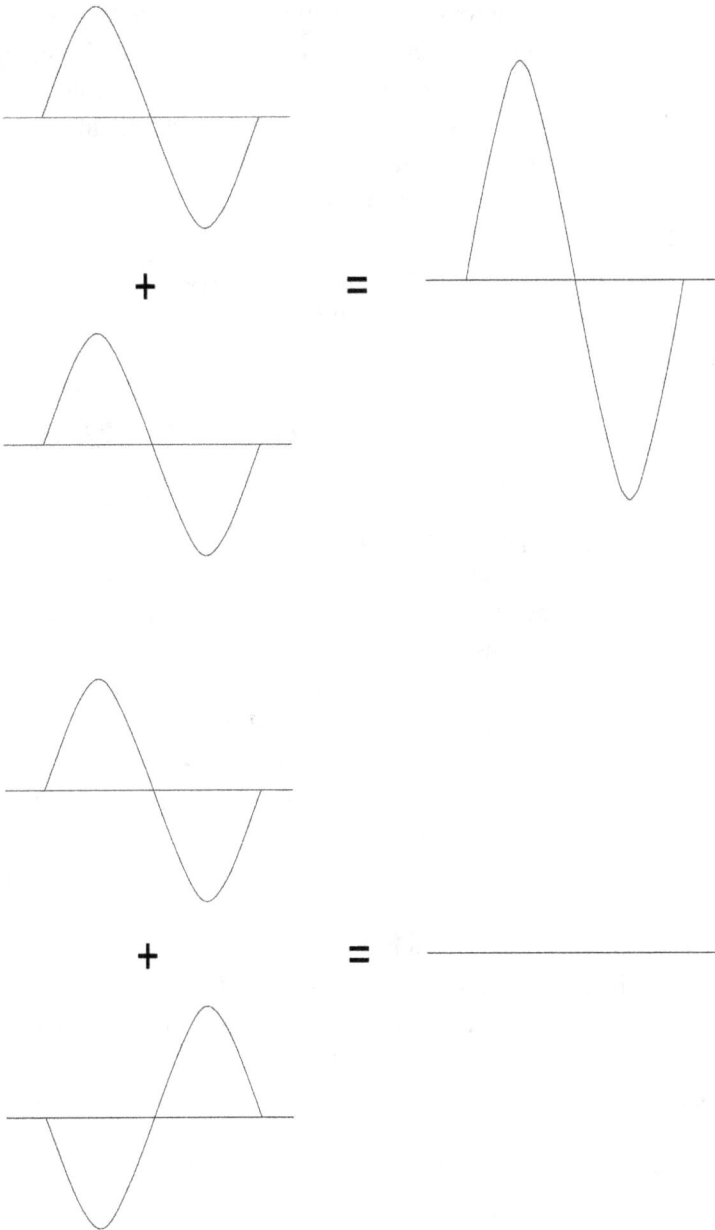

NOTE RELATIVE TO A GIVEN ORIENTATION IN PHASE WAVES REINFORCE
AND OUT OF PHASE WAVES CANCEL

FIG 2.7.1 THE WAVE REINFORCEMENT AND CANCELLATION PROCESS

In particular from our love of music we have an instinctive response to the harmonic resonances and standing waves forming the basis of musical instruments such as guitars, cellos and pianos.

Turning to the hologram and the fractal, the principles involved are perhaps less familiar to many readers.

In respect of the hologram, a laser beam used to illuminate any relevant three-dimensional structure when simultaneously directed on to a photographic plate produces an interference pattern on that plate in two dimensions but appearing three-dimensional to the eye. The visual impression of the three-dimensional structure is in effect enfolded on to two dimensions, a principle fundamental to an understanding of the unfolding and enfolding processes referred to in the Reference Model.

For a practical and detailed explanation of holograms an excellent source is that by Kasper and Feller (Ref 2.7.1).

Kasper and Feller compare the process of making and observing photographs with the making and observing of holograms. In the latter case it is important to ensure the coherence of the light beam and since that beam is also very narrow the use of a lens to ensure divergence of the beam to permit illumination of substantial areas.

They go on to explain the importance of the creation of a reference beam and an object beam. Part of the divergent light from the lens termed the reference beam falls on a mirror and is reflected on to a photographic plate while the remainder of the divergent light termed the object beam bypasses the mirror, falls directly on the object concerned and is reflected from it on to the plate.

Now whereas the photograph created by a camera gives just one representation of the scene, in the case of the hologram two images are created one virtual and one real, both three dimensional. The virtual image appears on the opposite side of the plate from the viewer and the real image on the viewer's side.

When viewing the developed plate the viewer's eye on one side of that plate sees the virtual image as a diverging beam having been acted on by the interference pattern on that plate but appearing to emanate from that image. However some of the light from the hologram is a converging beam which comes to a focus producing the real image, an optical replica of the object.

LASER BEAM
BEAM SPLITTER
MIRROR
OBJECT OF INTEREST
PHOTOGRAPHIC PLATE

NOTE   Diagrams showing the principle of the hologram
can be found on the internet.

FIG 2.7.2   COMPONENTS OF THE HOLOGRAM

The result is that whereas a photograph cut in to pieces cannot be seen as one whole if only a single piece is examined, in the case of a hologram a single segment allows a realistic three-dimensional view of the whole object. In other words the hologram records information about the whole scene in each small piece of the plate.

Fig 2.7.2 shows the basic components involved and numerous diagrams demonstrating their use in creating holograms can be found on the internet.

The principle of the hologram is also described by Bohm (Ref 2.5.1), in particular that the hologram contains an ordering that determines the arrangement of points that appear in the images referred to above when the object is illuminated, i.e. the enfolded implicit order. The order in the object as well as in the image is the unfolded explicit order.

Indeed it may be noted that the writing and associated figures for this book are like all other books presented in two dimensions when it is clear that much of the discussion is concerned with the world viewed in three or more dimensions. Holography plays a significant role in modern physics, but another ingredient is necessary.

The essence of this additional ingredient is to combine the unfolding and enfolding process of holography with the image of a pattern or sequence constantly repeating itself, sometimes in a chaotic manner. There is no limit in general terms to how simple that pattern can be, or any limit to its complexity.

The principles of the fractal and self similarity are described in detail by Schroeder (Ref 2.7.2) arising from the original discovery by Mandelbrot (Ref 2.7.3). Other relevant sources can be found on the internet and the mathematical principle of the Mandelbrot Set is given in Section 7.3.2.

Self similarity is simply an endlessly repeating pattern, patterns mind you with endless variations. Two simple patterns are shown in Fig 2.7.3, one with straight lines only and one with circles to illustrate the principle. In fact fractals give rise to patterns of incredible variety and complexity as the references mentioned above reveal.

The Model incorporates an order distribution curve described in detail in section 4.0 to represent such self similarity and suggests it is universally applicable to both real and imaginary worlds and furthermore gives rise to the infinite variety evident to both our physical senses and to our imagination.

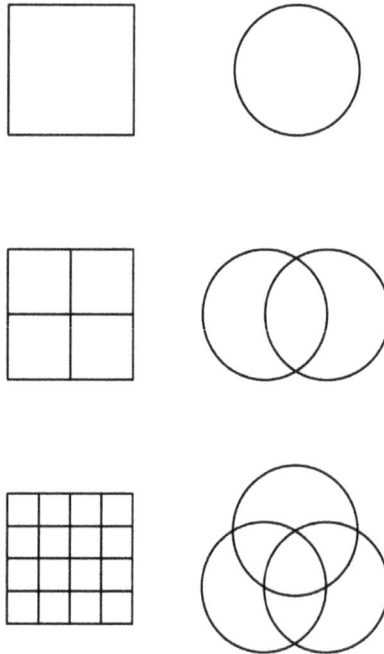

FIG 2.7.3   DIAGRAMMATIC REPRESENTATION OF A FRACTAL

The holographic and self similar principles are united in the Model, that unification being described as a holographic fractal. It combines the sense of ordering inherent in wave motion and the unfolding/enfolding holographic process, with the chaotic element associated with the fractal.

In addition the Model must relate to our understanding of the human brain. *'The Self and Its Brain'* is the title of a book by Popper and Eccles attempting to comprehend the body mind problem (Ref 2.7.4). They were dualists who believed in the existence of mind stuff and material stuff and that mind interacted with matter. Associated with this dualism the book developed Popper's original concept of different worlds integrating it with Eccles' neurological expertise.

In particular Eccles describes the relationship between the left and right hemispheres of the brain in various circumstances both normal and in cases where global lesions of the cerebrum have occurred for whatever reason.

In this description the left hemisphere is regarded as dominant and the right hemisphere as minor which may be associated with the lateralisation involved whereby the left/right sides of the brain interact primarily with the right/left sides of the body.

This again may relate to the fact that over 90 per cent of the human population is right-handed, an example of asymmetry termed chirality, used by the Reference Model to establish a model of consciousness.

While it is known that specific areas of the left side of the brain confirm a degree of lateralisation associated with language, the dominance of the left compared with right hemisphere is considered illusory by other scientists, who do not distinguish between mind and body, pointing to evidence that when the human brain is stimulated no evidence of dominance by either hemisphere is apparent. They consider that the brain acts as a unit in general terms.

Irrespective of the mind body issue studies involving the split brain by the Nobel prize winner R.W. Sperry and by M.S. Gazzaniga (Ref 2.7.5) have tended to imply that at least on average the left hemisphere is analytical and interpretive (epistemological) to the point of making up stories to fit what it regards as the facts in its efforts to answer how things are; whereas the right hemisphere tends to ask what is the literal (ontological) truth and why it is?

Adding another layer of argument to all this, are the efforts of Karl Pribam (Ref 2.7.6) who developed a holonomic brain model of cognitive function, supporting the idea that processing in the brain occurs in a non-localised manner.

The Reference Model resolves these contrary views with explanatory figures which accompany the text in Section 4.0. It becomes evident that the principles of both locality and non-locality are necessary to resolve the mind/body problem.

In particular the Model provides an alternative basis for the development of the three different worlds as proposed by Popper and Eccles, grounded in the principles of both the explicit order and its inherent symmetry and the implicit order and its inherent asymmetry as explained in this book. This alternative removes the limitation to three worlds and expands them to potentially an infinite set. Furthermore it provides a basis for resolving the argument concerning brain lateralisation referred to above.

The Model therefore assumes a limited degree of lateralisation favouring the left hemisphere of the (local) brain in terms of language, but additionally a correlation with the processes of differentiation and integration respectively conducted by the (non-local) mind inherently asymmetric in character. This correlation is implicit as opposed to explicit and therefore not measurable in scientific terms, giving rise directly to the argument concerning lateralisation.

In any case the combination of these processes is holonomic or holistic in principle.

Reversing the correlation with difference and similarity makes no difference in principle to the validity of the Model.

2.8   Hierarchical and holarchical (or holoarchical) structures.

Hierarchical structures are those for which the key term is level, the structure being divided in to levels on the basis of degrees of complexity in the case of inanimate structures or position or rank in terms of status, power or seniority in animate structures.

The principal emphasis is on separation in to parts, each part distinguishable from other parts. There is a further emphasis on absolutism as opposed to the relative, the objective as opposed to the subjective, differentiation as opposed to integration.

A hierarchical structure in human terms can be visualised as a pyramid or triangle with the height above the base representing status and the width representing the numbers of people at that level; typically one chief at the top and many workers at the bottom.

The primary linkages in the pyramid are horizontal and the secondary linkages vertical.

Holarchy as originally defined by Koestler (Ref 2.8.1) in his 1967 book, *'The Ghost in the Machine'*, as a connection between Holons which represent both wholes and parts rather than simply parts.

Although this structure retains the concept of level, the patterns at any one level are similar to those at another level. In other words the holons at one level are made up of the holons at another level. An example is the ordering constituting particles, atoms, molecules, macromolecules, organelles, cells, tissues, organs, organisms, communities, societies. Each holon is therefore ultimately a description of the whole set, a fractal pattern in effect.

The principle emphasis is on the dilution of parts in favour of the whole, the relative as opposed to the absolute, the subjective as opposed to the objective and integration as opposed to differentiation.

Holarchical structures are also like a pyramid but with the primary linkages being vertical and secondary linkages horizontal.

In human affairs both hierarchical and holarchical structures and combinations of both can apply.

The Reference Model supports Koestler's holarchical principle by relating it to implicit order and the hierarchical principle in turn to explicit order.

3.0 THE STANDARD REFERENCE MODEL; THE OBJECTIVE AND SUBJECTIVE PERSPECTIVES.

3.1 The Reference Model and the objective philosophies

Advances in scientific knowledge in the last few hundred years and particularly in recent decades prove to be of enormous value in clarifying the relationship between that knowledge and our comprehension of order and disorder.

Science has already extended Newton's original supposition of absolute three-dimensional space to Einstein's four-dimensional space-time but the Model must seek to establish a relationship between the apparently incompatible positions taken by that non-probabilistic theory of General Relativity and the probabilistic theory which relates the quantum state to the classical level of physics.

The book evaluates the principles of Relativity against the Model and finds no reason to dispute them in terms of the explicit. Indeed the principle of background independence so important in Relativity is fully compatible with the Model.

Where the Model deviates from Relativity is in its stance with regard to the quantum world.

There are over a dozen theories associated with what has been termed the quantum dilemma, three of which are the original theory associated with Bohr, the many worlds theory of Everett and its developments (i.e. parallel universes) and the particle and carrier wave theory of Bohm and de Broglie. Ten of those theories including these three are discussed in Section 7.0.

The book relates these in turn to the Model and contends they can all be construed as looking at the same problem from different perspectives.

As originally proposed by Bohm (Ref 2.5.1) the Model replaces absolute space (or space-time) and the associated Cartesian (or curvilinear) coordinates by an entirely different non-Cartesian concept of order, where at any given level space-time can be unfolded in the form of a dimensional continuum (the explicit order in four-dimensions) or enfolded again in the form of a multidimensional non-Cartesian ensemble (the Implicate Order). More specifically the Model incorporates the multidimensional mathematical concepts of phase space and Hilbert space and relates these in turn to non-Cartesian space referred to as Implicate space. It is demonstrated that the unfolding of animacy and consciousness is to be associated with that Implicate space, which is non-local in character.

If one chooses to carry out measurements at micro levels of scale the above concept of order will reveal itself. Should an observer choose to examine the micro world in particular experimentally the behaviour of particles as real (bounded) objects in a three-dimensional special case, he should not be surprised if they appear to function as a wave system intrinsically inseparable from an undivided whole.

Science may in principle be limited to the explicit but the difficulty in defining what is being measured in experiments at the quantum level has forced science, in the main unwillingly, to consider its relationship with the implicit.

This situation is further exacerbated by Bell's Interconnectedness Theorem (Ref 3.1.1), which demonstrates that local realistic theories which form the basis of classical physics are incompatible with quantum theory.

The same Theorem also brings into question the degree to which reality is independent of conscious entities, one which the Reference Model seeks to answer.

Many branches of science have made efforts to establish the relation between the inanimate and the living world; physics, chemistry and in particular biology and medicine come to mind.

The theory of natural selection for example becomes manifest at the boundary between inanimacy and animacy, where increasing chaotic disorder is in balance with increasing order.

It therefore best describes the ordering process when such balances apply.

However the explanation for life in general and consciousness in particular requires a robust model of ordering processes. It cannot be based on an insistence in taking measurability as the sole criterion of truth to the exclusion of imagination.

The Model on the other hand while maintaining the objective and subjective in a general state of balance associates the inanimate with that state and life with motion relative to it.

Further lines of scientific enquiry also play an important part in the construction of the Reference Model.

In particular the general principle of unfolding and enfolding is associated with the properties of the hologram as described in Section 2.7.

In addition the Model takes into account the self-similar properties which govern the behaviour of fractals, of equal importance to holographic properties.

Self-similarity makes it impossible to regard any point in the system as absolutely defined even at the macro levels of scale associated with classical physics.

Taking the simplest analogy and representing the relative truth as a scale and the absolute truth as a pointer on the scale, then the inclusion of fractal properties results in the principle of absolute truth breaking down when the level of detail on the scale is too fine for the absolute (unchanging) pointer to differentiate.

Coupled with the uncertainty related to the properties of systems at the micro or quantum level this makes defining absolute points in the system doubly absurd.

It is this problem and the relation between the objective and the subjective in particular that the Model seeks to resolve by relating symmetric, antisymmetric and asymmetric properties to an ordering/disordering system which has levels of coherence.

The word level is taken as key, since absolute truth is limited to the extent that it relates to a given level in the ordering sequence. The concept of level and the spreading out of ordered systems at every level is crucial to an understanding of the processes which in the form of the Reference Model offers a means of relating the animate and conscious implicit state to the quantum state and to the classical interpretation of everyday experience.

The sequence of order and the distribution (or spreading out) of differences and similarities at given levels in that sequence is incorporated by the Reference Model in what is termed the Ordering Centre all related to the big bang theory central to modern cosmology.

It is proposed that the classical space-time continuum unfolds at what is termed the mean level in the average state of the Ordering Centre when the formation of ordering, a holoarchical process matches the formation of levels in the ordering sequence, a hierarchical process.

It is further proposed that the explicit form of animacy and consciousness in the form of life emerges at this mean level.

The Model sets out the conditions that relate micro objects (particles), human scales and macro objects (galaxies). Our comprehension of micro-objects is grounded in Quantum Theory, of human scales in the classical physics of Newton and its extension to macro objects in the form of the General Theory of Relativity. The Model goes on to relate the quantum state of Hilbert space to that of classical physics and both to the additional concept of Implicate space necessary to explain consciousness.

The reason for the apparent incompatibility between Quantum Theory and General Relativity is considered to be due to the insistence on regarding the big bang as an absolute condition termed a singularity and in general to viewing the ordering sequence from different perspectives.

The Reference Model further implies that the difficulties in relating the scientific Standard Cosmological Model and the Standard Particle Model are due to the inability of the bulk of the scientific community to understand the relationship between local realism and non-local imagination.

The more recent modification to the Standard Cosmological Model, namely inflation, involves relating the Model to cosmological models of fractal universes and to the condition referred to in scientific parlance as space-time foam. This proposal is opposed by those who prefer to fine tune the Standard Cosmological Model. The Reference Model takes the view that chaotic inflation models correspond to more disordered conditions towards the base of the Ordering Centre or outside its influence. Models which rely on fine tuning the Standard Cosmological Model are more related to looking upwards from the mean level in the direction of increasing order, ultimately to a unified structure termed the Whole Ordering Centre.

If the latter contains the holy grail that scientists and theologians seek, it will be a theory which explains the relationship between the objective and subjective.

As has been stated earlier, the book finds the theory of evolution is the ordering process made manifest at the mean level of implicit order in the Ordering Centre referred to above.

Accordingly natural selection is a general description of the ordering process seen in terms of the explicit as opposed to the implicit.

The mean level by definition implies there exist, levels of implicit order higher and lower than that mean, of which natural selection, when seen as an explicit process, takes no account, ignoring imagination which it regards as essentially irrelevant.

The concept of survival of the fittest and what the fittest might be is then transformed when account is taken of the implicit ordering process.

The same principles apply in relation to the debate between human and artificial intelligence, the resolution of which simultaneously explains why the universe appears silent to us, without explicit evidence of conscious life elsewhere.

The Model demonstrates that existence has an explicit description inseparable from its corresponding implicit description which ties the whole sequence of orderings together as one whole. The point being made here is that science has ultimately to be related to the implicit as well as the explicit. Our sense of a singular explicit space-time with absolute validity is to be associated with viewing existence as having explicit properties only, a situation limited to a given level in the ordering sequence and one incompatible with human imagination.

When the effectively infinite range of possibilities is taken in to account neither the individual nor collective human brain for that matter is capable of comprehending the degree of compatibility available to the Implicate Order.

3.2   The Reference Model and the subjective philosophies.

The general relationship between the Reference Model and the subjective philosophies will now be briefly considered using religious and spiritual belief of a non-spatiotemporal nature in place of science as the primary subject matter. By religion is meant mainstream religious belief associated with large groups of people in a highly structured dimensional arrangement as defined by Smart (Ref 3.2.1) and discussed in some detail in Section 7.5.4.

By spiritual is meant belief held by individuals or small groups of individuals of a more personal nature and a more fluid nature in the sense that it changes more readily than that held by large groups. They commonly refer to themselves as spiritual but not religious.

Whereas science is in a constant state of flux with new or updated theories appearing frequently, religion tends to show little or no interest in change. There are exceptions of course when a new religion appears on the scene but in general the participants are even more inclined to absolute viewpoints than those scientists take about scientific principles.

Where religion is concerned about what is termed belief and more importantly imagination, science in principle is viewed as a state of non-belief or at least suspended belief.

The main elements of religion vary somewhat across the range of belief systems but many common elements are apparent from studies of comparative religion e.g. Smart (Ref 3.2.1).

The belief in good is associated in most cases, but not all, with belief in a non-personal deity (or deities) or in a personal God who creates or maintains that process.

An associated belief is in a rising upwards towards what is regarded as a heavenly state of being or a place called heaven.

A further associated belief that this state of being or heaven is to be associated with what is termed the afterlife either in the form of a one off occurrence following death or as a continuous process of alternate life and death, termed reincarnation, with a view to ultimately escaping this process and attaining heaven.

A further belief in the power of prayer or meditation as an assistance to achieving goodness on behalf of oneself or on behalf of others.

Alternatively the inverse or opposing process to the above, ultimately nihilistic in character, involving ever increasing degrees of disorder; where belief in order of any kind is abandoned to a greater or lesser extent. In particular cases nihilism may take the form of evil associated in some cases with belief in an entity referred to as the devil and a further associated state of falling into hell or being in hell as opposed to heaven.

These states of heaven and hell are generally considered to be different states of being from the earthly one, but in some cases not, hence the terms heaven/hell on earth.

Corresponding to the belief in prayer as an assistance in achieving heaven, there is a more primitive belief in incantations of one sort or another with a view to achieving the opposite effect.

In general the impression is widespread in the human population that though the forces of evil win frequent battles the forces of good win the war as it were.

The degree of commonality between the main religions constitutes in effect what may be termed a Standard Model of Religion, though this description is not used in practice, an interesting fact in itself.

Of course the individual religions add liberally to these main elements in the form of rituals, narratives, doctrines etc in addition to the practical aspects of providing places of worship and hierarchical structures associated with the human organisation of religion.

In addition there are large numbers of people with beliefs of a non-spatiotemporal nature who tend to dissociate themselves from organised forms of religion. Their spiritual beliefs are of a more personal nature related primarily to individuals or small groups of individuals.

There are again large numbers of people who are atheists/ secularists with no spiritual beliefs, or agnostics who consider these matters unknowable, or combinations of these with varying degrees of belief in matters spiritual.

Just as the Ordering Centre and its extension the Whole Ordering Centre, referred to in section 3.1, is considered to explain the objective philosophies, so it explains the subjective philosophies including religion and the spiritual.

The author will contend that as with the sciences, the entire range and content of human belief, non-belief and suspended belief can be explained in terms of the Reference Model.

In particular it will be shown the belief in heaven and ascent to it is to be associated with the ordering process and changing levels of order. Likewise hell and descent to it is to be associated with the disordering process and changing levels of disorder.

The other aspects together with the various degrees of belief are too complicated to discuss in this section but are examined in detail in subsequent sections of this book.

Just as the Model relates the theory of evolution to concepts forming at a particular level in the ordering sequence so the alternative religious belief in intelligent design can be attributed to the concept of unfolding order at many levels.

In particular the Model exposes religion's fatal insistence and reliance on the absolute, similar to the scientific insistence on measurement as absolute. The removal of this concept and its replacement with infinite regress establishes clarity in place of confusion and enhances the intelligent design thesis rather than the opposite.

In terms of the Whole Ordering Centre the intelligent design thesis is a general description of the implicit ordering process, while natural selection is a general description of the corresponding explicit process, each applying dependent on the perspective taken.

The Model implies that ignoring fluctuations, over time theists, spiritualists, agnostics and atheists will be found in roughly constant proportions in the human population, since all views can be related to the perspective of the individual within the ordering structure.

It should be noted furthermore that although religion and the spiritual commonly shows scant regard for science it is very much concerned with the practical realities of the explicit world. The main religions may view the world in terms of people and their feelings, nevertheless if they ignored the practical problems of their flocks to the exclusion of reality there would be few worshippers. Generally this illustrates the more practical nature of belief compared with imagination which is unbounded.

Again the point is being made, that the generally subjective philosophies in principle implicit, are as with scientific philosophies, and for the same reasons, spread across the entire range of the manifold ordering/disordering distribution referred to earlier. As before the Model shows that neither conscious analytical reasoning on its own or emotional awareness on its own is capable of comprehending a theory which relates the explicit and implicit, only versions with limited validity.

Similar assessments will be applied in the course of the book to philosophies other than those associated directly with mainstream science and religion including politics, economics, mathematics, the arts and literature, with the same results namely that their contents can be explained with equal facility in terms of the Reference Model.

Fig 3.2.1 illustrates, admittedly in a simplified way but sufficient for general purposes, the main difference between science, any given subjective philosophy and the Reference Model in their treatment of the total sum of human records and the range of philosophy in general.

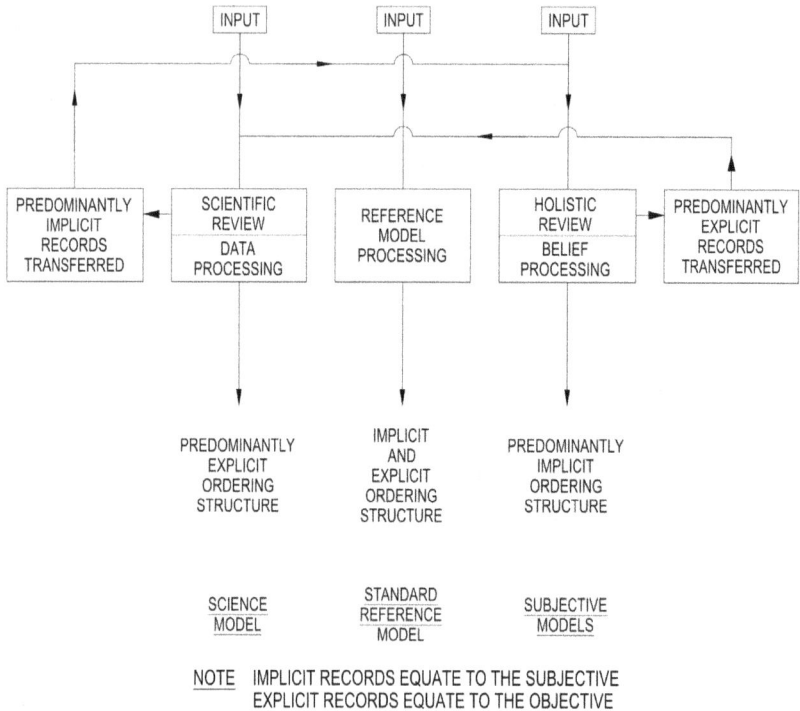

FIG 3.2.1   COMPARISON OF SCIENCE, THE SUBJECTIVE AND THE REFERENCE MODEL IN THEIR TREATMENT OF HUMAN RECORDS

4.0　THE STANDARD REFERENCE MODEL.

4.1　General.

The Reference Model is an overall description of the processes whereby low degrees of ordering are transformed into ever higher degrees of order and the inverse process of disordering, whether random or chaotic in character.

No initial assumptions are used in the compilation of this description as our concepts of time and space merely form part of the processes involved. In other words there are no absolute beginnings or ends from an overall perspective.

All of the aspects involved can be considered therefore to have primarily relative validity or primarily absolute validity depending on the viewpoint. To the extent that relative validity applies there is no limit to the degree of disorder or to the degree of order involved, i.e. existence as a whole has no boundary and accordingly no boundary condition. Correspondingly to the extent that criteria in an absolute sense are accepted appropriate boundaries and boundary conditions apply.

The Model has to put together in one coherent and integrated whole the elements listed in the earlier section 2.0 under the heading, 'The basic principles of ordering and disordering processes'.

The reader is therefore encouraged to think of a rich picture when trying to comprehend the Model and to realise that individual opinions are solely that, as opposed to the overall perspective for which the closest description we have is that held by the human race as a whole.

4.2　Randomness, chaos and the path to order.

The relationship between random, chaotic and ordered states has already been discussed in general terms in Section 2.0, but now a more detailed description is necessary.

Scientific comprehension of the random, stochastic, indeterminate state is that of unstable quantum fluctuations associated with the scientific concept of space-time foam; the corresponding religious concept is that of the formless void. The unstable quantum fluctuations represent a state of negligible energy density which the Model views as being directionless such that the notions of order and disorder are rendered meaningless.

In this case the random state swamps the formation of both order and disorder. In fact ordered and disordered states require the corresponding random state to exist and bear the mark of its influence, as will become evident.

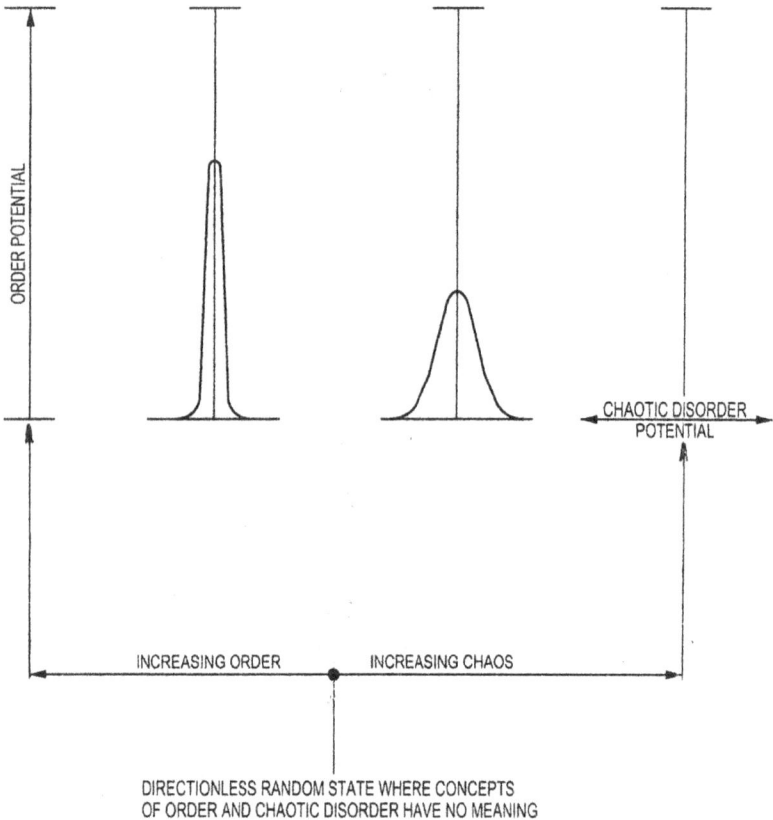

ORDER POTENTIAL

CHAOTIC DISORDER POTENTIAL

INCREASING ORDER   INCREASING CHAOS

DIRECTIONLESS RANDOM STATE WHERE CONCEPTS
OF ORDER AND CHAOTIC DISORDER HAVE NO MEANING

NOTE   THE REVERSE OF THESE PROCESSES ALSO APPLIES IN THIS AND SUBSEQUENT
FIGURES WITH THE ARROWS IN THE OPPOSITE SENSE TO THAT SHOWN.

FIG 4.2.1   THE BALANCED GROWTH OF ORDER AND CHAOTIC DISORDER
RELATIVE TO THE RANDOM STATE

The balanced growth of order and chaos and by implication the opposite processes, all relative to the random state is shown in Fig 4.2.1. The space-time foam conditions are represented in the figure by a random discrete point centred so that the tendency to create order is in balance with that creating chaos. The directionless unbalanced states correspondingly can be represented in the figure by random oscillations of the order and chaotic potential arrows in any direction about the random point.

The path to order from the order potential involves states of ever-increasing energy density, shown as a vertical line in this figure, and its dissipation forming a distributed, i.e. indiscrete state represented in the figure by a horizontal straight line, the potential for chaotic disorder.

The completely distributed state corresponds to the concept of a uniform scalar field.

The relationship between order, chaos and the random state is discussed in more detail in Section 7.1. The apparently unstable nature of the interactions between the fluctuations associated with space-time foam conditions and the essentially uniform scalar field is the subject of debate in cosmological circles and will be discussed in relation to the Reference Model in Section 4.7.

However it is agreed that the situation converts in whatever manner to the more stable conditions described by what is termed the Standard Cosmological Model.

The next state to which we can refer to and shown in Fig 4.2.2 (next does not refer to passage in time) is that where the random state interacts with the order and chaotic potentials giving rise to wave characteristics, which we would regard in everyday terms as high amplitude short wavelength and low amplitude long wavelength undulations, the former sharp and discrete, the latter in comparison distributed and non-discrete. In this connection we can regard discreteness as relating to the present time, which will be termed the present state and non-discreteness as relating to difference and similarity, forming a level which will be termed the variance or distributed state.

In effect we are unfolding the present state from any given state of distribution (non-discreteness) and conversely unfolding distribution from any given present state (discreteness) as portrayed in Fig 4.2.3.

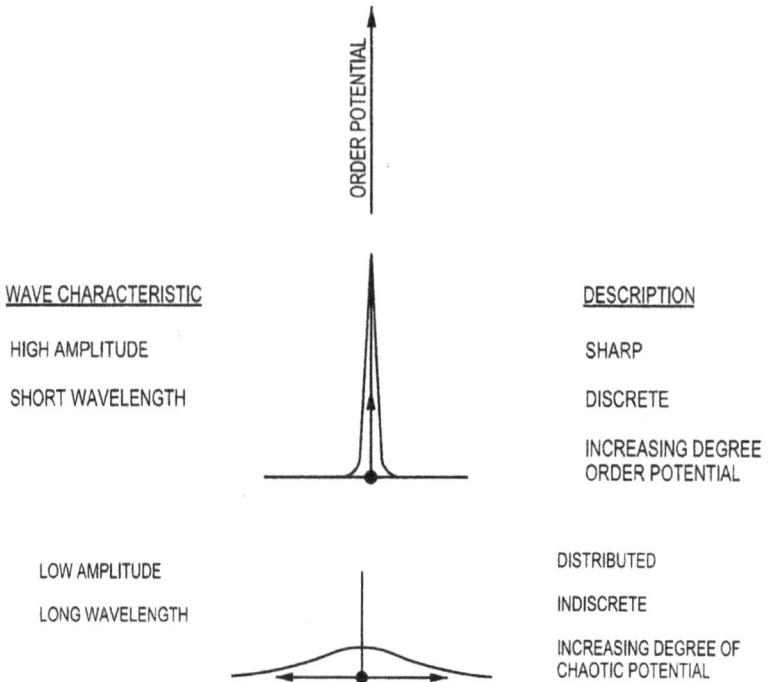

FIG 4.2.2   DEVELOPMENT OF WAVE CHARACTERISTICS RELATIVE TO THE RANDOM STATE

The distributed state forms a base for what becomes an ordering structure while the present state elevates that structure.

Considering that we are dealing with a multiplicity of waves in all possible orientations we can remind ourselves that relative to a given orientation in phase waves reinforce and out of phase waves cancel, as shown in Fig 2.7.1 (Section 2.0). Waves can of course also interfere resulting in partial reinforcement or partial cancellation.

Most waves are out of phase and therefore to a greater or lesser degree cancel, this can be considered to represent the disordering processes since there are many more ways of disordering existence than there are ways of ordering it.

Such waves transfer energy in a manner analogous to transverse waves which occur when a flexible string under tension is displaced or plucked from a static condition and quickly released. Transverse waves on a string consist of individual travelling waves which propagate

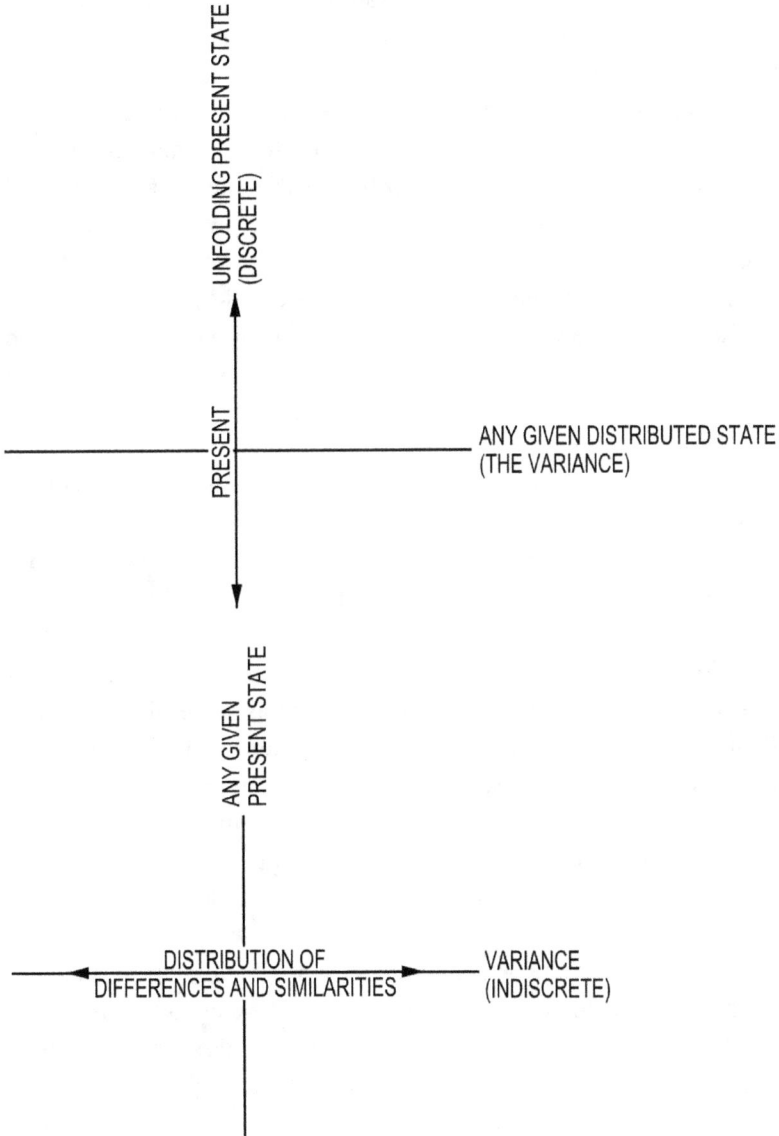

FIG 4.2.3   THE DEVELOPMENT OF THE PRESENT STATE AND THE VARIANCE
(THE DISTRIBUTION) FROM WAVE CHARACTERISTICS

along the string and twin waves of identical amplitude and frequency travelling simultaneously in opposite directions on the string, termed standing waves since the wave pattern does not advance along the string. Standing waves create conditions of zero amplitude on the string, namely nodes and maximum amplitude namely antinodes, involving resonant frequencies or harmonics from the first fundamental mode upwards, all of which forms the basis for musical instruments. Wave forms centred on the wave peaks or antinodes are symmetrical, those centred on the wave minima or nodes are anti-symmetrical.

The relation between sinusoidal travelling waves on a stretched string and wave disturbances representing standing waves on that string is described for example by Elmore and Heald (Ref 4.2.1) and by other freely accessible references on the internet.

Standing waves contribute to the present state while travelling waves contribute to the distributed state.

The wave reinforcement and cancellation processes relative to a given peak energy density, are then as shown in Fig 4.2.4, which is a region of stability in comparison to the unstable phase where quantum fluctuations are predominant. The distributed base state is the condition where waves effectively cancel out. Waves may have all amplitudes and all wavelengths but those with low amplitude and long wavelength are positioned nearer the base state while those with high amplitudes and short wavelength take positions nearer the discrete peak of the overall waveform. This process can be related to the formation of the normal (Gaussian) distribution as discussed in more detail in Section 7.3 of this book.

Fig 4.2.5 shows the ordering structure or what may be termed the ordering envelope for a given energy density. The envelope indicated is made up of the lines joining the peaks of the sub waves which reinforce each other to form the waves on the central axis at each level as shown in the earlier Fig 4.2.4. The centreing line of reinforcement at the peak energy density represents the present state.

The convention in this and further figures is to show order increasing to the top of the page and disorder vice versa.

Summarising what has been said so far is that there is a balance between the ordered state, the random state and the distributed state. For every coherent state of order there is an incoherent state of disorder.

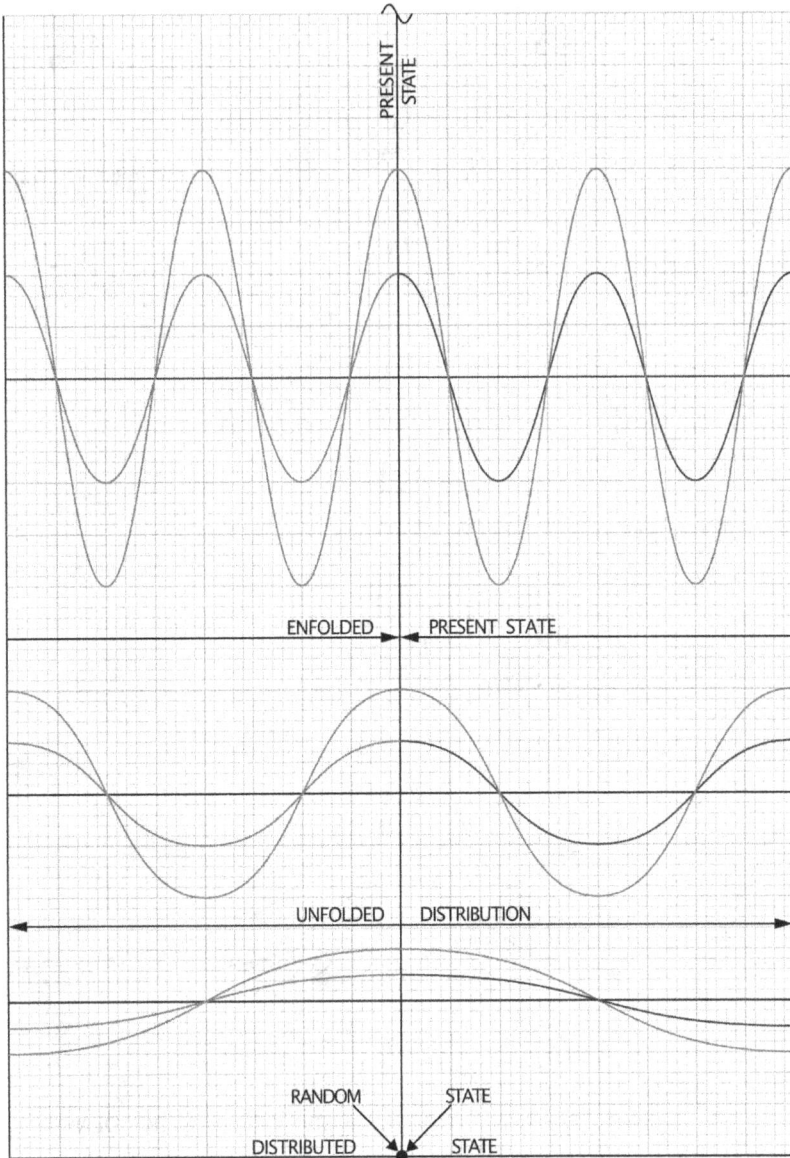

FIG 4.2.4   WAVE REINFORCEMENT AND CANCELLATION PROCESS IN THE
FORMATION OF THE PRESENT STATE AND DISTRIBUTED STATE

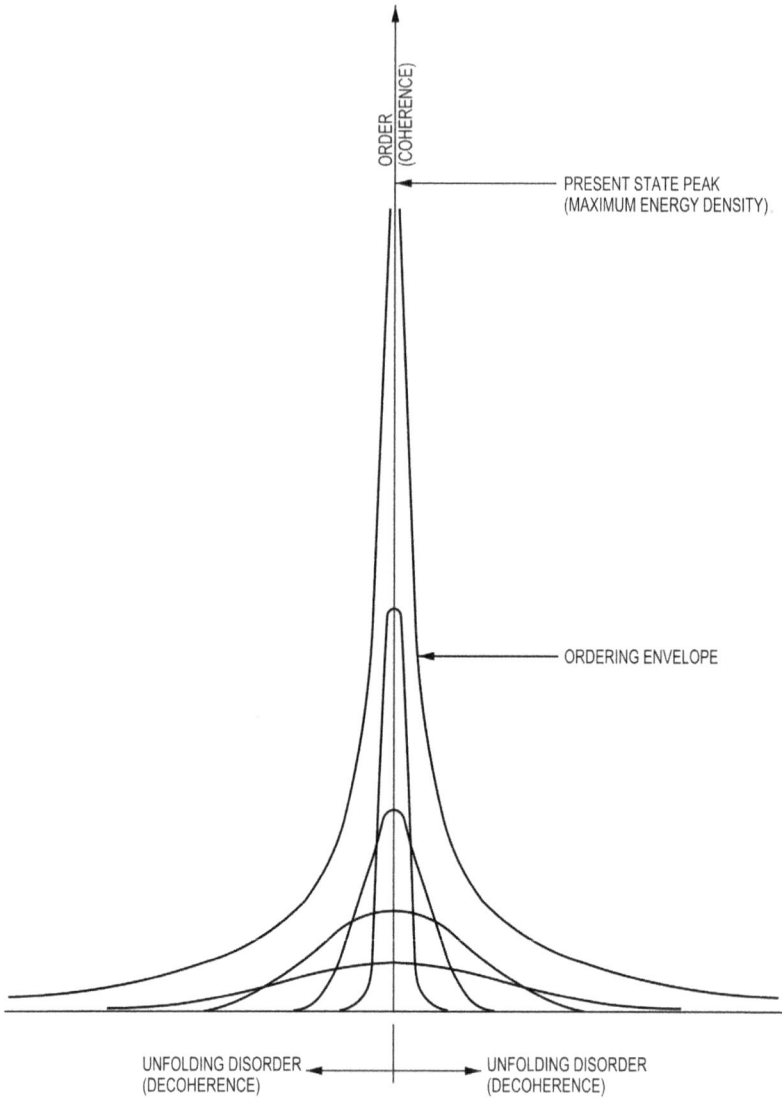

FIG 4.2.5   THE ORDERING ENVELOPE FOR A GIVEN PEAK ENERGY DENSITY

When this balance collapses there is no order and the random state rules the roost.

4.3 The ordering envelope and the Standard Cosmological Model.

The ordering envelope is clearly related to the familiar concept of the Big Bang state.

So it is useful at this point to consider some basic aspects of our understanding of spatial dimension and its relationship with time, as shown in Fig 4.3.1.

Fig 4.3.1 (a) shows an unfolded space S in two dimensions at a given time T. The disc can be replaced with a sphere in three dimensions. The space S can be enfolded to a point also at time T as shown in Fig 4.3.1 (b), the two states being equivalent. For such an enfolded

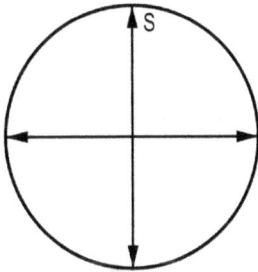

(A) UNFOLDED SPACE S IN TWO
DIMENSIONS AT A GIVEN TIME T

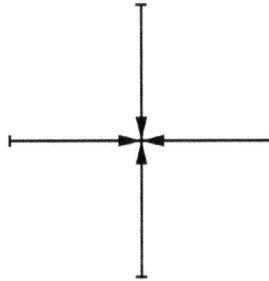

(B) ENFOLDED SPACE S
AT A GIVEN TIME T

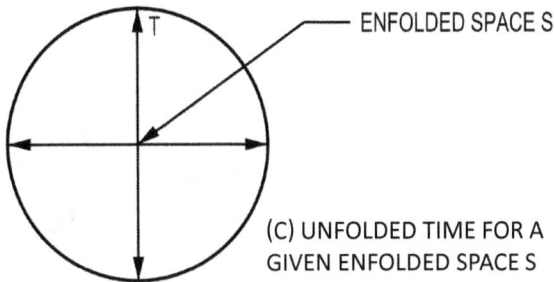

ENFOLDED SPACE S

(C) UNFOLDED TIME FOR A
GIVEN ENFOLDED SPACE S

NOTE   FOR 3 DIMENSIONS REPLACE DISC WITH SPHERE

FIG 4.3.1   UNFOLDED SPACE VERSUS UNFOLDED TIME (SPACE DECOUPLED FROM TIME)

space S we can show unfolding time as shown in Fig 4.3.1 (c). In each case time and space are to be considered independent.

However the Theory of Relativity emphasises that space and time are interrelated such that there is only one stuff namely space-time. So the process shown in Fig 4.3.1 can be repeated on this basis as shown in Fig 4.3.2.

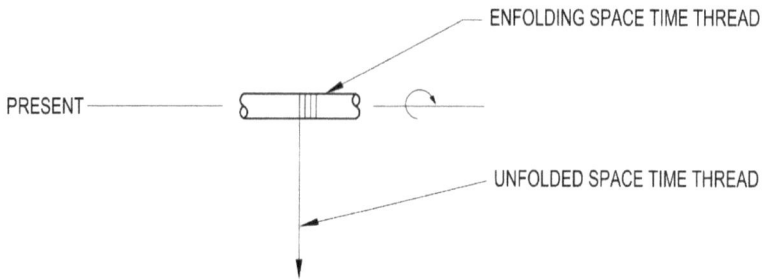

(A) UNFOLDED AND ENFOLDED DIMENSION RELATIVE TO THE PRESENT

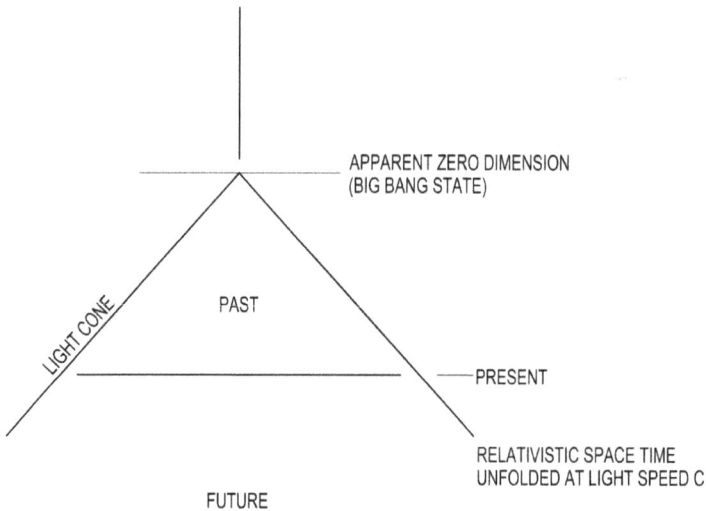

(B) UNFOLDED OBSERVABLE DIMENSION RELATIVE TO THE BIG BANG STATE

FIG 4.3.2   UNFOLDING AND ENFOLDING OF OBSERVABLE DIMENSION
            RELATIVE TO THE PRESENT AND APPARENT BIG BANG STATE
            USING THE ANALOGY OF A THREAD WOUND ON A REEL (SPACE
            COUPLED WITH TIME)

Fig 4.3.2 (a) shows the process of unfolding relativistic space-time from an enfolded state or indeed the reverse, a state of enfolded four-dimensional space-time, like winding a thread on to a reel. Bohm (Ref 2.5.1) used a mechanical stirring device in to which ink droplets were inserted to demonstrate this folding/unfolding process.

Fig 4.3.2 (b) portrays the unfolding of observable (explicit) four dimensional space-time from what is referred to as the Big Bang state, shown as a cone which expands at the speed of light in accordance with General Relativity and what scientists refer to as the Standard Cosmological Model, which incorporates in addition the cosmic expansion of space. The Big Bang state implies a singularity, but this is an absolute assumption which the Reference Model shows to be inconsistent with the principle of no absolute boundaries. It is therefore regarded as only an apparent state of zero dimension in the Model.

Fig 4.3.3 uses the principle of the ordering envelope in terms of the observed unfoldment, a process which commenced approximately 13.8 $\times$ 10^9 years ago. The Standard Cosmological Model is described in some detail in Section 7.4 of this book, together with relevant references to assist the reader.

Fig 4.3.3 is important to the thesis of this book since it introduces the concept of the mean level of explicit order in this (explicit) ordering process. The mean level corresponds to the standard normal Gaussian distribution as shown in the figure with which the reader may be familiar.

In accordance the known physics mass diffusion increases as order increases (entropy decreases) while mass aggregation increases as order decreases (entropy increases) to the point where black holes with enormous entropy form at extremely low temperatures ultimately evaporating and, as some argue, potentially destroying the information associated with the aggregation. Also in accordance with known physics increasing distribution is to be associated with increasing particle diffusion as entropy increases. Increasing order corresponds to increasing coherence and vice versa increasing disorder corresponds to a process of decoherence. Dark matter and attractive gravity dominate towards the present state peak (the apparent big bang state) while dark energy and repulsive gravity dominate towards the distributed base state, all as further discussed in later sections.

This mean level correlates with the dark matter/dark energy balance which is now the subject of cosmological debate. So the figure should be interpreted as applying to cosmological scales.

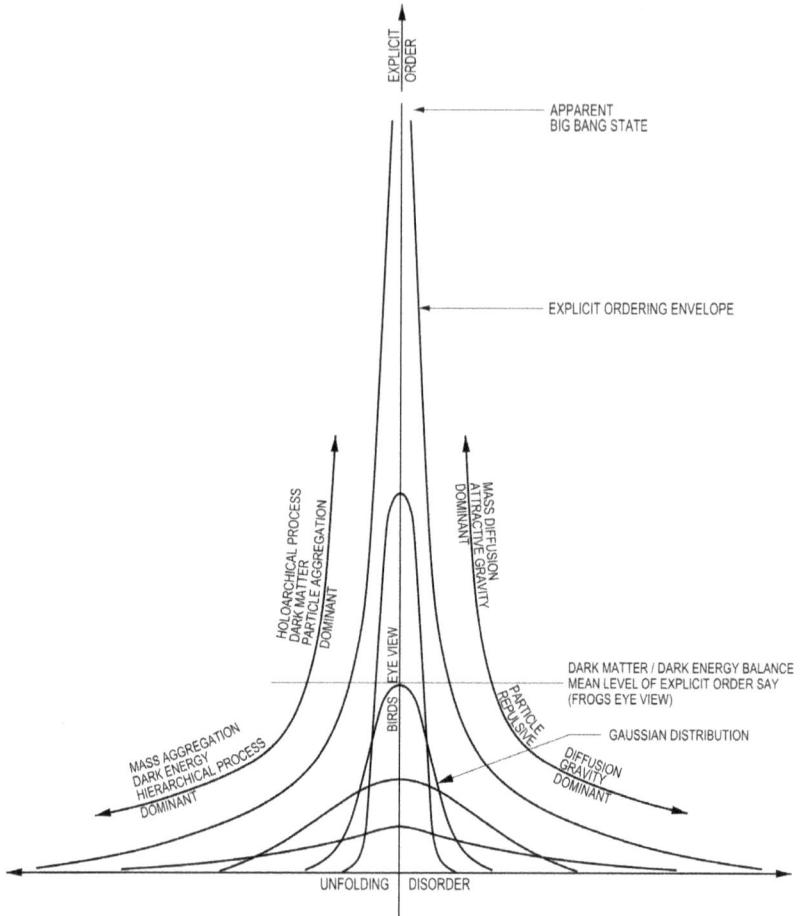

FIG 4.3.3   THE BIG BANG STATE AND THE MEAN LEVEL OF EXPLICIT ORDER (FROGS EYE VIEW) VERSUS MULTIPLICITY OF LEVEL (BIRDS EYE VIEW). COSMOLOGICAL SCALES APPLY

In particular the ordering envelope is regarded as being primarily holoarchical in character above the mean level and primarily hierarchical below that level.

This is further illustrated in Fig 4.3.4 where the mean level of explicit order corresponds to the condition when the hierarchical process associated with distribution is in balance with the holoarchical process associated with the present state. This represents what may be

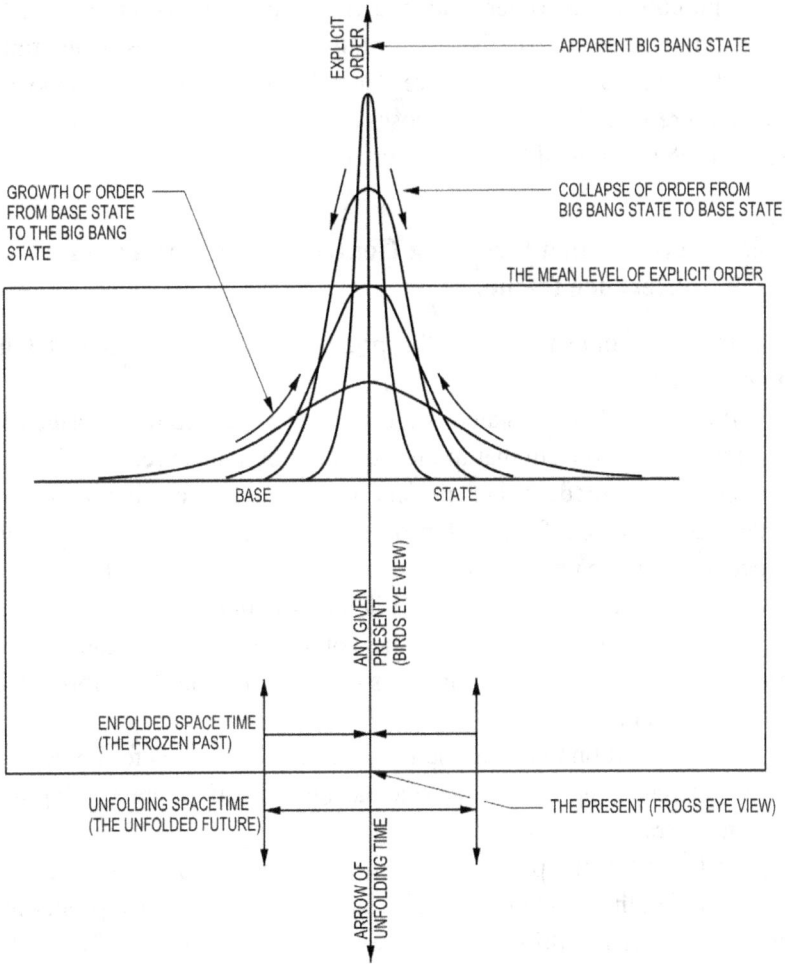

FIG 4.3.4   THE MEAN LEVEL OF EXPLICIT ORDER AND THE ARROW OF
             UNFOLDING TIME

termed the average state of the ordering distribution under the envelope
previously referred to.

The figure shows the enfolded past and unfolding future more
disordered than the past all relative to the average state of the Ordering
Centre which represents the present.

The figure also shows the arrow of unfolding time in accordance
the above description.

The collapse of order from the Big Bang state to the base state represents the action of the laws of thermodynamics as space-time unfolds from the past to the future. The growth of order in the opposite sense represents that which opposes the laws of thermodynamics as space-time is enfolded from the future to the past.

4.4   The unified Ordering Centre and the preferred reference frame.

We are now going to restate the principles given in section 4.3 in another way.

It has already been stated that order can be related to the principle of sequencing and from that in turn to a distribution curve.

Order is better described as consisting of a discrete pair of enfolded orders the one a differential process and the other an integration process, each separated from a centred line representing the path to greater or lesser order as shown in Fig 4.4.1(a) and (b).

The centred line represents the condition for which there is in relative terms no dimension in terms of space or time. It performs the function of a mirror.

The separation between the two waves corresponds to the degree of decoherence applicable, ever-increasing separation representing the path to the random state.

In Fig 4.4.1, the primarily differential (continuous) properties are shown as full lines and the primarily integral (discontinuous) properties as dotted lines. The differential curve Fig 4.4.1(a) shows the frequency distribution with the norm or condition of greatest similarity in the centre and the conditions of greatest difference from the norm to either side. In this case difference is said to be unfolded from the centre while similarity is enfolded at the centre.

It is important to note that the differential curve should be associated with continuity and the integral curve with discontinuity. The integral curve Fig 4.4.1(b) shows that the condition of greatest difference moves to the centre with that of greatest similarity to either side. In this case similarity is said to be unfolded from the centre while difference is enfolded at the centre.

(A) UNFOLDED DIFFERENTIAL DISTRIBUTION CURVE

(B) UNFOLDED INTEGRAL DISTRIBUTION CURVE

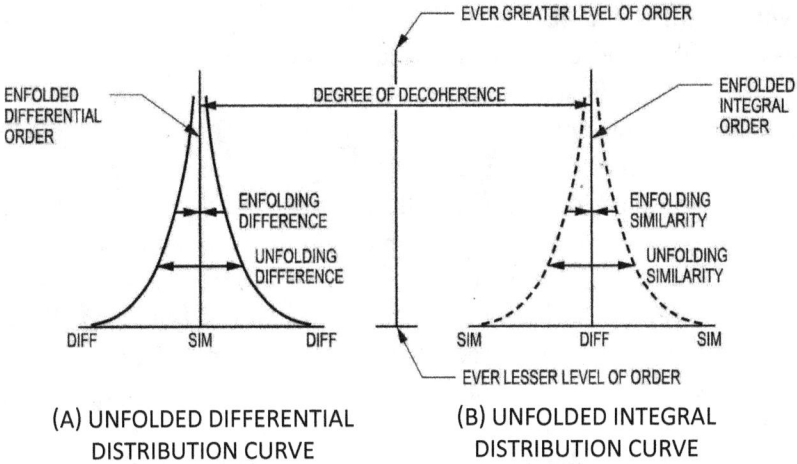

UNREINFORCED (DISCORDANT) DISTRIBUTION CURVES

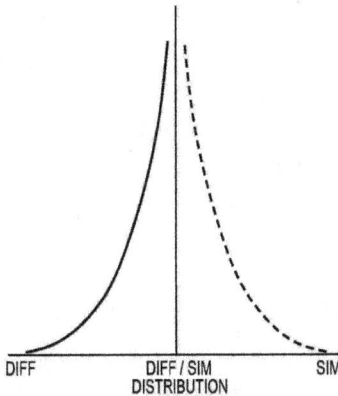

(C) REINFORCED (SUPERIMPOSED) DISTRIBUTION CURVES

NOTE 1   ORDER GROWS WHERE PROCESS OF DIFFERENTATION IS IN BALANCE WITH PROCESS OF INTEGRATION

NOTE 2   CONVENTION USED IS TO SHOW ORDER / DISORDER TO TOP / BOTTOM OF PAGE ALSO DIFFERENCE / SIMILARITY TO LEFT / RIGHT OF PAGE

FIG 4.4.1   THE COMPOSITION OF THE UNIFIED ORDERING CENTRE

Fig 4.4.1(c) shows the how a unified Ordering Centre is formed with the differential and integral curves superimposed. This figure is based on the concept of a path to order based on a preferred reference frame whereas Figures 4.4.1 (a) and (b) relate to states where that concept does not apply.

The convention as indicated by the figures is to use the left-hand side of the Centre to represent differentiation and the right-hand side integration, for reasons introduced in Section 2.7 and reinforced by the theory of consciousness put forward in this book.

The primary thesis being put forward here is that in terms of Fig 4.4.1 differentiation is a process of unfolding difference while enfolding similarity and inversely that integration is a process of unfolding similarity while enfolding difference.

It should be noted that Fig 4.4.1 also indicates a means of addressing the second dichotomy in that the peak of the Centre implies an association with the concept of absolute truth with relative truths lying to either side.

The Reference Model uses the verticals in the Ordering Centre to represent a set of ordering sequences made up of differences and similarities and by using the horizontals to represent the spread or distribution of the difference/similarity sequences. The system of vertical and horizontal lines so formed within the Ordering Centre can be likened to a grid when related to the coordinates in a Cartesian sense.

The peaks in the vertical and horizontal wave motions are considered to give rise to our sense of the absolute while the lines to either side of those peaks give rise to our sense of the relative.

The intersection of vertical and horizontal axes and the distribution curve can be viewed explicitly as a Cartesian (absolute and measurable) system. Alternatively the fractal aspect (allowing each line to become pairs or pairs to become a single line again) can be added to the holographic nature of the unfolding and enfolding of opposite characteristics about such lines, together with the curvature imposed by unfolding and enfolding wave motion.

Then the description alters to a non-Cartesian (relative and immeasurable) system with the characteristics of a holographic fractal as implied in Fig 4.4.2. The fractal description implies there is no

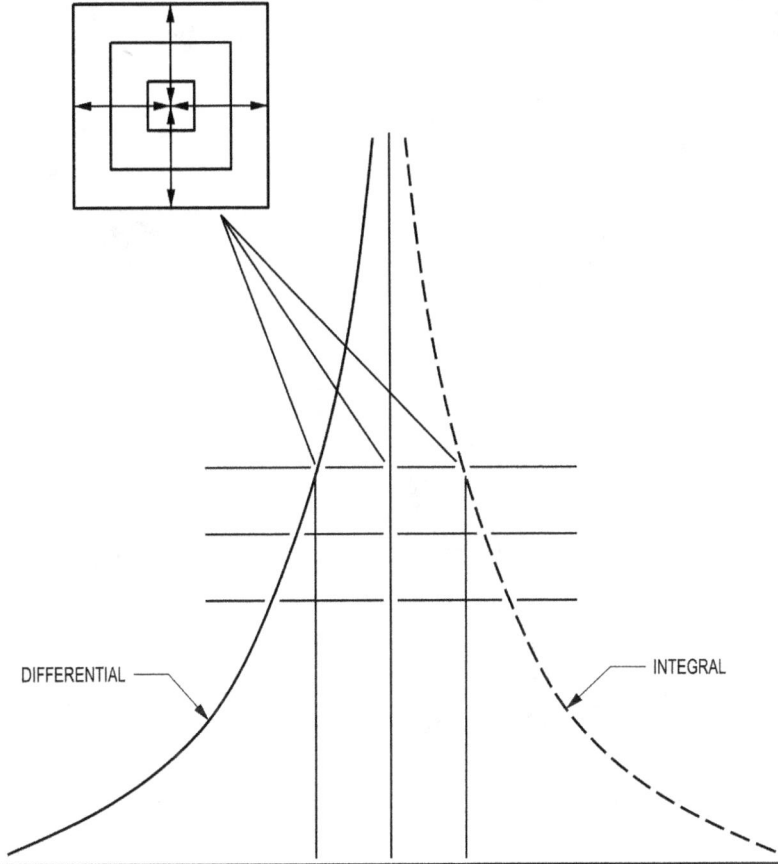

NOTE ALL VERTICALS REPRESENTING PRESENT STATE,
ALL HORIZONTALS REPRESENTING LEVELS OF ORDER AND
ALL POINTS IE. INTERSECTIONS OF PRESENT STATE AND
LEVEL HAVE AN EXPLICIT CARTESIAN DESCRIPTION AND
AN IMPLICIT FRACTAL DESCRIPTION

FIG 4.4.2   IMPLICIT FRACTAL NATURE OF THE UNIFIED ORDERING CENTRE
COMPARED TO THE CARTESIAN DESCRIPTION

absolute peak to the distribution, which allows the order potential to
take other values, as discussed later in section 4.7. In general terms the
Ordering Centre can be described as an entirely simple unfolding and
enfolding repeating pattern, herein termed a holographic fractal, which
gives rise to an ever more complex ordering of difference and similarity.

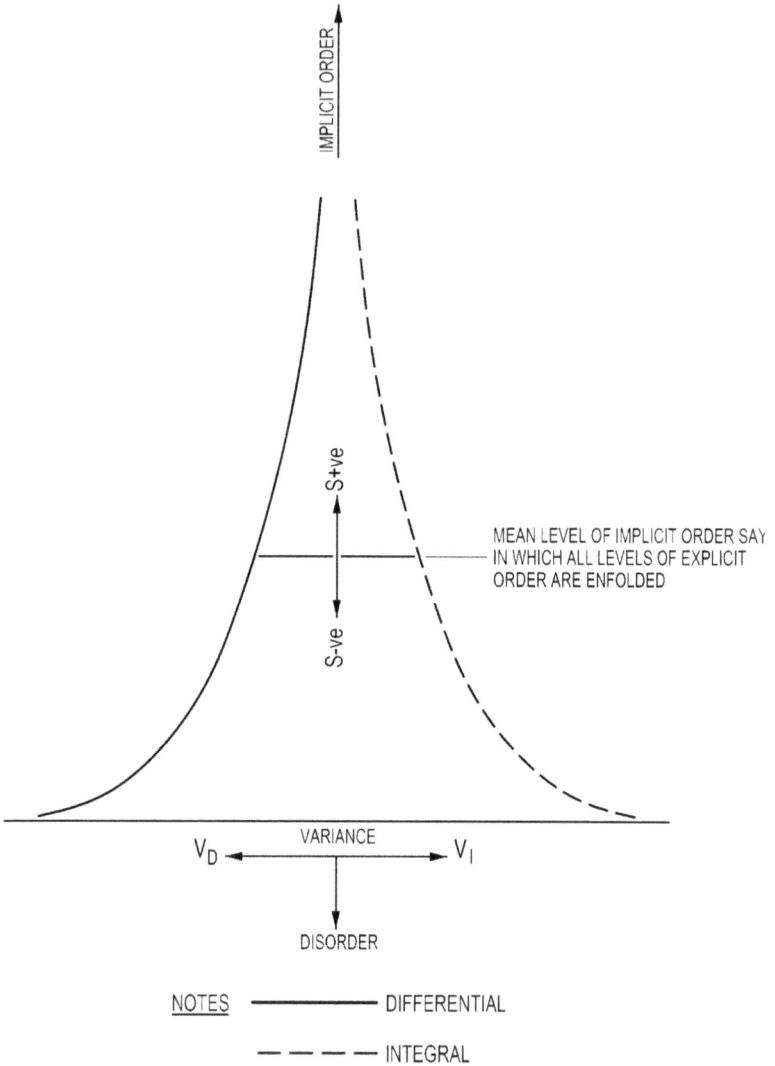

FIG 4.4.3   UNIFIED ORDERING CENTRE SHOWING ORDER AND VARIANCE
RELATIVE TO MEAN LEVEL OF IMPLICIT ORDER

The primary definitions relating to the Ordering Centre are given
in Fig 4.4.3.

Vertical lines (ordinates) under the distribution curve can be
regarded as measures of implicit order (s for sequence) relative to the

base of the figure. Horizontal lines (abscissae) from side to side of the curve can likewise be regarded as measures of chaotic or explicit disorder relative to the peak of the curve. The horizontal distribution is termed the variance (v).

The s and v axes form the Ordering Centre reference frame.

So s + ve and s − ve correspond to implicit order and implicit disorder respectively; increasing variance corresponds to explicit disorder.

Decreasing variance towards a state of zero variance corresponds to explicit ordering.

Implicit and explicit order corresponds to coherence while implicit and explicit disorder corresponds to decoherence.

Explicit disorder equates to the operation of the laws of thermodynamics, while explicit ordering equates to the increasing complexity of the material substrate from the Big Bang state to the present time apparent to human observers. That increasing complexity in turn equates to the process of evolution referred to as natural selection and explained further in Section 7.2.2.

It will be shown in the following sections that implicit ordering is associated with processes leading to animacy and consciousness.

The measures of order on the differential side are shown under the full curve while those on the integral side are shown under the dotted curve.

The overall structure reveals unfolded order can be regarded as enfolded disorder and inversely unfolded disorder as enfolded order, order being the inverse of disorder.

The variation of wave peaks in the vertical direction distributes order and disorder, while wave motion in the horizontal direction away from the central axis distributes differential variance to the left and integral variance (similarity) to the right. These properties of variance shown in Fig 4.4.3 can be related to Fig 2.5.1. Differential/integral variance corresponds to the negative/positive values of x in the earlier figure.

In particular the unified Ordering Centre has a mean level of implicit order, whereas previously we have referred to the mean level of explicit order. The explanation from the fractal description is that all levels of the four-dimensional explicit order including the mean level as described in Section 4.3 are enfolded in the mean level of implicit order.

We have now moved from the Big Bang model of the ordering process entirely local and explicit in character, to a new non-local model both explicit and implicit in character.

Nevertheless we can still describe the characteristic shown in Fig 4.4.1(c) in an explicit manner as follows. In the case of curve (c) the greatest degree of difference lies on one side, greatest similarity on the other, while complex combinations of difference and similarity lie towards the centre. When the two waves combine the resulting wave peaks correspond to a state with a very high degree of potential energy, in a state of thermodynamic instability; when this order loses potential energy the kinetic energy so released distributes or unfolds its differential order content in the form of difference and correspondingly unfolds its integral order content in the form of similarity. Together these unfoldments take the form of complex wave patterns reinforcing, cancelling and interfering. Where reinforcement by super positioning occurs manifestly ordered sequencing is established giving rise to ever higher degrees of sequential order. Inversely where the wave patterns tend to interfere or cancel out, which is by far the most probable, increasingly chaotic disorder applies to the point where the kinetic energy, initially free energy i.e. unequally distributed, converts to a state of thermal equilibrium with the energy uniformly distributed, the so-called heat death. The peak order is a state of very low entropy and the heat death very high entropy.

Fig 4.4.4 shows order and variance associated with the implicit order unfolding and enfolding in an Ordering Centre, in which all levels of the four-dimensional explicit order are enfolded in the mean level of implicit order referred to simply as the mean level of order.

The enfoldment and unfoldment of the Ordering Centre to and from the vertical and horizontal axes of the preferred reference frame can be related to the principle of the holographic fractal.

Fig 4.4.5 illustrates clearly the expansion of distribution to give level of order and the corresponding contraction to give the present state, all on the Implicit Ordering Centre envelope shown in the previous Fig 4.4.3.

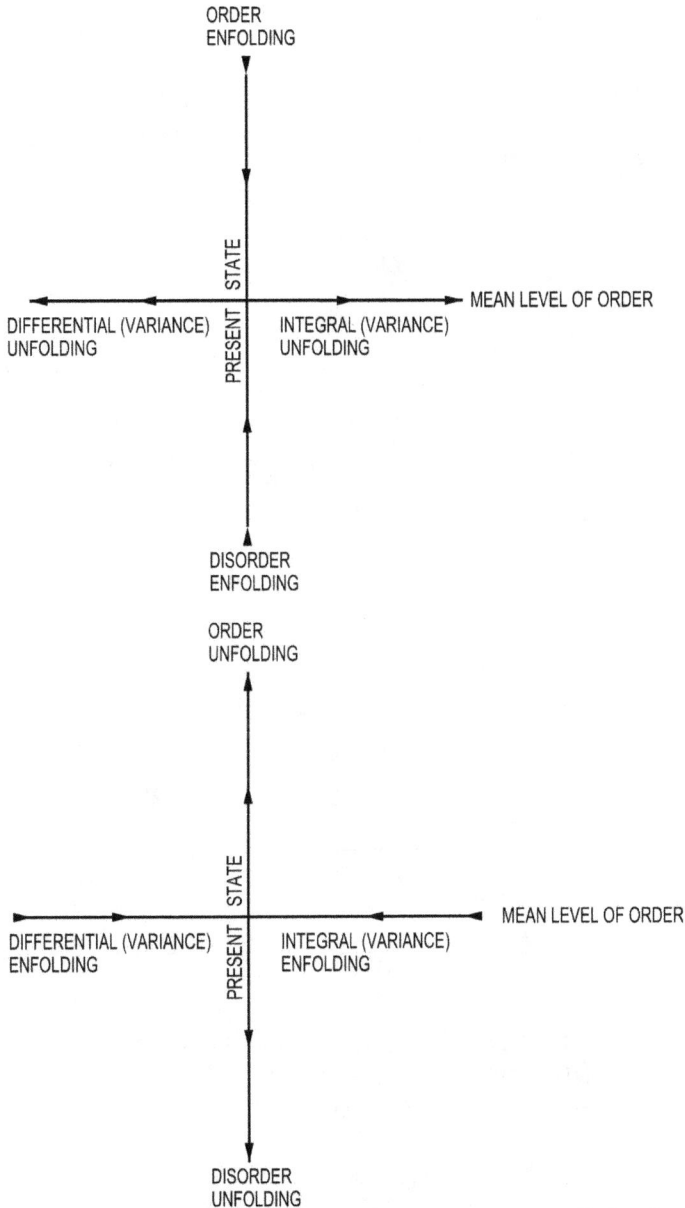

NOTE ALL LEVELS OF EXPLICIT ORDER ARE ENFOLDED IN
MEAN LEVEL OF ORDER IN ORDERING CENTRE

FIG 4.4.4 THE ORDER / VARIANCE UNFOLDING / ENFOLDING PROCESS IN
AN ORDERING CENTRE

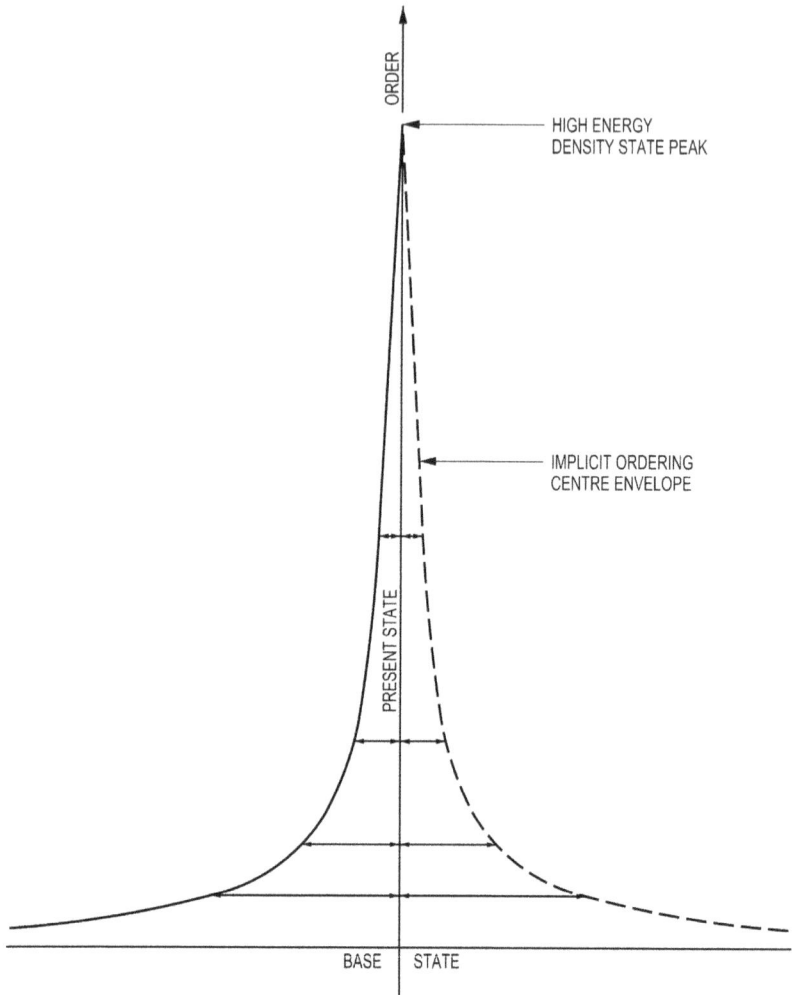

FIG 4.4.5   EXPANSION OF DISTRIBUTION TO GIVE LEVEL OF ORDER AND
CONTRACTION TO GIVE THE PRESENT STATE ON THE IMPLICIT
ORDERING CENTRE ENVELOPE

Fig 4.4.5 should also be compared to the explicit Cartesian envelope shown in Fig 4.3.3.

Fig 4.4.6 shows the ordering/disordering process in three different states corresponding in principle to those shown in the earlier Fig 4.2.4.

A represents the degree of ordering and B the degree of distribution.

In the first case the present state is dominant, with A clearly greater than B.

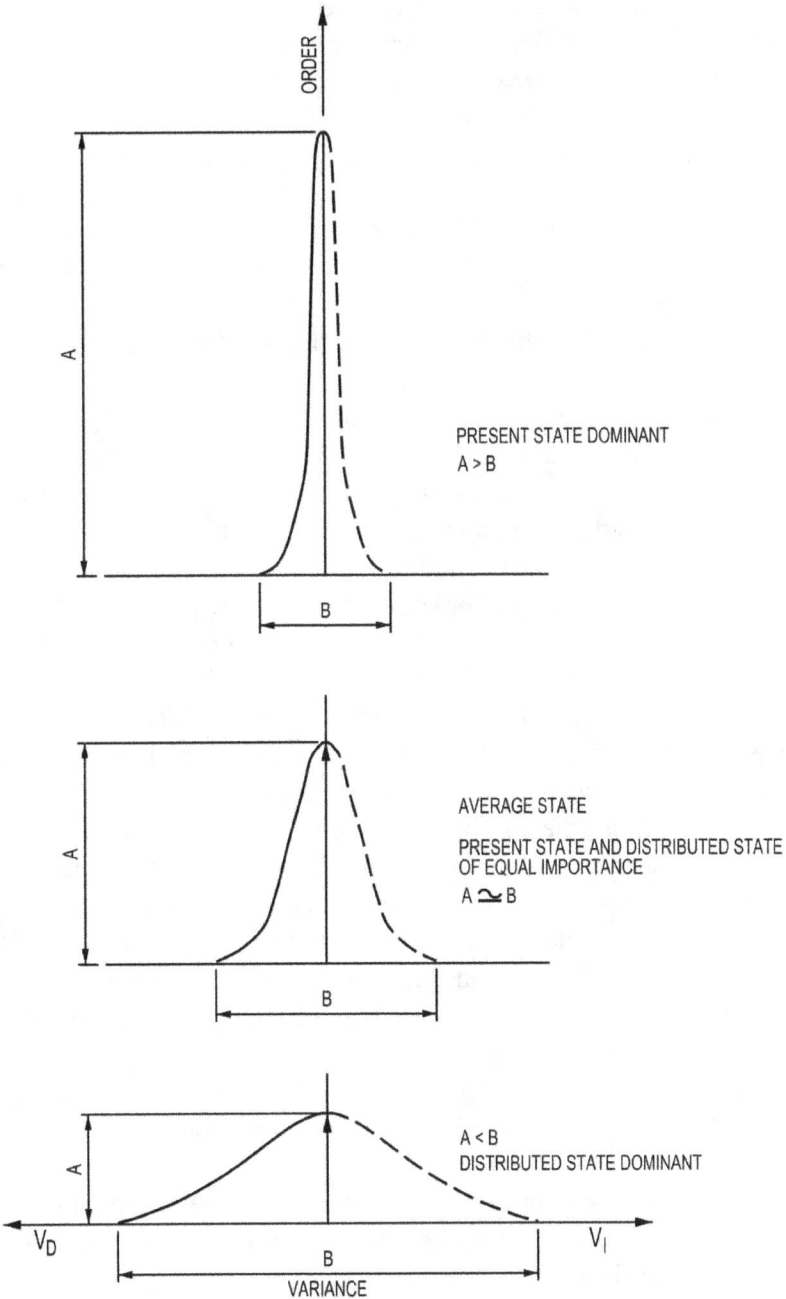

FIG 4.4.6   THE RELATIVE IMPORTANCE OF THE PRESENT STATE AND
DISTRIBUTED STATE WITH DEGREE OF ORDER AND VARIANCE

This favours imagination in the form of the present state at the expense of reality and large vertical components imply movement by enfoldment towards the vertical symmetric axis.

The second case shows the average state with the present state and distributed states of equal importance, A and B being approximately equal.

This favours reality and imagination acting in concert.

The third case shows the distributed state dominant, with area A clearly less than B.

This favours reality in the form of level (distribution) at the expense of imagination and large horizontal components imply movement by unfoldment towards the horizontal axis.

It is therefore contended that as with mathematics, ordering and disordering stuff has real and imaginary components and that at the mean level of order these components are in a state of balance.

This is the level at which the unfolding of variance has the same importance as unfolding of ordering i.e. where the formation of level by the hierarchical process matches the formation of present state through ordering, the holoarchical process.

Above the mean level imaginary components dominate at the expense of reality and vice versa below the mean level the real components dominate at the expense of imagination.

It is clear that the distributed state emphasises not only reality but also the transformation of waves from those of high amplitude and short wavelength to those of low amplitude and long wavelength.

It is equally clear that the present state emphasises not only imagination but also the transformation of waves from those of low amplitude and long wavelength to those of high amplitude and short wavelength.

It is the reductionist process from the above to the classical state which in an immediate sense is responsible for our common sense view of reality.

The reductionist process related to reality increases in influence towards the base of the Ordering Centre and vice versa the imaginary process increases in influence towards its peak.

It will be shown that Figures 4.4.5 and 4.4.6 provide keys essential to unlocking the nature of animacy and consciousness as opposed to the inanimate, involving imbalance between difference and similarity. It is this imbalance which provides the basis for the divergence from the inanimate as explained in the next Sections 4.5 and 4.6.

The unified description given above is based on the principle of a preferred reference frame, which can be traced back to the random state when difference and similarity are maintained in balance. Where this is not the case increasing decoherence as shown in Figure 4.4.1 provides the myriad of alternative unbalanced states which collapse order in an increasingly directionless manner culminating again in a return to the random state.

The reader should note all these alternatives remain in play.

4.5   Phase space v Hilbert space.

Antisymmetry v symmetry.

The mean level of implicit order and classical physics.

Symmetry v asymmetry.

The description thus far of the Big Bang explicit ordering and distribution all as shown in Figures 4.3.1 thru 4.3.4, must now be related to the Implicate order in a more detailed manner by relating the four-dimensional cosmological model which extends Newtonian to Einsteinian physics allowing for the curvature of space-time, to multidimensional concepts in both classical terms and the non-classical quantum state all in terms of the Reference Model.

Multidimensional classical physics, derived in particular from the work of the mathematicians, Lagrange and Hamilton, is based on the atomistic concept of individual particles each having their own three-dimensional space. Physics therefore allows for 3n dimensional space termed configuration space and 6n dimensional space termed phase space when the momentum of the particle is included, where n is the number of particles involved. When dealing with a very large number of particles the number of dimensions grows accordingly up to infinity. This can be considered the equivalent of viewing all the explicit levels in the ordering envelope as one continuous whole.

A basic mathematical description of the above is given in Section 7.3 with appropriate references to aid the reader.

There is a further multidimensional concept to be considered, this time based on waves and wave motion namely the quantum theory, all housed in what is termed Hilbert space.

The background to the discrepancies between observations by experimental scientists in the late 19th and early 20th centuries and that of classical physics which precipitated the formulation of quantum mechanics and its development through the work of theoretical physicists including Bohr and Schrödinger is discussed in Section 7.4 of this book, for the benefit of the reader, again with appropriate references.

The reader will discover that phase and Hilbert spaces are closely related in that Hilbert space is not limited to the multidimensional since it includes finite dimensional Euclidean space; also that when Hilbert space has such a classical analogue then that analogue may be considered a phase space. Basically Hilbert and phase spaces are equivalent when viewed as measurable spaces. In terms of the Reference Model phase space is the quantum state seen in terms of particles in unfolded space, whereas the quantum state is phase space seen in terms of waves in enfolded space. Hilbert space is associated with symmetry and phase space with anti symmetry (or skew symmetry) in mathematical terms (see Section 7.3.7).

There is nevertheless a dichotomy between classical physics and the quantum theory which continues to this day, also discussed in some detail in Section 7.4, with relevant mathematical aspects in Section 7.3.

In particular the relation between quantum theory and reality arising in particular from Bell's Interconnectedness Theorem, previously mentioned in section 3.1, was reviewed by B. d'Espagnat (Ref 4.5.1) and followed up by experiments in the early 1980s by A. Aspect and colleagues (Ref 4.5.2) which brought in to question the extent to which real objects are independent of human consciousness. B. d'Espagnat gives an excellent description of the background against which to assess the problem, which the reader would find informative.

An assessment of this description in terms of the Model is made in Sections 7.2.4 and 7.2.8.4 of this book.

The relation between our sense of four-dimensional reality and the above multidimensional worlds as seen from the perspective of the

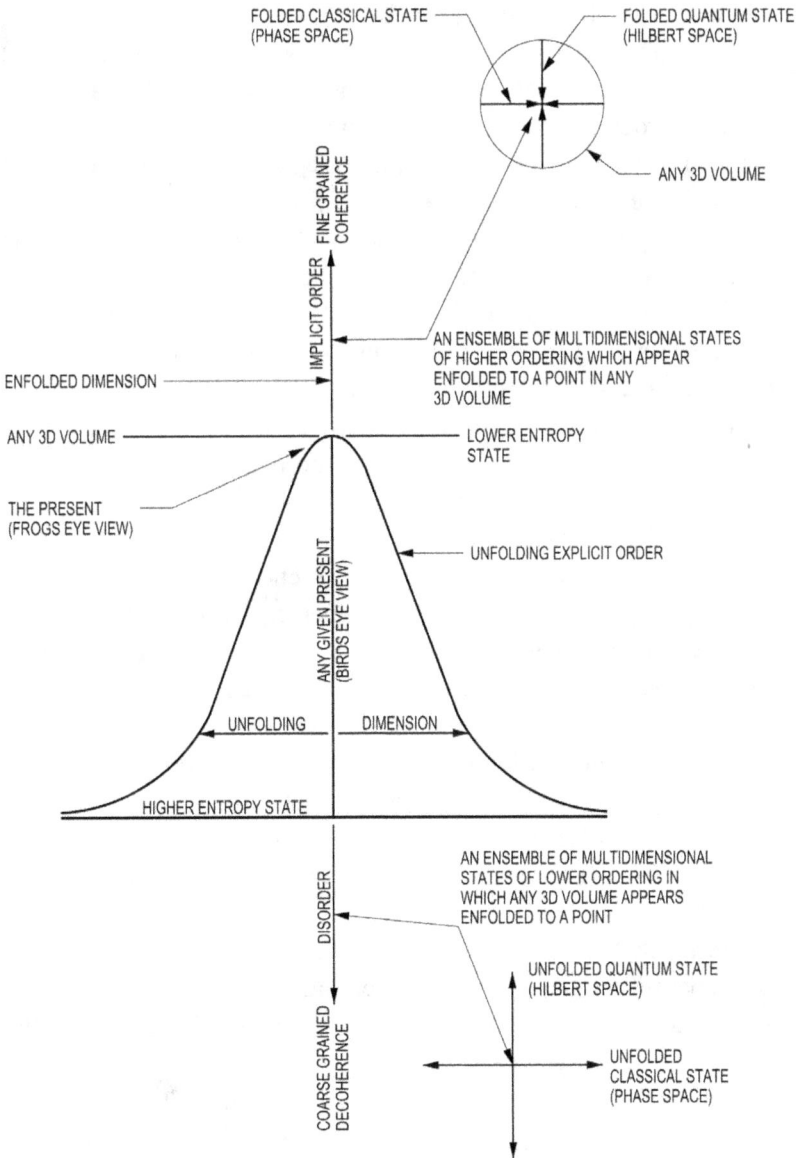

FIG 4.5.1   THE RELATIONSHIP BETWEEN THE MULTIDIMENSIONAL
CLASSICAL AND QUANTUM STATES AND ANY 3D VOLUME
APPLICABLE TO THE MEAN LEVEL OF EXPLICIT ORDER

Reference Model is illustrated in Fig 4.5.1 which relates the multidimensional classical (phase space) and quantum (Hilbert space) states and any three dimensional volume applicable to the averaged state of the ordering distribution or mean level of explicit order.

Relative to any 3D volume the ensemble of multidimensional states of higher ordering appear enfolded to a point in that volume. Correspondingly that 3D volume appears enfolded to a point in the ensemble of multidimensional states of lower ordering.

To the extent that we can think of the ordering distribution as being represented by a grid then the higher degrees of ordering can be considered fine grained in comparison with the lower degrees of ordering which are coarse grained. As has previously been stated those increasingly higher degrees of ordering correspond to a process of coherence and vice versa the increasingly lower orderings (disorder) to a process of decoherence.

Now taking all the statements in this section together, it will later become apparent firstly that our comprehension of the either/or state that represents classical physics i.e. that a given statement is either absolutely true or absolutely not true can be related in the Model to the explicit local, continuous and causal condition represented by the mean level in the ordering distribution.

Secondly that our comprehension of phase space associated with the concept of particles representing an enormous range of possibilities, can be related to the totality of distribution on the determinate variance axis.

Thirdly that our comprehension of the quantum state as a wave like probability distribution representing an enormous range of possibilities, can likewise be related to the totality of ordering on the determinate present state axis.

To these three aspects must be added a fourth namely the nature of the enfoldment process from the multidimensional states to the four-dimensional explicit reality with which we are familiar. Accordingly the figure portrays explicit four-dimensional space-time as the frog's eye view of dimension at the mean level in the ordering distribution with phase space in parallel, allowing for the extension to an unlimited number of dimensions but still in terms of classical physics. Hilbert space, associated with the quantum state, corresponds to the bird's eye view of dimension acting along the present state imaginary axis. The bird's eye view corresponds to a state of zero dimensions relative to the unfolded distribution or variance. The central origin represents the present.

It is a primary thesis of the Reference Model that there is an interaction between phase and Hilbert space which gives rise to the above process of dimensional enfoldment.

Our ability, as unfolded conscious entities, to see three large spatial dimensions only and to relate them to a large time dimension has been associated with the fact that alternative numbers of large space and time dimensions are either (a) unstable or (b) incapable of forming complex structures at least in terms of the physics we can comprehend. In this connection while string theories popular among modern physicists allow additional spatial dimensions they do so by assuming that particles can be interpreted as being vibrating strings which does not in principle render the conclusions derived from the Reference Model inadmissible.

In support of all this the subject of fractals was introduced in section 2.0 and discussed again in section 4.4, in which it was made apparent that all the axes referred to are only relative axes with fractal characteristics, only appearing absolute when viewed from a stationary or fixed perspective but otherwise representing lines of discontinuity.

So the process previously illustrated in Fig 4.3.2 could be repeated in an explicit sense for other orientations of an xyz reference frame, or alternative origins for that frame. Each orientation produces a different sequence of individual points as the volume of space being considered is traversed relative to that reference frame.

Other interpretations of the relation between classical physics and the quantum state have been referred to in the scientific literature, see Everett (Ref 4.5.3) and De Witt and Graham (Ref 4.5.4), namely the various scenarios of parallel universes and many worlds.

This has led to the bizarre supposition that all possible events occur in a vast range of possible universes (see Section 7.4.8).

The Model gives an explanation for the above (see Section 7.4.10).

While we have the ability to increasingly separate, in fact fragment one level from another, this has to be on average in balance with our ability at a deeper level of comprehension to enfold these levels in to one comprehensive whole. This conclusion is supported by Bohm and Hiley (Ref 4.5.5) and additionally in the following sections 4.5 and 4.6 where it will be shown that the resulting balance of reality and imagination is to be associated with our level of consciousness.

What is implied is that there is a multiplicity of unfoldments or threads and that they are ordered in a subtle manner, in effect interwoven in an increasingly complex pattern as the degree of ordering increases, not unrelated to the principle of weaving patterned carpets.

The concept of reaching an absolute centre in Cartesian terms must be replaced therefore by a concept of never ending non-Cartesian movement towards an Ordering Centre.

Movement away from such an Ordering Centre implies an increasing number of differentiated (separated) sets of worlds.

Ultimately expanding dimension (variance) leads to increasingly chaotic explicit disorder while the collapse of dimension is a reduction process which defines the mean level of explicit order as explained later in this section.

The Ordering Centre contains an ensemble of both classical and quantum states, with many potential futures and pasts as portrayed in Fig 4.5.2. This ensemble consists of what the physicist refers to as the unitary quantum state (U) which is determinate in character and develops in accordance Schrödinger's wave equation, acting along the imaginary axis and the equally determinate classical state which develops in accordance the Hamiltonian framework of phase space acting along the real axis.

In particular the quantum state represented by this wave equation in so far as it relates to particles is called the wave function of the particle (see Sections 7.3 and 7.4).

Associated with that concept is that of the wave packet in Hilbert space, comprising of groups of waves of slightly different wave lengths, as described for example by Bohm in his original conventional textbook on the subject (Ref 4.5.6).

Hilbert space contains waves in motion, the whole being described as a complex vector field.

Likewise fundamental to classical physics is the concept of the particle in phase space.

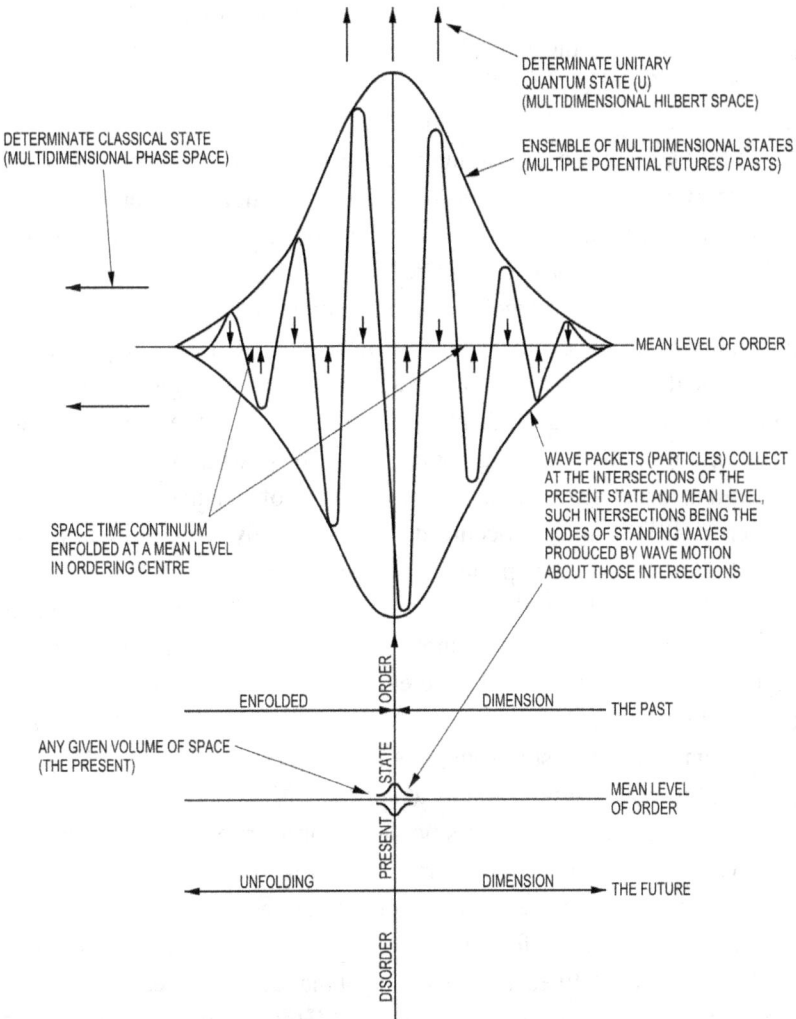

FIG 4.5.2   COMPARISON OF THE ENSEMBLE OF DETERMINATE
MULTIDIMENSIONAL STATES AND THE SPACE TIME CONTINUUM
AT A MEAN LEVEL OF ORDER (FROGS EYE VIEW)

Phase space contains particles in motion, each particle having position and momentum, the whole being described as a vector field where the vector space is represented by a continuous and differentiable manifold down to ever smaller dimension.

From the above ensemble it is now proposed that such wave packets collect at the intersection of the present state and mean level as particles, such intersections being the nodes of standing waves produced by wave motions about those intersections. Accordingly the Reference Model regards the main axes of the Ordering Centre to be determinate and therefore inanimate.

The space-time continuum expands from the present time corresponding to the intersection concerned, explicit disorder increasing on average with that expansion; in the opposite sense that continuum is enfolded as explicit order is restored.

The explicit mean level has so far only been considered in terms of cosmological scales, but it is now necessary to relate this state to human scales 40 orders of magnitude below cosmological scales and some 40 orders of magnitude above scales relative to the quantum state.

Following the Big Bang some 13.8×10^9 years ago the mean level of explicit order defined earlier as the state of balance between dark matter and dark energy occurred, approximately 5×10^9 years ago. How therefore can it be applied to human scales?

Well simply that following that time, 4.6×10^9 years ago the solar system formed including planet Earth. Liquid water, an essential ingredient for life, may well have existed on the surface of the earth at least as long as 4×10^9 years ago and the first life forms in the form of microorganisms at least as long as 3.8×10^9 years ago. It has taken that period of time for Homo sapiens to appear.

It is the main thesis of this book that order rises and disorder falls in an Ordering Centre, so the increase in complexity of life, an ordering process has matched the disordering process required by the laws of thermodynamics. Therefore it is contested by the author that the mean explicit level as defined by the energy balance can be equated to the mean implicit ordering level about which Homo sapiens operates and furthermore other intelligent entities throughout the visible universe. In other words that balance precipitated over the following 5×10^9 years the emergence of life forms on rocky planets in solar systems of which our own is an example. All of this will be supported and reinforced in the remainder of this book, in particular Sections 7.2.6 and 7.2.8.

Fig 4.5.3 show how these processes are applied to form our common-sense notions of continuity. The indeterminate reduction process (R) non-local, non-causal and non-continuous in character, converts the determinate quantum state to the level of classical physics

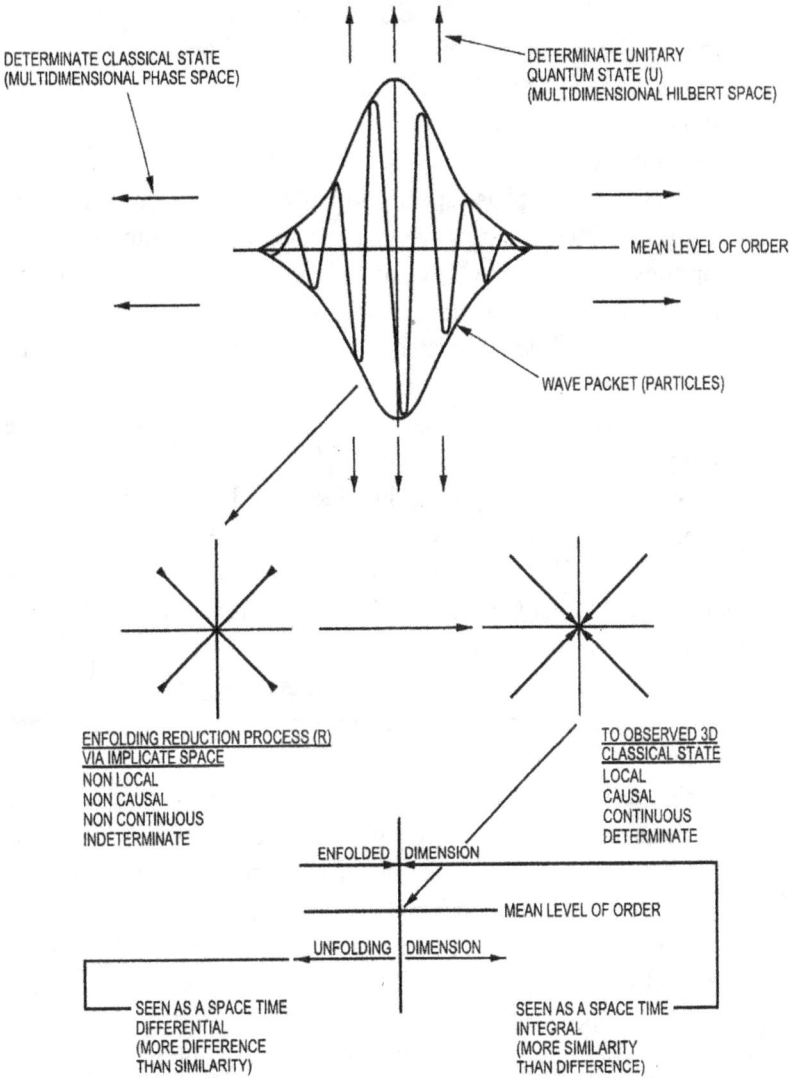

DETERMINATE CLASSICAL STATE
(MULTIDIMENSIONAL PHASE SPACE)

DETERMINATE UNITARY
QUANTUM STATE (U)
(MULTIDIMENSIONAL HILBERT SPACE)

MEAN LEVEL OF ORDER

WAVE PACKET (PARTICLES)

ENFOLDING REDUCTION PROCESS (R)
VIA IMPLICATE SPACE
NON LOCAL
NON CAUSAL
NON CONTINUOUS
INDETERMINATE

TO OBSERVED 3D
CLASSICAL STATE
LOCAL
CAUSAL
CONTINUOUS
DETERMINATE

ENFOLDED | DIMENSION

MEAN LEVEL OF ORDER

UNFOLDING | DIMENSION

SEEN AS A SPACE TIME
DIFFERENTIAL
(MORE DIFFERENCE
THAN SIMILARITY)

SEEN AS A SPACE TIME
INTEGRAL
(MORE SIMILARITY
THAN DIFFERENCE)

FIG 4.5.3   THE INTERACTION BETWEEN THE ENSEMBLE OF DETERMINATE
MULTIDIMENSIONAL STATES AND THE SPACE TIME CONTINUUM VIA
AN INDETERMINATE REDUCTION PROCESS. HUMAN SCALES APPLY

in this Implicate space (when an act of measurement is carried out) that
conversion being portrayed here as arrows converging to a point. Those
arrows act at an angle to the determinate axes the justification for which
will become further apparent in this section 4.0.

The horizontal component of R in the figure is of great importance as it identifies the mean level of the order potential on the y-axis of the Ordering Centre relative to the base state.

In doing so it also identifies the chaotic dimensional distribution of the mean explicit order along the x-axis of the Ordering Centre.

It is concluded that phase space and Hilbert space pull the relevant vector fields in different directions, an incompatibility similar to that which applies between the Theory of General Relativity and the quantum theory. The resultant motion gives rise not only to the indeterminate process associated with the animate state, but also the inanimate core of that motion namely the four-dimensional classical state, local, causal and continuous in character, portrayed as the converged point like state shown in the figure.

Now while this may appear to involve what is generally termed a collapse of the wave function when the act of measurement is made, that appearance is a matter of perspective; the quantum state is how the classical state sees other possibilities while the classical state is how the quantum state sees a particular case.

The wave function collapses in a real sense along the mean level of order, but no such collapse applies to the imaginary y-axis since explicit dimensional limitations do not apply.

By this means the Ordering Centre keeps a balance between the ordering processes in the form of wave reinforcement and the disordered processes in the form of wave cancellation as previously explained in Section 4.2 and summarised in Fig 4.4.4.

The reduction process (R) separates the explicit order from the mean level of implicit order. This process can therefore be considered to have two components: an explicit particle orientated component, from that perspective seen by the explicit human brain as real (objective) and an implicit wave orientated component, the wave function from that perspective seen by the implicit human mind as imaginary (subjective).

The differing emphasis placed on these components gives rise to the diversity of human belief as illustrated in the following sections of this book. It also gives rise to the range of interpretations of quantum mechanics discussed in Sections 7.4.8 and 7.4.10.

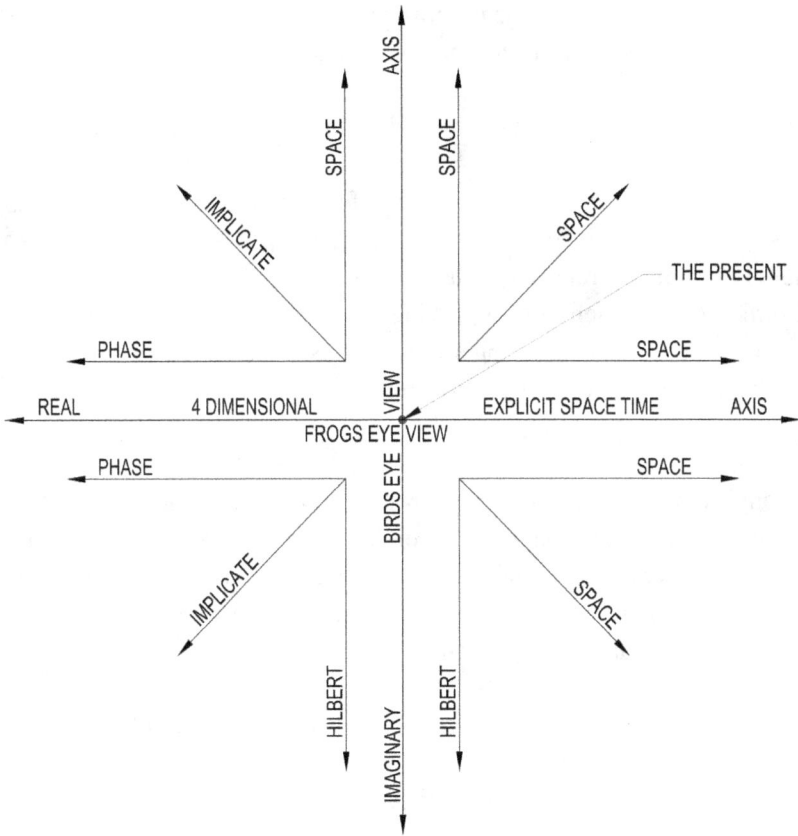

FIG 4.5.4   4 DIMENSIONAL / PHASE / HILBERT SPACE COMPARED WITH
IMPLICATE SPACE

The figures in this book distinguish between variance associated with difference and variance associated with integration or similarity.

The ensemble of multidimensional states referred to above on this basis contains wave packets (particles) and aggregations of particles seen as a differential associated with unfolded dimension and as an integral associated with enfolded dimension.

The interaction between the vector field in phase space related to anti-symmetry and the complex vector field in Hilbert space related to symmetry gives rise to what the Reference Model refers to as a general state of asymmetry and asymmetric motion in what is termed Implicate space.

Fig 4.5.4 illustrates the proposed relation between implicate space, phase space, Hilbert space and explicit space-time. Implicate space emphasises asymmetry, an essential ingredient of animacy, as opposed to the symmetrical or anti-symmetrical inanimate character of Hilbert space and phase space.

In particular the frog's eye view referred to earlier in Figures 4.3.3 and 4.3.4, is provided to us by 4D explicit space-time i.e. the mean level of explicit order. Explicit order is enfolded in the mean level of implicit order which in turn provides the bird's eye view also referred to in those figures. The frog's eye view is therefore to be associated with what we perceive as reality and the bird's eye view with what we perceive as imagination.

Asymmetrical states become evident when these processes occur at angles to the vertical ordering process and the horizontal distribution process as shown in Fig 4.5.5 which illustrates the various asymmetric order/disorder vectors and their components operating at the mean level and present state in an Ordering Centre. In particular the core state at that level in an Ordering Centre is an inanimate condition as shown in the figure.

Fig 4.5.6 in accordance Figures 4.5.1, 4.5.2 and 4.5.3 illustrates the relation between the asymmetric vectors, 3D classical and multidimensional physics, all relative to the present state at the mean level in the Ordering Centre.

The application of these asymmetric vectors and their components become evident with the help of Fig 4.5.7 (a) thru (d) which illustrates the process of unfolding differential and integral ordering sequences while enfolding disorder and in the opposite sense the unfolding of disorder while enfolding order, again all relative to the mean present state and level.

From the viewpoint of the Ordering Centre as a whole the vectors act from points at many levels and displacements in the ordering structure bearing in mind the enormous range of scales applicable as was discussed in section 4.4.

Figures 4.5.5 thru 4.5.7 reveal that the changes in ordering level can be interpreted as asymmetric movements across the vertical and horizontal lines constituting what in a Cartesian sense was previously referred to as the grid in Section 4.3.

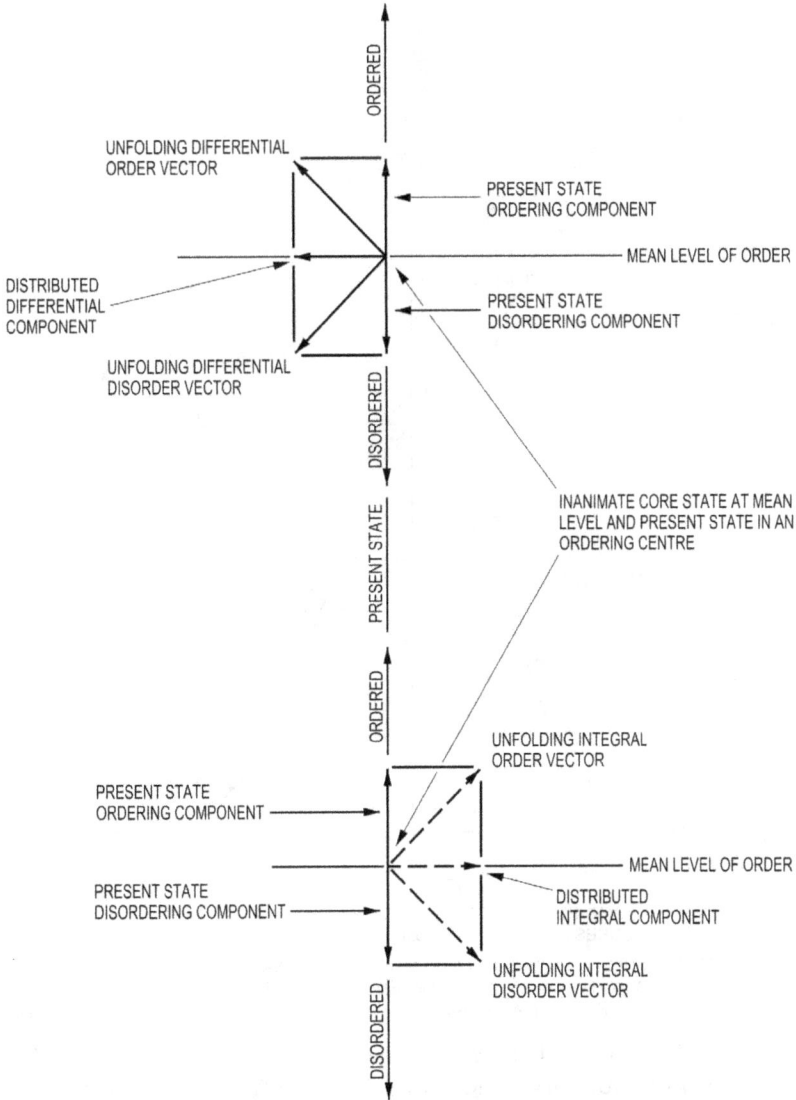

FIG 4.5.5   ASYMMETRIC VECTORS AT A MEAN LEVEL IN THE ORDERING CENTRE
COMPARING PRESENT STATE AND DISTRIBUTED STATE ORDERING

Furthermore each vector has components parallel to the grid both vertically and horizontally. The vertical components represent the present state and the horizontal components represent the distributed state.

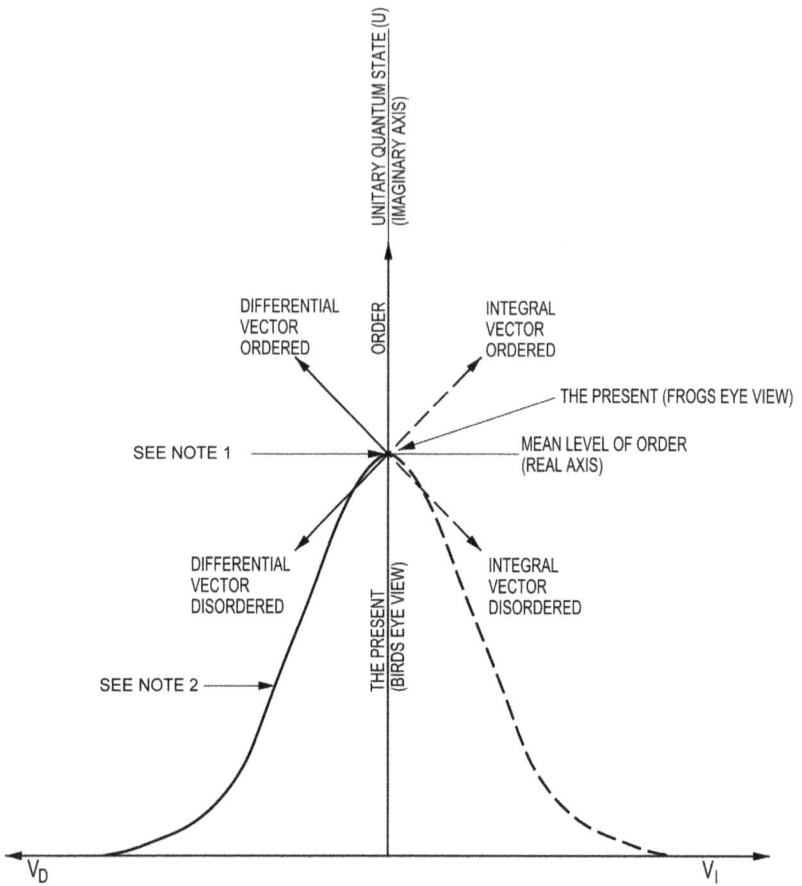

NOTE 1   OBSERVED 3D CLASSICAL STATE AS PER
          FIGURES 4.5.1, 4.5.2 AND 4.5.3

NOTE 2   UNFOLDING ENSEMBLE OF MULTIDIMENSIONAL STATES
          WHICH CONTAIN UNFOLDING EXPLICIT ORDER AS PER
          FIGURES 4.5.1, 4.5.2 AND 4.5.3

FIG 4.5.6   THE RELATION BETWEEN ASYMMETRIC VECTORS, 3D CLASSICAL
             AND MULTIDIMENSIONAL PHYSICS RELATIVE TO THE PRESENT
             STATE AT THE MEAN LEVEL OF THE ORDERING CENTRE

Another essential feature the Reference Model emphasises, is that
unfolding and enfolding processes are in a state of balance about a
centreing condition, namely the relevant present time or present state at
the mean level of order, a condition about which all such unfoldments
and enfoldments may be said to occur.

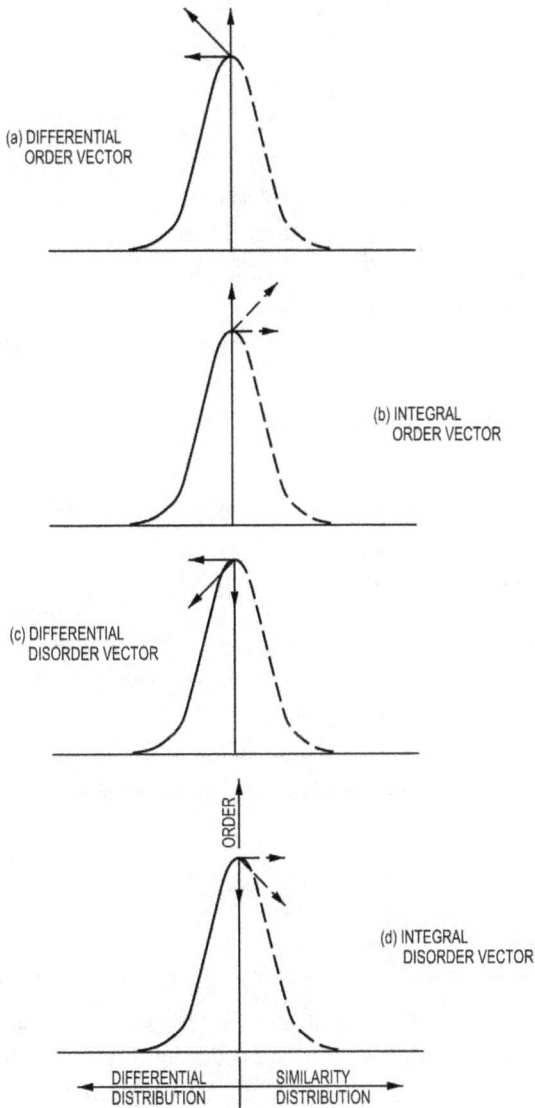

FIG 4.5.7   ASYMMETRIC VECTORS IN THE ORDERING CENTRE PROMOTING
THE GROWTH AND DISSIPATION OF ORDER

Reduction processes in each of these figures can be portrayed by reversing the direction of the arrows. Enfolding variance towards the inanimate core state on the real axis establishes and reinforces the explicit order in the form of four dimensional space-time reality.

It is the contention of the Model that it is the above asymmetric motion which positions the wave shown in Fig 4.5.7.

The resultant of all such asymmetric motion must be such that it represents the Ordering Centre as a whole. To understand this imagine the sv-axes to be capable of rotation and translation, so a given vector will try to move the Centre and rotate it to coincide with itself and similarly for all the other vectors. So the vector sum must be such as correspond with the ordering direction and any applicable movement of the Ordering Centre as a whole.

The comparison between symmetric or anti-symmetric motion and asymmetric motion relative to given present states and levels in an Ordering Centre will be considered in greater detail in Sections 4.6 and 4.7.

4.6   Inanimacy, animacy and the emergence of consciousness associated with non-Cartesian properties of the Ordering Centre.

The Model now equates the Ordering Centre in its balanced symmetric or anti-symmetric state with the inanimate and its unbalanced asymmetric state with the animate.

Asymmetric vectors change the level of an animate entity in the ordering system by reason of its movement from the inanimate balanced state.

Inanimate (non-living) states are regarded by the Model as balanced conditions, meaning balanced about a centreing condition, or unbiased.

Conscious/aware (living) entities move in all directions with respect to the intersections of what are regarded as neutral vertical and horizontal states, unbalancing the ordering system from the inanimate state, while endeavouring to tune in to the system as a whole.

Vertical components are associated with the present state and horizontal components with the distributed state relative to the mean level in the Ordering Centre as shown in these figures.

Animate living states are considered to be those which move purposefully in the direction associated with unfolding order or disorder (the present state), while also unfolding the variance (the distributed state). Such states are regarded as unbalanced conditions or biased.

It is the animate motion which ultimately can see and know anything as opposed to the inanimate motion which sees and knows nothing.

In support of this we will return to the concepts of wave motion, reinforcement, cancellation and interference discussed in section 4.2 and the further comments made in Section 4.3 concerning multi-dimensional phase space.

The inanimate system as already stated is represented by the vertical and horizontal axes or grid making up the envelope of the Ordering Centre and the distribution within it, all related to the principles of travelling waves and standing waves referred to in Section 4.2.

Accordingly the animate system is represented by asymmetric wave motion, all relative to this grid as shown in Figs 4.6.1 and 4.6.2,

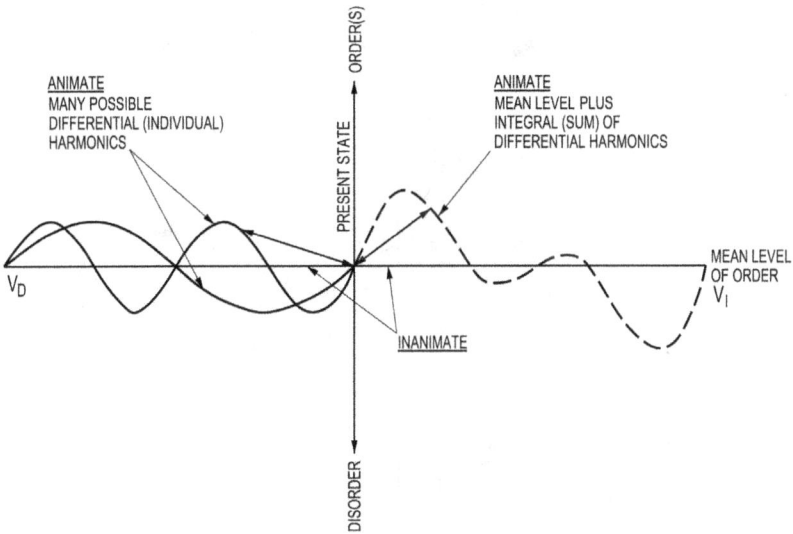

ANIMATE
MANY POSSIBLE
DIFFERENTIAL (INDIVIDUAL)
HARMONICS

ANIMATE
MEAN LEVEL PLUS
INTEGRAL (SUM) OF
DIFFERENTIAL HARMONICS

ORDER(S)

PRESENT STATE

V_D

MEAN LEVEL
OF ORDER
V_I

INANIMATE

DISORDER

NOTE 1   THE WAVE PATTERNS SHOWN CAN BE ROTATED THROUGH 360 DEGREES
TO REPRESENT ANIMATE / CONSCIOUS VARIABILITY

NOTE 2   ANIMATE WAVE MOTION CAN BE PRESENTED BY UNFOLDING
VECTORS WITH ORIGIN AT THE INTERSECTION OF
THE S AND V AXIS AND TERMINUS AT ANY
GIVEN POINT ON THE WAVE AND ENFOLDING VECTORS
IN THE OPPOSITE SENSE AS SHOWN

FIG 4.6.1   NOTIONAL SCHEMATIC OF THE ANIMATE WAVE MOTION COMPARED WITH THE INANIMATE WHILE TUNING IN TO A MEAN LEVEL OF ORDER

these being notional schematics which model the means whereby the animate tunes in to the mean level of order (the v-axis) and the mean present state (the s-axis) respectively.

These schematics show the reduction process illustrated in Fig 4.5.3 in reverse, creating waves as opposed to the reduction process creating particles.

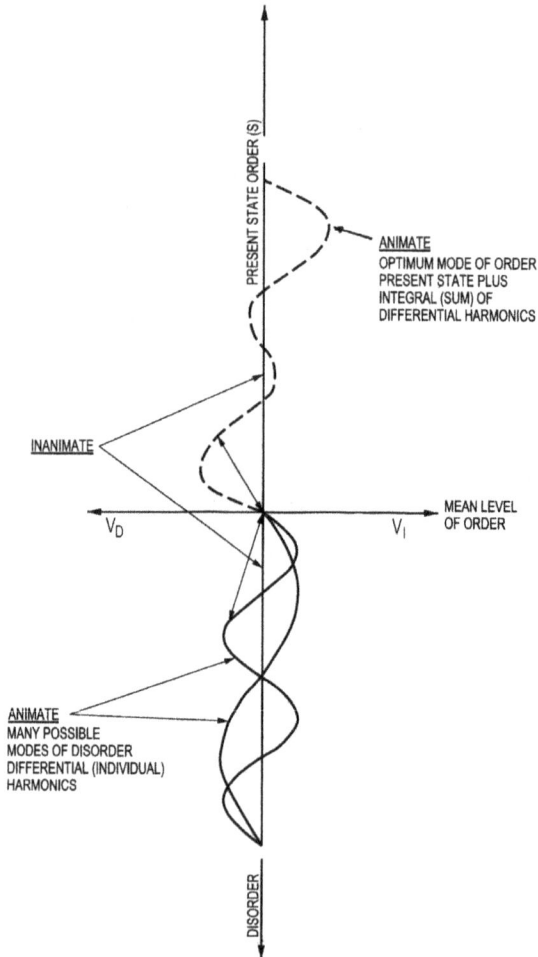

FIG 4.6.2   NOTIONAL SCHEMATIC OF THE ANIMATE WAVE MOTION COMPARED WITH THE INANIMATE WHILE TUNING IN TO THE PRESENT STATE
NOTES 1 & 2 AS FIG. 4.6.1 APPLY

In Fig 4.6.1 the animate in the case of unfolded dimension is represented by different individual harmonics and in the opposite sense by a single wave corresponding to the inanimate variance, plus the integral or sum of those harmonics.

In Fig 4.6.2 the animate in the case of unfolding disorder is represented by different individual harmonics corresponding to the many possible modes of disorder and in the opposite or ordering sense by again a single wave motion corresponding to the optimum mode of order made up of the inanimate present state plus the integral or sum of those harmonics.

In each of these figures the wave motion is non-local, non-causal and non-continuous in character and can be represented by unfolding vectors with origin at the intersection of the s- and v-axes and terminus at any given point on the wave and enfolding vectors in the opposite sense. The variance and present state axes represent the inanimate state, local, causal and continuous at the mean level of order.

Animate and conscious variability implies the s- and v-axes must rotate relative to (or be displaced from) the preferred reference frame so as to form the perspective of a given individual entity as opposed to an increasing range of individuals up to that of the human race as a whole; this implies in turn extra-terrestrial intelligence associated with 'our' big bang (see Sections 7.2.6 and 7.2.8).

There is of course no limit to the complexity of wave motion or number of harmonics involved in the growth and dissipation of an Ordering Centre but the same underlying principles associated with the holographic fractal continue to apply.

The works of M.C. Escher (Ref 4.6.1) include prints which can be related to these principles and in some cases illustrate the emergence of animacy from the inanimate state.

Fig 4.6.3 shows in principle how the vectors referred to above can be related to individual conscious entities such as human beings and to the human race as a whole, when all individuals are taken in to account, the latter being represented by the resultant vector shown in the figure which uses the familiar Cartesian principle to ease initial comprehension.

As defined earlier the vertical and horizontal axes in the figure represent the preferred reference frame for the Ordering Centre. If the human race was alone in the explicit universe then the resultant vector would on this basis align itself with the vertical axis.

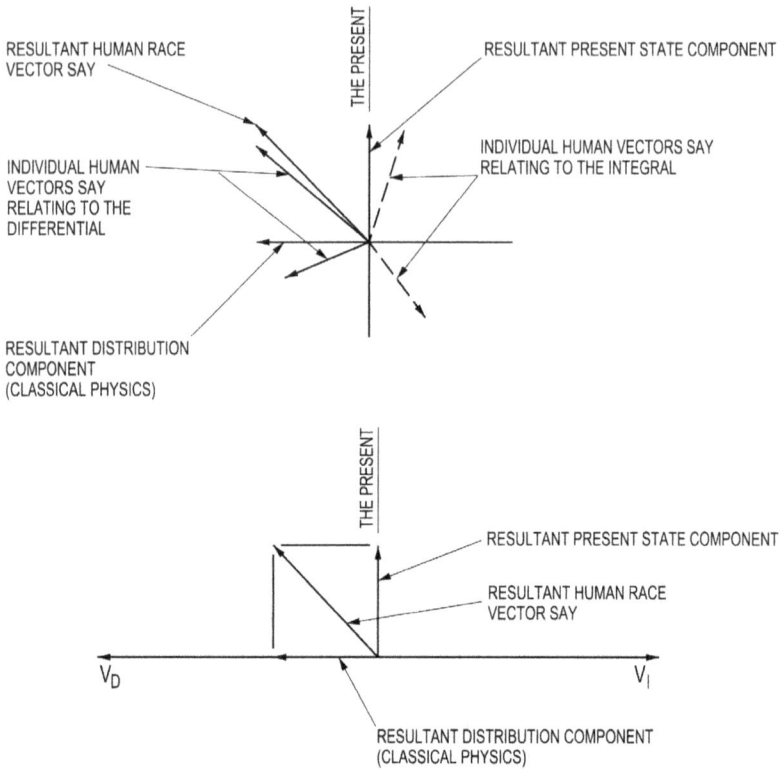

RESULTANT HUMAN RACE
VECTOR SAY

THE PRESENT

RESULTANT PRESENT STATE COMPONENT

INDIVIDUAL HUMAN
VECTORS SAY
RELATING TO THE
DIFFERENTIAL

INDIVIDUAL HUMAN VECTORS SAY
RELATING TO THE INTEGRAL

RESULTANT DISTRIBUTION
COMPONENT
(CLASSICAL PHYSICS)

THE PRESENT

RESULTANT PRESENT STATE COMPONENT

RESULTANT HUMAN RACE
VECTOR SAY

V_D

V_I

RESULTANT DISTRIBUTION COMPONENT
(CLASSICAL PHYSICS)

FIG 4.6.3   THE RELATION BETWEEN INDIVIDUAL HUMAN VECTORS AND THE
HUMAN RACE VECTOR

However consciousness cannot be explained using Cartesian principles.

The holographic fractal nature of ordering and distribution properties must be incorporated, together with the asymmetric motion associated with Implicate space, the combination offering at the very least a gateway to the non-Cartesian concepts necessary to provide that explanation.

Fig 4.6.4 illustrates the effect of including for non-Cartesian properties in an Ordering Centre in more detail.

The present state ordering component for the mean level of order in the earlier figures is vertical, but allowing for the enfolding present state, as shown in Fig 4.6.4 (a), reveals that same component implies asymmetric motion at higher (or lower) levels of order in a holoarchical

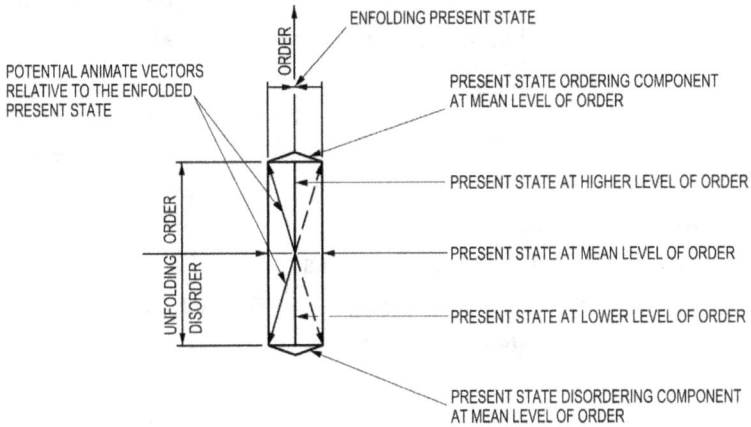

(A) THE HOLOGRAPHIC FRACTAL UNFOLDING OF ORDER / DISORDER ARISING FROM THE ENFOLDING PRESENT STATE

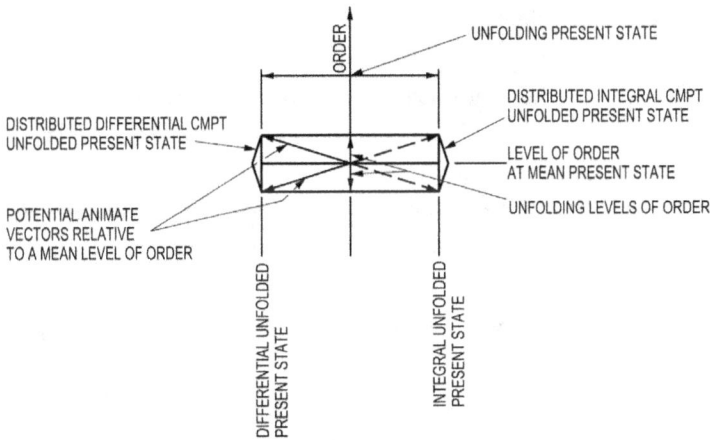

(B) THE HOLOGRAPHIC FRACTAL UNFOLDING OF THE DIFFERENTIAL / INTEGRAL ARISING FROM THE UNFOLDING PRESENT STATE

FIG 4.6.4   THE NON CARTESIAN PROPERTIES OF THE ORDERING CENTRE

sense in the Ordering Centre. It is the enfolding present state which gives rise to human animacy and intelligence.

The same situation applies to distributed components when allowing for the unfolding present state as shown in Fig 4.6.4 (b). It follows that the unfolding present state in an Ordering Centre unfolds levels of order in a hierarchical sense, enhancing the distributed state and the influence of chaotic disorder. This has implications for the concepts of artificial animacy and intelligence (See Sections 7.2.5 and 7.2.8).

Again the reader is reminded that while all the figures in Sections 4.5 and 4.6 are based on a preferred reference frame, the directionless states corresponding to the cases where there is no preferred reference frame remain in play.

While the holographic fractal nature of the Reference Model preserves the principle of background independence so important to General Relativity and dimensional concepts generally, the non-spatio-temporal character of the present state axis requires a preferred direction in terms of order, while dimensional concepts associated with variance diverge from that direction.

This applies to both the inanimate motion of the preferred reference frame and the animate motion of the vectors relative to that frame.

The overall implications of the above, for both the objective and subjective philosophies, are profound indeed.

It is clear that these non-Cartesian properties are responsible for the confusion between the discontinuous state and that of both classical physics and the quantum theory.

The same properties explain the debate between natural selection and intelligent design, the further debate between human and artificial intelligence and last but not least the lack of explicit evidence of extra-terrestrial consciousness, organic or otherwise, in the universe, as discussed in later sections.

4.7   The formation of the Whole Ordering Centre and opposing decoherence processes.

The Holomovement.

The previous sections have considered in detail only one peak energy density associated as it were with 'our' apparent Big Bang, but some consider the standard cosmological model has too many deficiencies. A split has developed in the physics community between those who favour fine tuning the cosmological model and those who favour the alternative concept of inflation (See Section 7.4.4).

The Reference Model offers clear reasons for this divergence as explained below.

The space-time foam conditions involving minimal quantum fluctuations of the otherwise uniform vacuum state scalar field all associated with the random state, have been referred to earlier in Section 4.2. In terms of the Reference Model these fluctuations come under the influence of ordering processes, where they ultimately coalesce to more unified structures with the overall wave function promoting order and demoting disorder.

This process is modelled in Figure 4.7.1 which illustrates for what is termed the Whole Ordering Centre, the unfoldment and enfoldment of inanimate and animate states without limit, from and to the random state. It is responsible for defining the preferred reference frame, lesser Ordering Centres having frames rotated/displaced from it, in the same manner as the s- and v-axes in Figures 4.6.1 thru 4.6.3 allow for animate/conscious variability.

The Whole Ordering Centre extends the principles discussed in relation to a given Ordering Centre, as described in Sections 4.4 thru 4.6, to include for a vast range of energy densities with the growth and dissipation of order occurring on the same basis.

The base line representing the minimal degree of order corresponds to the totally indiscrete state previously discussed in Section 4.2. The growth and dissipation of order for the Whole Ordering Centre is therefore also similar to that of 'our' apparent big bang for example.

Each potential for order corresponds to the condition where the potential for differentiation is matched by the potential for integration. The greater the capacity for both difference and similarity, the greater the order potential.

So the peak representing the order potential for 'our' apparent big bang and which appears absolute to us is in fact only a relative peak in the context of the range of order potentials implied in the figure. If the Whole Ordering Centre is generated by a super big bang say, then lesser big bangs such as 'our own' merge, making contributions to form lower levels of order in that Whole Centre, similar to the processes illustrated in Figures 4.2.4 and 4.2.5.

The Whole Ordering Centre will also have wave motion at intermediate scales similar to the wave motion at human scales at the

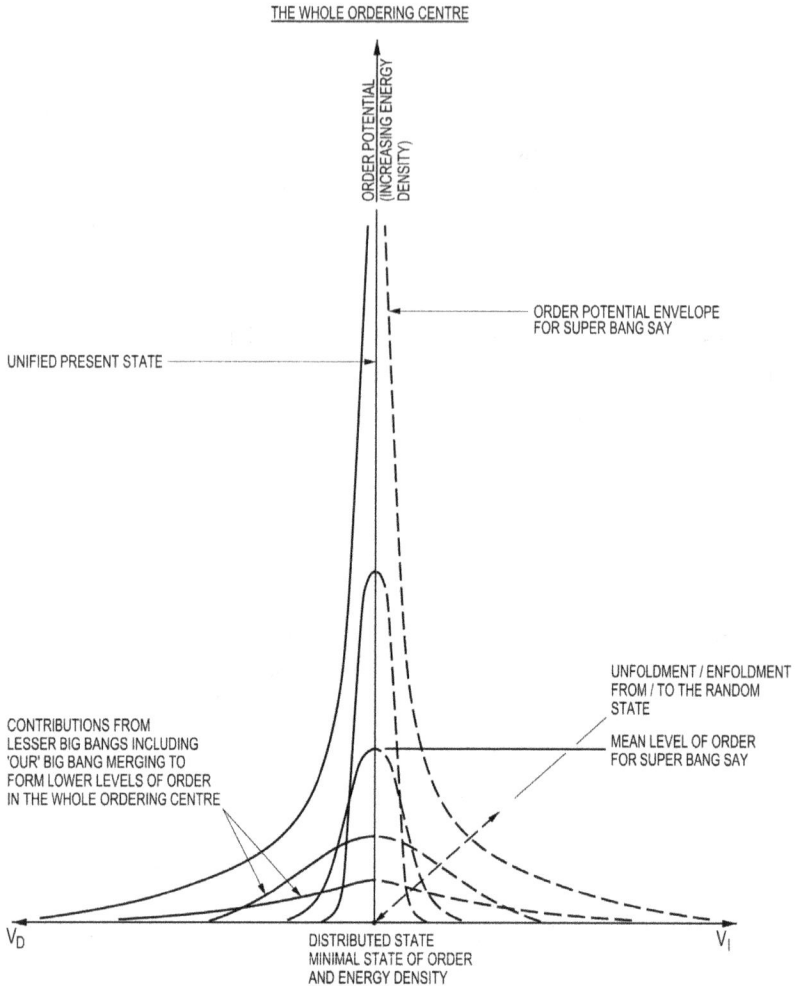

THE WHOLE ORDERING CENTRE

NOTE THE GREATER THE ENERGY DENSITY THE GREATER THE CAPACITY FOR EVER HIGHER
LEVELS OF CONSCIOUSNESS ARISING FROM ASYMMETRIC MOTION WITHIN THE ORDERING
CENTRE ABOUT THE PRESENT STATE WHICH INCREASES ORDER.
IN THE OPPOSITE SENSE CONSCIOUSNESS DECREASES AS ASYMMETRIC MOTION IS
LIMITED TO THAT ABOUT THE DISTRIBUTED STATE WHICH DISSIPATES ORDER.

FIG 4.7.1   ORDER V VARIANCE FOR THE WHOLE ORDERING CENTRE
INCORPORATING A RANGE OF ENERGY DENSITIES

level applicable to our big bang, (i.e. those which were intermediate between the quantum scale 40 orders of magnitude below and the cosmological scale 40 orders of magnitude above).

The reduction process described in Section 4.5 also acts at the mean level of order in the Whole Ordering Centre, that (whole) mean level having the mean levels of aligning lesser Ordering Centres enfolded within it.

The emphasis so far has been placed on the formation of Ordering Centres but their collapse or decoherence to the random state must also be explained to complete the description referred to in Section 4.2.

Wave reinforcement unfolds order and the distribution of dimension, while wave cancellation processes collapse order and dimension. Total decoherence of the Whole and lesser Ordering Centres is completed by the return to discordant distribution curves shown in Figures 4.4.1 (a) and (b), followed by the collapse of any form of distribution culminating in the return to minimal quantum fluctuations.

To the extent that lesser energy densities do not contribute to the Whole Ordering Centre they lose order relative to it, to the point where they collapse to this random state. The latter condition seems to approach, ever more closely, but without ever reaching a state of non-existence about which by definition nothing can be said.

It is the peak of the Whole Ordering Centre that is the source and explanation for energy, attractive gravity, mass diffusion and particle aggregation. It is the base distributed state that is the source and explanation for the dissipation of energy, repulsive gravity, mass aggregation and particle diffusion. Finally, it is the collapse of the distribution leading to the random state where the concepts of wave, particle, space and time have no meaning, which is in turn responsible for the collapse of order to the state of minimal quantum fluctuations.

It can now be seen that both the peak of the Whole Ordering Centre and the random state on the base line represent opposite poles or great attractors, conscious 'stuff' being attracted to the pole of greatest order and unconscious 'stuff' to the pole of least order.

It should be noted that the range of mean levels of order implied above lie on the centreline of Fig 4.7.1. This centreline represents the unified present state which is made up of inanimate complex vectors unfolding from each mean level of order to the next higher mean level.

The sum of all such vectors is directly analogous to the quantum theory description of Hilbert space, as a complex vector space with an inner product defined on that space, which is complete (see Section 7.3).

Figure 4.7.1 also appears to offer clues to the relationship between fine-tuned versions of the standard cosmological model and the inflationary hypothesis.

Fine tuning may be considered to relate to the state of the Ordering Centre which gives rise to 'our' apparent big bang and more ordered states with higher energy density above our level, while the fractal set of energy densities associated with inflation relate to the more fragmented state of the Ordering Centre at states of ordering below our level, leading ultimately to its collapse.

The figure also offers clues to the extraordinary specialness of the apparent big bang, even the roles played by ordinary visible matter and what is termed dark matter and dark energy, all of which will be referred to again in Section 7.0 of this book.

Since however there is no boundary to existence there is no absolute unified present state, no state of absolute completeness and therefore no absolute Whole Ordering Centre. There is only a never-ending progression towards greater wholeness, a statement entirely consistent with Godel's Incompleteness Theorems (See also Section 7.3).

The figure gives the impression of a cycling process between order, disorder and the random state as if it were a closed loop. However the removal of the concept of the absolute replaces that closure with an ever-widening spiral implying that indeed incompleteness rules the roost.

The Whole Ordering Centre is then responsible for the growth of consciousness and its dissipation. It has not only the greatest peak energy density but correspondingly the greatest base variance. The higher the energy density the higher the degree of consciousness at its particular mean level, with increases in the imaginary component of consciousness at the expense of reality above that level and vice versa below that level.

In particular the greater the energy density the greater the capacity for ever higher levels of consciousness arising from the asymmetric motion within the Ordering Centre about the present state which increases order. In the opposite sense consciousness decreases as asymmetric motion is limited to that about the distributed state which dissipates order.

There is therefore no limit to levels of consciousness since existence has no boundary.

Similarly there is no state of non-existence as has been said earlier.

The process described above is considered in principle to equate with what Bohm (Ref 2.5.1) referred to as the Holomovement, a form of overriding consciousness oscillating about the unified present state, as portrayed in Fig 4.7.2.

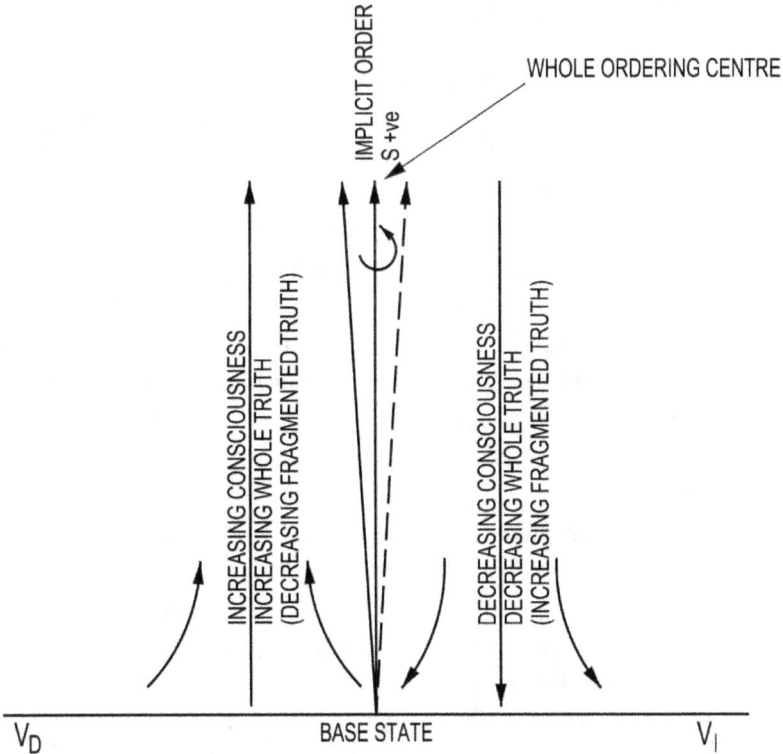

FIG 4.7.2   THE HOLOMOVEMENT

91

The distributed base state would evidently produce no ordering whatsoever and appear to take an infinitely long time doing so, conversely the Holomovement would seem to reach any given level above the base state effectively instantaneously.

The figure shows the whole truth and likewise consciousness increasing with the growth of order. In the opposite sense the whole truth and consciousness collapse towards an ever more fragmented state. This fragmented state can be associated with theories of cosmology involving the concept of inflation which implies a fractal set of many greater and lesser energy densities with varying physics giving the appearance of a spectrum as described by Linde (Ref 4.7.2).

The greater the wave peak in the figures he provides the greater the energy density and the greater that density the greater the strength of the scalar field. Such scalar fields are again considered to grow by wave reinforcement, providing the potential energy for the inflationary process preceding the more moderate rates of expansion associated with the standard cosmological model. In this scenario minimal energy density and minimal inflation dimensionally are represented by valleys and flat plains between the peaks.

It is the contention of the author that the Model demonstrates that the world we see explicitly in three dimensions varying with time is merely a special case in effectively an infinite range of possibilities.

The reader is reminded that the Model does not dispute that endless quantum fluctuations arise which do not in turn give rise to patterns of order and variance, but it is concluded that animacy and least of all consciousness, belief and imagination do not emerge in these circumstances.

What existence cannot have knowledge of, nor can its conscious constituents.

Bohm and Hiley (Ref 4.4.5) are almost alone in attempting to relate Relativity, Quantum theory and physics generally not only with the explicit but also the implicit.

Figures 4.5.2 and 4.5.3 in particular are intended to simulate their mathematical variation to Schrödinger's quantum state wave equation, which regards particles as real objects in four-dimensional space-time, while confirming the existence of the non-local Implicate order from which our understanding of reality unfolds.

While this is done at the cost of adding extra terms to that equation, such terms are considered to represent the tension or friction between classical physics and the quantum state, which gives rise to consciousness, all relative to a mean level in the Ordering Centre (see Sections 7.4.8 and 7.4.10).

It is not at all evident that a mathematical formulation of the Implicate Order is possible.

The Reference Model implies that rather than trying to unify reality and the imaginary in a mathematical sense, the formation of a coherent theory relating the objective and subjective of the type put forward in this book, is the practical way to proceed.

The efforts referred to above are not in any case generally supported by those who hold sway in scientific circles, for reasons the Reference Model makes all too clear.

Nor can the Model expect to receive support from the contradictory mix of belief systems.

The Model in fact predicts that no individual will agree with it but that it nevertheless corresponds to the average of their summated opinions and beliefs.

With regard to the validity of Implicate space in the absence of a mathematical basis, it is overwhelmingly apparent that the Model correlates with not only the diverse range of human philosophy but also the degree of convergence or commonality which is also evident.

That divergence is expressed in terms of distribution and convergence in terms of the present state, the inevitable consequence being the conflict between reality and imagination, which gives rise to Implicate space. This conflict reflects the fact that existence is in motion and needs to keep stirring the pot.

It is nevertheless the refusal to accept a coherent model of the kind put forward in this volume that maintains the divisions so apparent in human society.

So we must be prepared to appreciate we are dealing with a Model and what matters is whether at the end of the day it adequately describes the ordering process, at least in averaged terms, so as to give guidance on a very difficult subject in place of the current vacuum in comprehending the relation between reality, belief and imagination.

4.8 The relation between the Ordering Centre, the human level of consciousness within it and the arrow of time.

The Reference Model described to date in this Section 4.0 uses figures which assume a common base state with varying mean levels of order above that base and nothing as it were below that base. However the figures can be redrawn to show instead increasing implicit order above the mean and increasing implicit disorder below the mean but where the former is a mirror image of the latter. This is a horizontal mirror as compared to the vertical mirror implied by the central ordering line shown in Fig 4.4.1.

We now have increasing order in the s +ve direction and increasing disorder in the s -ve direction at right angles to the mirror and increasing distribution along the v axis in the plane of the mirror. The plane of the horizontal mirror then represents a single mean level as opposed to the varying levels implied in Fig 4.7.1.

The random state is no longer apparent as by definition it is inconsistent with the Ordering Centre concept. When that concept collapses the random state reappears.

The figures in the following Section 4.8 can be interpreted in terms of either convention.

Up to now emphasis has been laid on reaching a degree of comprehension of the Ordering Centre as a whole, but this section 4.8 is concerned primarily with understanding its relation to human levels of both scale and comprehension and also the relation between human consciousness and the inanimate state. That inanimate state is defined by what may now be termed the horizontal line representing a mean level of order and by the vertical centre line representing zero variance with the growth of order and disorder formed by wave motion relative to those lines.

The Reference Model is based on the definition of order increasing in the direction common to all the figures shown in this book, regardless of the orientation of asymmetric vectors, including that for the human race as a whole as indicated in Fig 4.6.3.

Accordingly Fig 4.8.1 reveals the explicit and implicit domains relative to the implicit lifespan of a conscious entity.

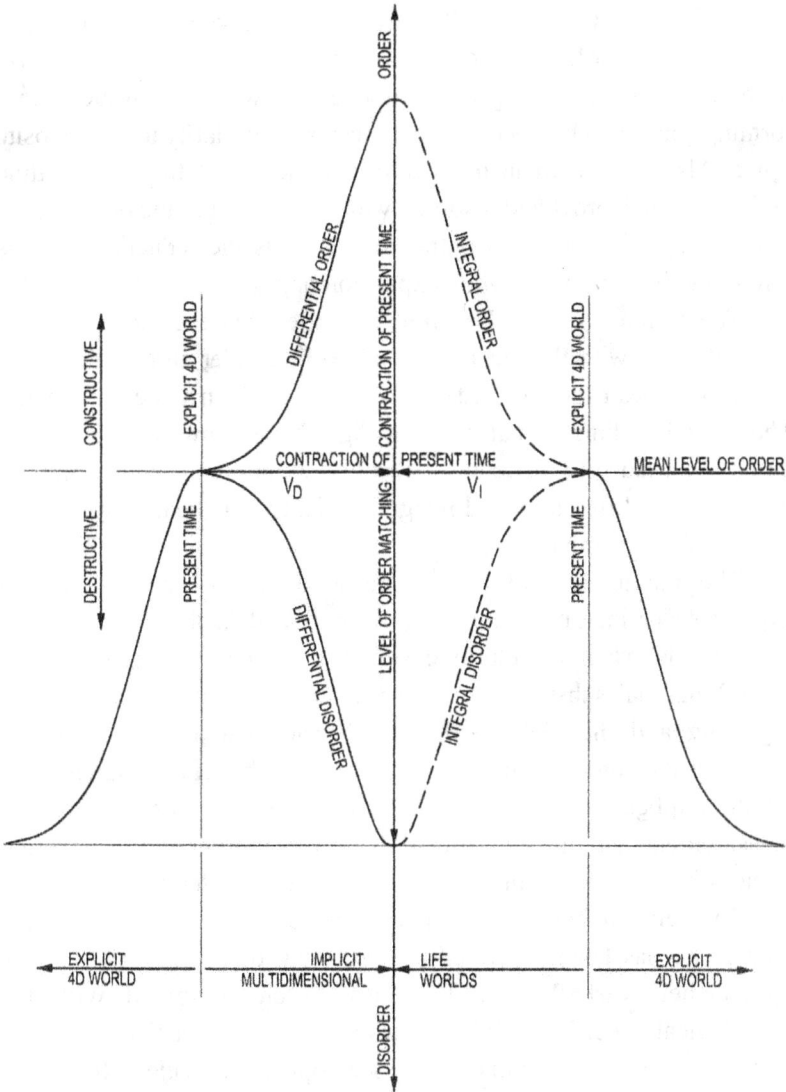

NOTE CONTRACTION OF PRESENT TIME UNFOLDS THE IMPLICIT
EXPANSION OF PRESENT TIME UNFOLDS THE EXPLICIT

FIG 4.8.1   EXPLICIT AND IMPLICIT DOMAINS RELATIVE TO IMPLICIT LIFESPAN
OF CONSCIOUS ENTITY

Fig 4.8.1 can be directly related to Fig 4.7.1 representing the Whole Ordering Centre, by allowing the mean level of order to move up or down the ordering column. The implicit at any given mean level becomes explicit at the higher mean level, a new implicit wave pattern forming above and below the higher mean and similarity in the opposite sense. Also, as shown in the figure, contraction of the present time unfolds implicit order and disorder, while expansion of the present time unfolds explicit disorder. Contraction supports the reduction process referred to in Section 4.5 while expansion opposes it.

The Reference Model therefore implies that the mean level of implicit order with the mean level of explicit order enfolded in it, or simply the mean level of order, is fixed by a given Ordering Centre. The explicit 4-dimensional world in Fig 4.8.1 is shown as full lines in terms of variance to distinguish it from the multidimensional implicit world, where differential and integral variance is indicated by full and dotted lines respectively.

The inanimate real axis corresponds to the mean level of explicit order in four dimensions, all enfolded in the mean implicit level of order which is multidimensional. The latter therefore represents all the material substrate from present human existence back to the big bang and this defines the real world. The multidimensional domain is the home of animacy, consciousness, belief and imagination, i.e. the implicit worlds enfolded in that we refer to as real. Scientific measurement and the interpretation of the present time by Homo Sapiens is that which applies at the mean level of order or real world.

The relation between the inanimate, consciousness and human knowledge has been described as consisting of three worlds by the philosopher Karl Popper. This description combined with the neurological expertise of John Eccles was further enunciated by them in the latter part of the 20th century; see Popper and Eccles (Ref 2.7.4). It is primarily a scientific perspective adapted to include the dualist interpretation of mind as opposed to matter. It is therefore not a reductionist viewpoint.

The Reference Model provides an alternative description using the entire range of human thought matched to the principle of a universal ordering sequence as described in previous sections, where increasing order represents the holistic perspective and decreasing order the reductionist perspective. A potentially infinite number of worlds form the background to this alternative scenario. This infinity of worlds is reduced in human terms to five main zones or worlds as shown in Fig 4.8.2. These are the real (physical/biological/autonomic) world or material substrate, two animate world levels pre and post the mitosis/meiosis transition (see Sections 7.2.3, 7.2.8.3 and 7.6.8.5), the belief world of coarse grained consciousness associated with the explicit and implicit senses and the imaginary world of fine-grained consciousness associated with intelligence. Each related by folding and unfolding processes.

In the Reference Model the real world unfolds from the mean level of order, initially a physical inanimate (abiotic) state it develops in to the biological (biotic) state, then in to the animate states, from these into the coarse-grained conscious world of the senses, and in turn the fine-grained world of knowledge, imagination and ideas including mathematics.

It does so under the influence of the multidimensional Ordering Centre as opposed to the explicit world which gets left behind as it were, all as exposed by procedures involving the basic scientific method as discussed in Section 7.4.1.

It is important to understand that the inanimate state is represented by the vertical and horizontal grid structure parallel to the s and v axes and the animate and conscious states by motion asymmetric to the grid.

Fig 4.8.3 illustrates the above in another way. The real and animate worlds unfold from the horizontal inanimate axis by wave motion into a coarse-grained conscious world and by further wave motion into the fine-grained world of ideas (knowledge and belief). All this is divided into the conscious (differentially unfolded) aspect, reductionist and analytical in character and the aware (integrally enfolded) aspect, holistic and emotional in character. The ordered wave motion is constructive and the disordered motion inversely destructive. The order and disorder levels at which the average human intelligence will apply should correspond to some intermediate level which includes a degree of fine graining in addition to the coarse-grained conscious level.

The inanimate, the animate and the coarse- and fine-grained conscious states should not be regarded as separate or mutually exclusive, rather they are complementary and intimately related.

ZONES APPLY TO EACH QUADRANT

A   REAL WORLD (PHYSICAL / BIOLOGICAL SUBSTRATE / AUTONOMIC)
B   ANIMATE WORLD (AUTONOMIC / ANIMATE TRANSITION)
C   ANIMATE WORLD (ANIMAL / CONSCIOUS TRANSITION)
D   BELIEF WORLD (COARSE GRAINED CONSCIOUSNESS / SENSES)
E   IMAGINARY WORLD (FINE GRAINED CONSCIOUSNESS / INTELLIGENCE)

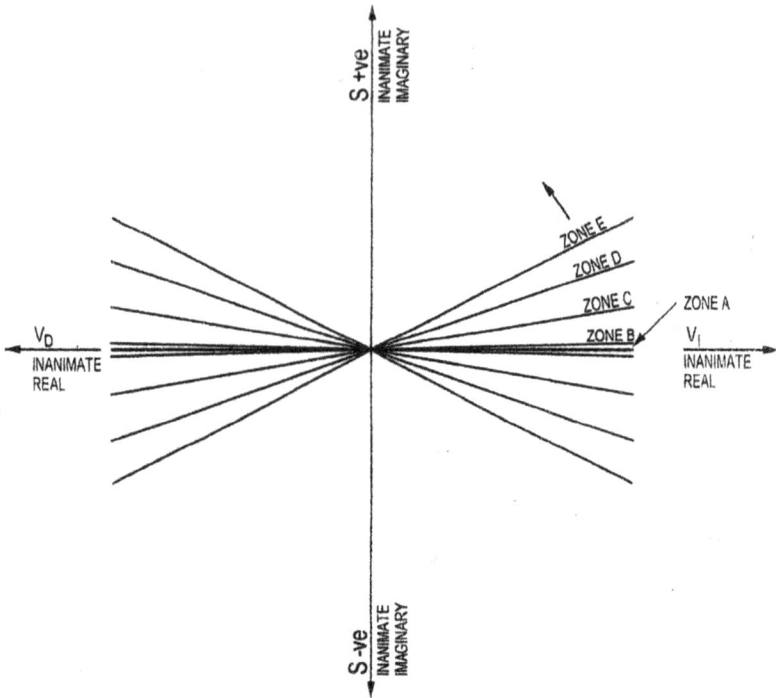

NOTES
WORLDS A THRU E ARE DERIVED FROM THE REFERENCE MODEL USING THE ENTIRE RANGE OF
HUMAN THOUGHT MATCHED TO THE PRINCIPLE OF A UNIVERSAL ORDERING SEQUENCE.
ALTHOUGH THE INDIVIDUAL WORLDS ARE SIMILAR TO WORLDS 1, 2 AND 3 AS DERIVED BY
POPPER AND DESCRIBED BY POPPER AND ECCLES (REF 2.7.4), THE LATTER ARE BASED ON
A MORE LIMITED RANGE OF THOUGHT AND SEQUENCING.
DETAILS OF POPPER'S WORLDS 1, 2 AND 3 CAN BE FOUND ON THE INTERNET.

FIG 4.8.2   THE CONCEPT OF DIFFERENT WORLDS RELATIVE TO THE
            INANIMATE AXES

Consciousness is enfolded in the animate that is in turn enfolded in the
inanimate from which unfolds the animate which in turn unfolds
consciousness.

98

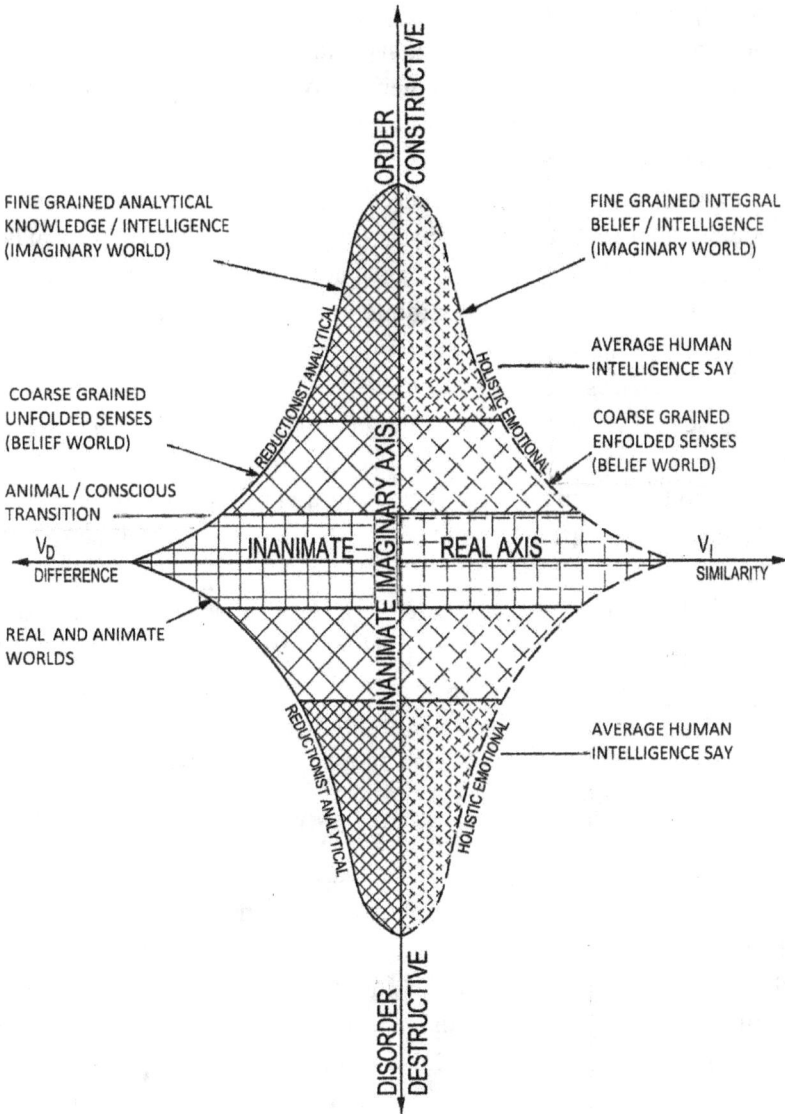

FIG 4.8.3   THE GROWTH OF CONSTRUCTIVE AND DESTRUCTIVE
CONSCIOUSNESS RELATIVE TO THE INANIMATE AXES

For example, we may initially take intelligent decisions in carrying out a given work pattern or activity, such as learning to drive but if the work is repetitive and certainly driving on a straight road without points of interest is in that category, then that which involved a high degree of concentration in the first instance becomes an autonomic condition to the point where no memory is retained of carrying it out. Some event or thought inevitably occurs in due course which stimulates return to the fully conscious state.

Figures 4.8.1 thru 4.8.3 relate the asymmetric vector motion of conscious entities relative to the motion of the symmetric inanimate axes i.e. the preferred reference frame, corresponding to that indicated in Figures 4.6.1 thru 4.6.3. The motion of individual human beings as opposed to the human race as a whole, the human race as opposed to other intelligences in the visible universe and so on is provided for in the Model by allowing for unfolding/enfolding vector motion in any direction through 360 degrees as described in earlier sections, all relative to the inanimate axes. This in turn allows for the inclusion of the range of opinions and beliefs of all those entities as will be shown below.

Fig 4.8.4 describes the contents of the different worlds in more detail, making the distinction between consciousness which is related to difference and awareness which is related to similarity. The increasing asymmetry of these worlds relative to the symmetric state on the inanimate baseline axis is also shown, that symmetric state being defined by the explicit 4D world.

Fig 4.8.5 demonstrates the asymmetry of the conscious/aware worlds in terms of difference and similarity. The differential aspect is primarily concerned with analysis and reductionist in character while the integral aspect is primarily concerned with synthesis and holistic in character. The Reference Model implies that the existence of the hemispherical split in the human brain and any consequent lateralisation of skills referred to in Section 2.7, should be associated primarily with the intrinsic requirement for difference and similarity; an asymmetry shown in the book to be directly related to the development of consciousness. The relation between consciousness, awareness and the Reference Model in general terms is discussed in Section 7.2.8.9.

Following the principles shown in Figures 4.8.1 through 4.8.5, Fig 4.8.6 relates the conscious world of unfolded and enfolded senses with the intelligent worlds of analytical knowledge and integral belief, whether used constructively or destructively. Action is then considered

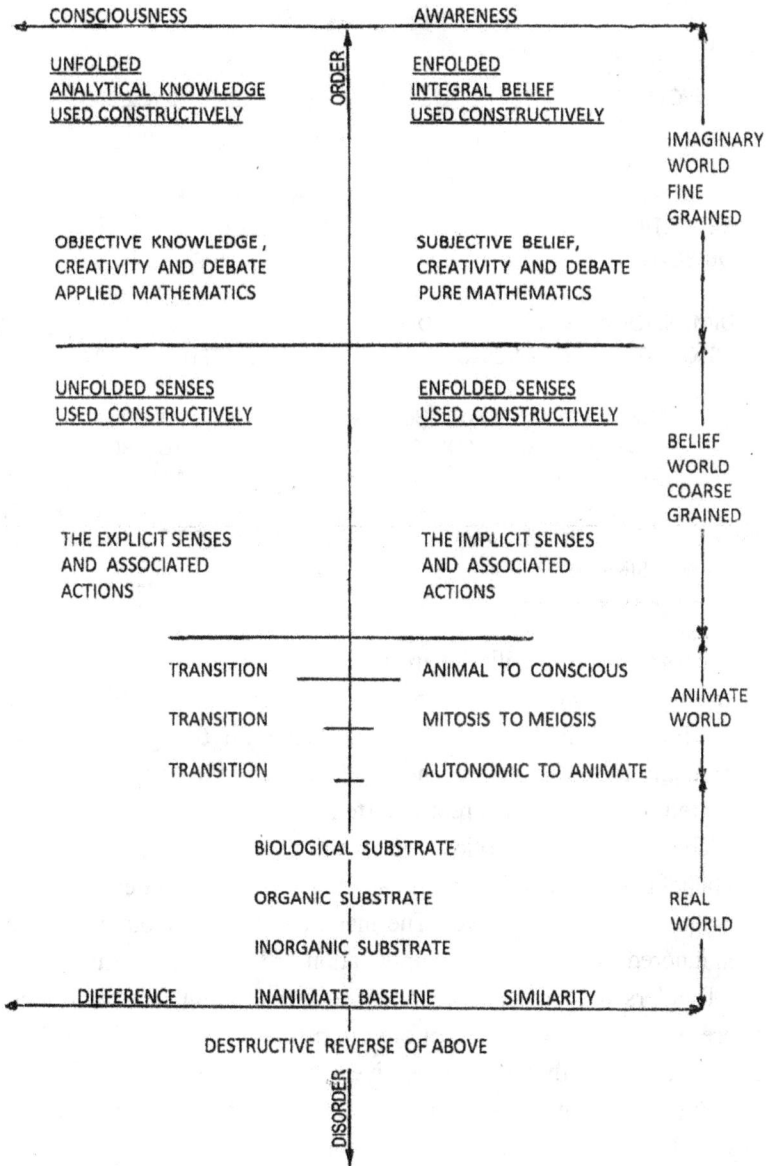

FIG 4.8.4   THE INCREASING COMPLEXITY AND ASYMMETRY OF
THE DIFFERENT WORLDS WITH GROWTH IN ORDER
RELATIVE TO THE INANIMATE BASELINE AXIS.

DIFFERENCE	SIMILARITY
CONSCIOUSNESS (UNFOLDED SENSES)	AWARENESS (ENFOLDED SENSES)
LINGUISTIC INNOVATIVE REDUCTIONIST OBJECTIVE	MUSICAL / ARTISTIC INTEGRATIVE HOLISTIC SUBJECTIVE
DIMENSIONAL ANALYSIS AND ASSOCIATED SEQUENCING	DIMENSIONAL SYNTHESIS AND ASSOCIATED INTEGRATION

NOTE   THE TABLE ILLUSTRATES THE ASYMMETRIC CHARACTERISTICS ASSOCIATED WITH CONSCIOUSNESS AS OPPOSED TO THE SYMMETRIC STATE ON THE INANIMATE BASELINE AXIS.

FIG 4.8.5   CONSCIOUS / AWARE ASYMMETRY RELATED TO DIFFERENCE AND SIMILARITY

a process associated with the growth of wave motion and the expansion of the animate world to the belief world as activated by the human cerebellum, while thought is associated with the growth of wave motion and the expansion of the belief world to the imaginary world as activated by the human cerebral cortex.

The figures which follow relate various aspects of human activities and belief to the main properties of an Ordering Centre, order and variance, all relative to the mean level. The inanimate and autonomic zones have been ignored for purposes of simplification and to ensure clarity.

Readers may wish to reverse the left to right orientation in the figures or indeed the top to bottom orientation if that is their preference or to suit commonly held perspectives.

Fig 4.8.7 relates firstly human philosophy to the above properties. Physical epistemology and the philosophical viewpoints corresponding to positivism, determinism and nominalism are in the first instance to be associated with the mean level of explicit as opposed to implicit order.

Neutral monism assumes the physical and metaphysical worlds are differing ways of ordering the same neutral stuff, which is neither physical nor mental. In this sense an intermediate state relative to other

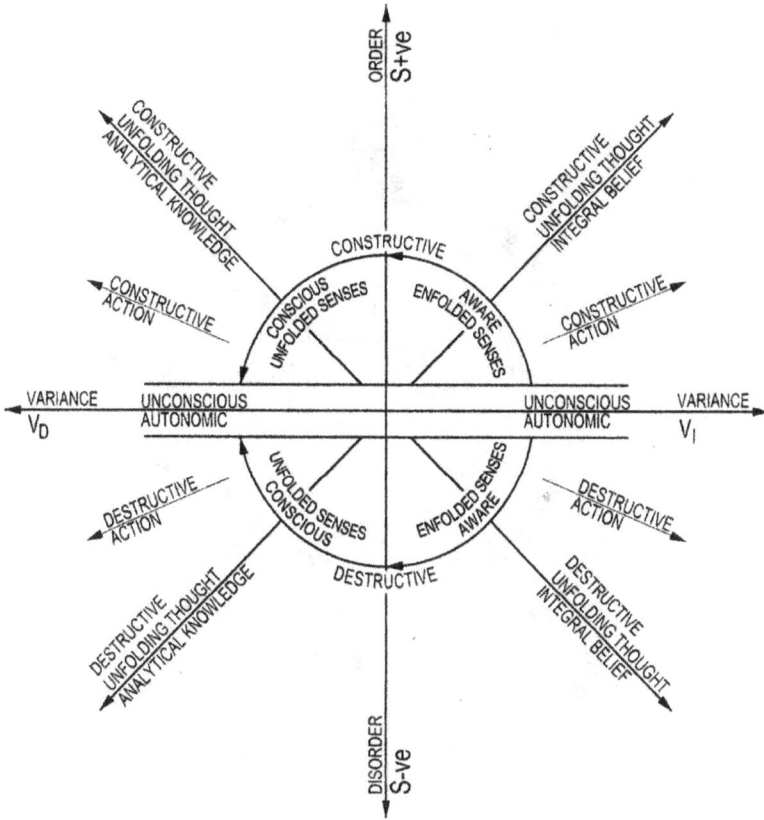

FIG 4.8.6   THOUGHT AND ACTION RELATED TO ORDER AND VARIANCE

forms of monism such as materialism and idealism. It is therefore to be associated with the mean level of implicit as opposed to explicit order. Since the mean explicit level is enfolded in the mean implicit level, positivism, determinism, nominalism and neutral monism are all shown as related to the variance axis.

The present state is associated with metaphysical ontology and the philosophical concept of freewill. Increasing order is associated with what may be termed positive absolutism, religious monism, morality and ethics, and in the opposite sense of increasing disorder with what may be termed negative absolutism, nihilism, immorality and the unethical.

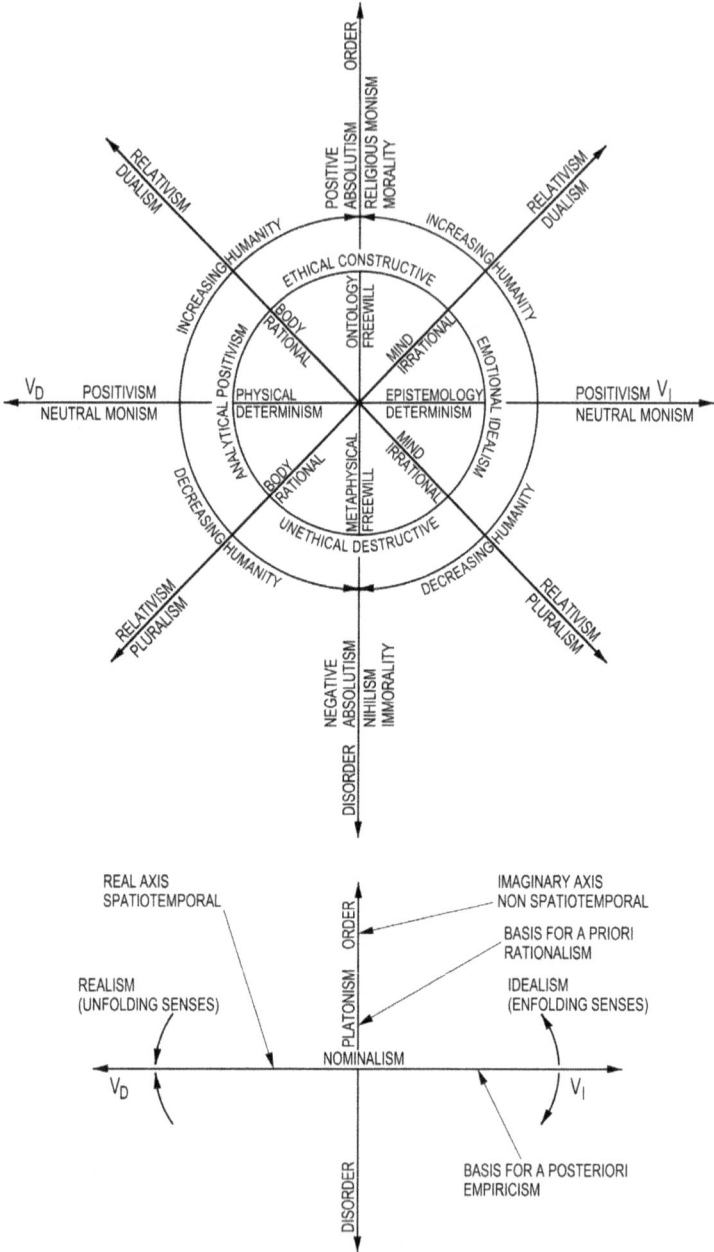

FIG 4.8.7   HUMAN PHILOSOPHY RELATED TO ORDER AND VARIANCE

Accordingly humanity increases as order increases, giving rise to the philosophy of humanism above the neutral axis of variance. Between these states lie forms of relativism and the philosophical concepts of dualism in the ordering direction which decays to pluralism in the direction of disorder.

The left hand side of the diagram is further associated with analytical positivism, the rational and concepts associated with the body, while the right-hand side is associated with emotional idealism, the irrational and concepts associated with the mind.

The lower half of the figure relates realism to idealism and rationalism to empiricism. Realism is related to the unfolding senses at the mean level of order and idealism to the enfolding senses.

A priori rationalism is related to the non-spatiotemporal axis and a posteriori empiricism to the spatiotemporal axis.

Nominalism is related to the spatiotemporal axis as opposed to Platonism (transcendental realism) which is related to the imaginary non spatiotemporal axis.

The reasons for the above are more fully discussed in Section 7.0.

Fig 4.8.8 relates politics again to the same properties. Difference is again associated in general terms with politics which support free enterprise and similarity with politics which favour state control. Again more specifically order is associated with stable democracy and disorder with unstable autocracy while difference is associated with egocentric individualism and similarity with radical socialism. Stable democracy splits further into moderate constructive conservatism and socialism as indicated by the radial arrows shown in the figure, essentially honest, law abiding and responsible in character. The mean level of order and variance corresponds to democratic liberalism, effectively middle of the road politics, which can in turn be related in psychological terms to tender mindedness.

At the opposite end of the political spectrum unstable autocracy may take the form of regimes which are simply corrupt, criminal and irresponsible in character or split into either extreme fascist or extreme communist forms of totalitarianism, which in turn can be related in psychological terms to tough mindedness.

Fig 4.8.9 relates economics again to the same properties. Difference is associated in general terms with private capital orientated economies

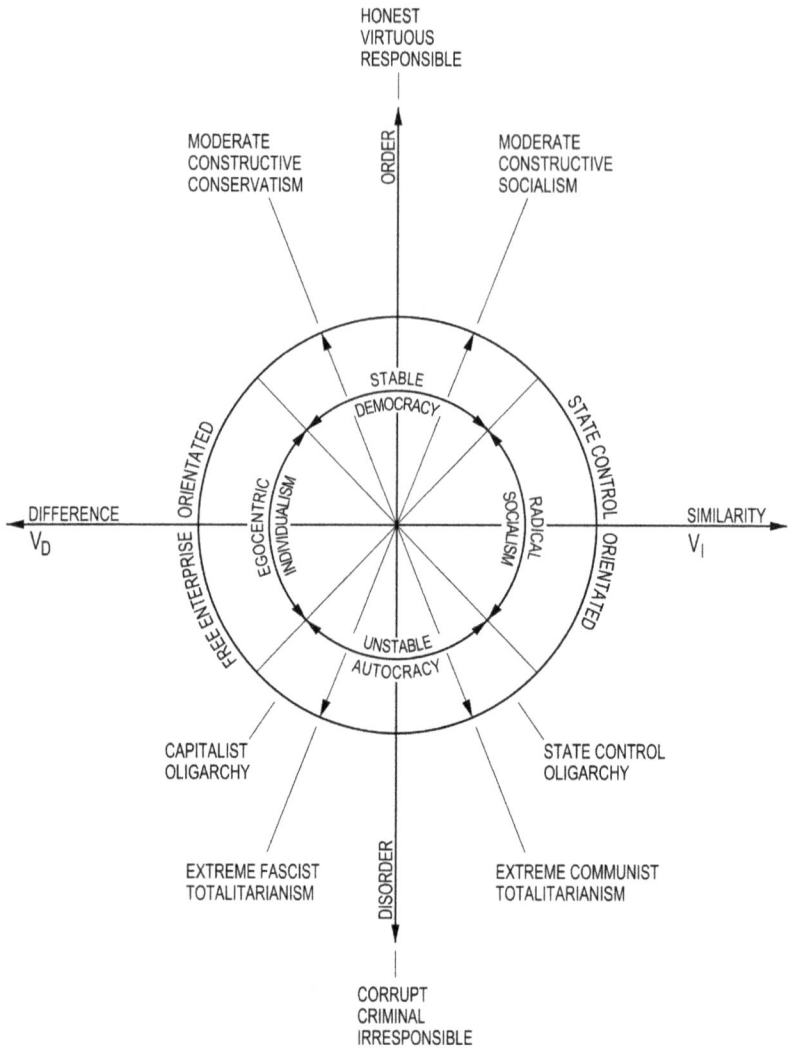

FIG 4.8.8   POLITICS RELATED TO ORDER AND VARIANCE

and similarity with state control orientated economies. More specifically order is associated with stable mixed economies and disorder with unstable corrupt (discordant) economies, while difference is associated with *laissez faire* economies and similarity with those centrally controlled. All relatively ordered economies, whether capitalist orientated or state controlled, may grow constructively or suffer at

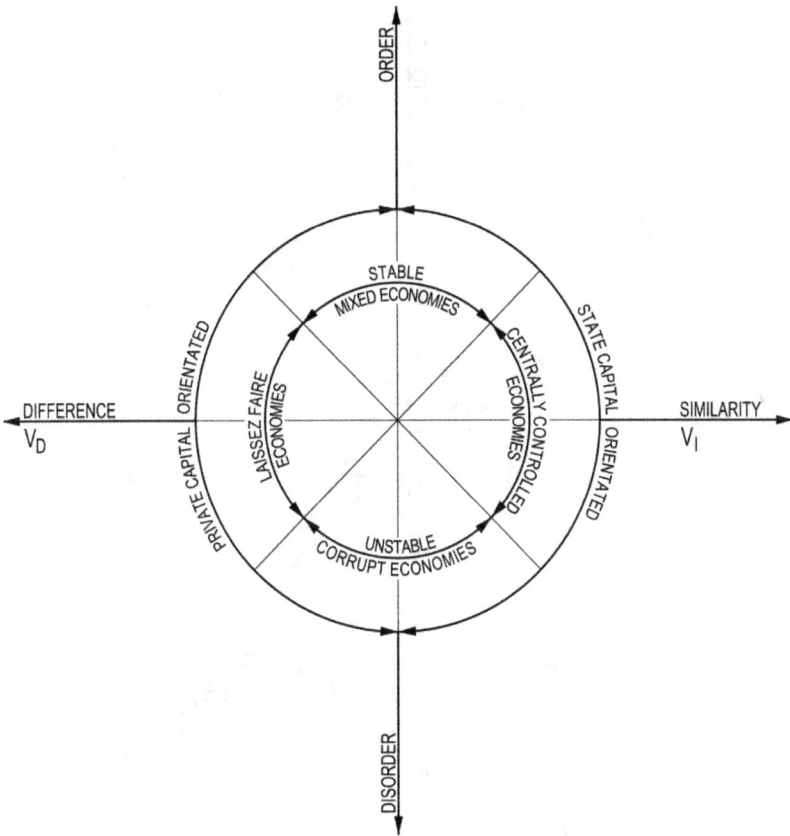

FIG 4.8.9   ECONOMICS RELATED TO ORDER AND VARIANCE

times the opposite, with varying degrees of disorder from relatively minor to calamitous.

Fig 4.8.10 gives an overall view of the relationship between reality, belief and imagination relative to order and variance. Unfolding differential variance and integral variance gives rise to spatiotemporal reality (the explicit, objective and measurable) on the real axis at the mean level in an Ordering Centre where order is in balance with variance as described earlier.

Belief and imagination (the implicit, subjective and immeasurable) being non-spatiotemporal are then related to animate vector motion across the grid.

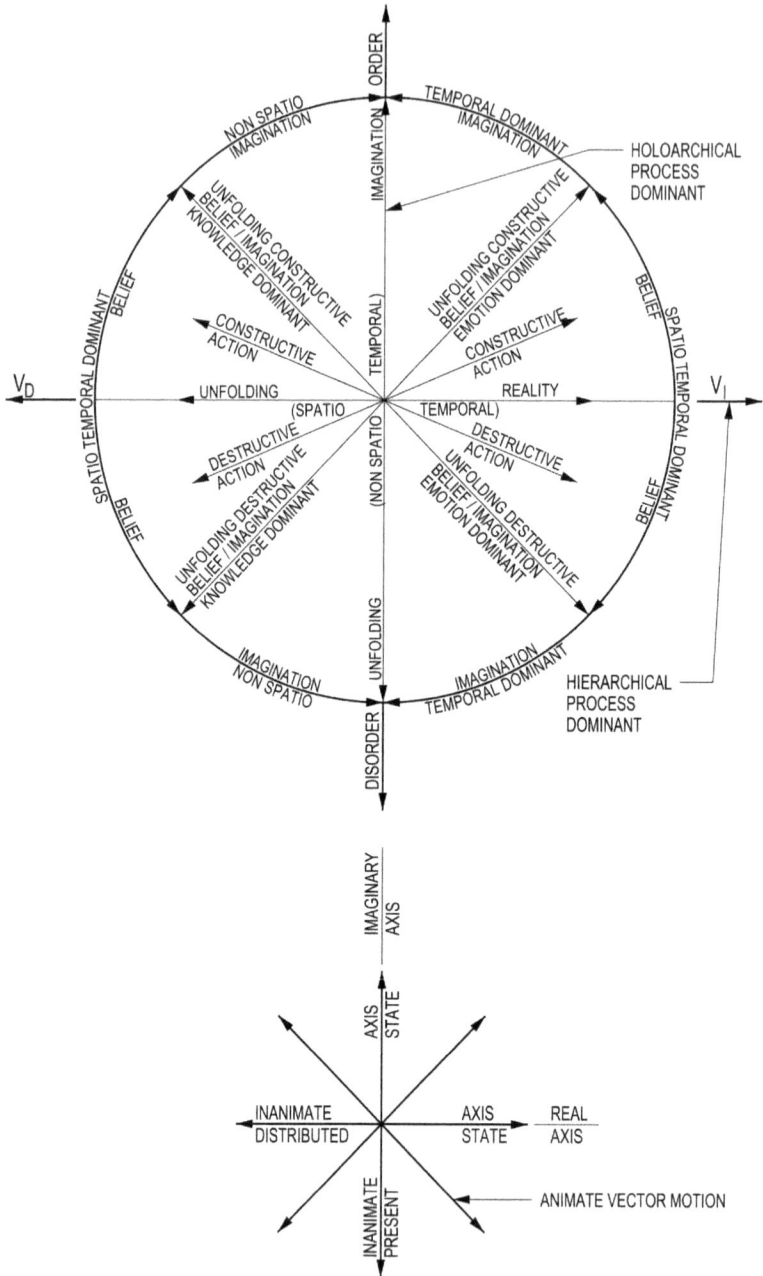

FIG 4.8.10   REALITY, BELIEF AND IMAGINATION RELATED TO ORDER AND
VARIANCE

Belief is very much concerned with practical affairs on a day-to-day basis and therefore tends to be dominated by the distribution of variance (the real axis); imagination being in effect other worldly is dominated by the present state (the imaginary axis). Again order is associated with constructive belief and disorder with destructive belief. Likewise differential variance is associated with belief that is knowledge dominant and integral variance with belief that is dominated by emotion.

It is this background which explains our common-sense notion of a fixed reality while also explaining the interaction of that reality with our ability to believe and imagine, a facility to which we are equally attached.

At the mean level a hierarchical structure predominates giving rise to atheistic/physical/reductionist philosophies in which the importance of measurement is stressed. Above that level imagination and a holoarchical structure predominate giving rise to theistic/metaphysical/holistic philosophies in which emphasis is laid on wholeness and the spiritual as opposed to measurement. Below that level order ultimately collapses to an essentially inanimate base giving rise to the concept of parallel universes and chaotic inflation referred to in Section 4.4.

Fig 4.8.11 relates secularism and religion to the properties of an Ordering Centre. Scientific records, observations and predictions are considered to relate to the mean level of order (the distributed state) and represent a state of non-belief, atheistic (non-God) in character, all tempered by the associated ethical philosophy of humanism, in mitigation of the cold world implied by the 'red in tooth and claw' description of nature corresponding to reality at this level. Belief in the secular equates to unfolding the present state as per Fig 4.6.4 (b) and swinging the bell curve of Figures 4.8.1 and 4.8.3 through 90 degrees from vertical to horizontal, emphasising reality. Agnostic states corresponding to suspended belief in a deity occupy sectors of the figure adjacent those of atheism, as explained in Section 7.5.2.

The figure is not intended to reflect the numbers of adherents in respect of the four sectors involved, which are shown as if equally distributed.

Religion on the other hand is concerned primarily with belief in wholeness (holism) which is concerned with other levels of order.

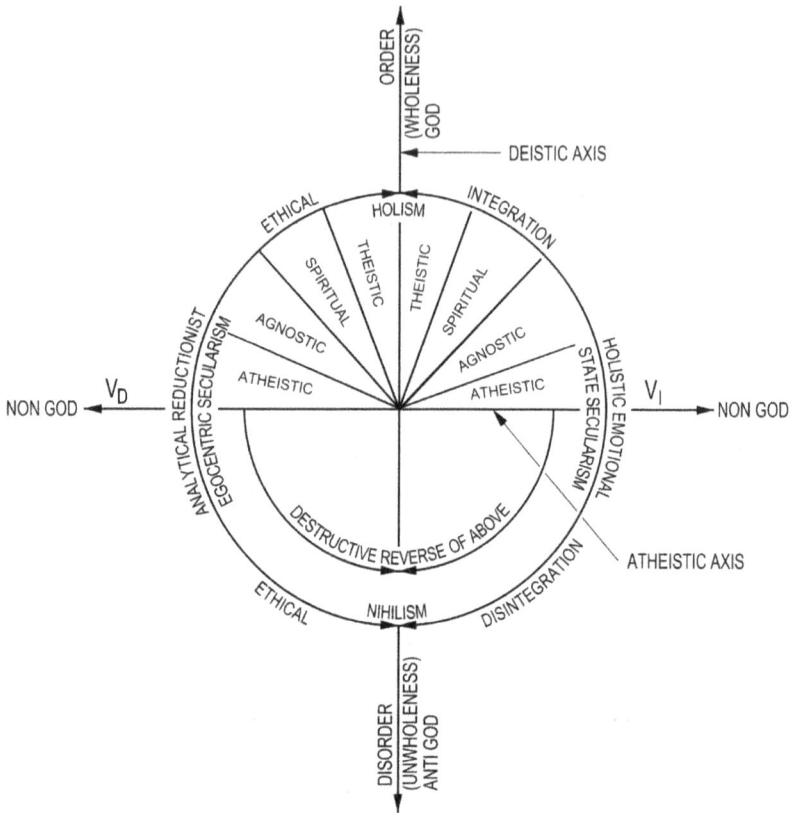

NOTE AGNOSTICISM IS A STATE OF SUSPENDED BELIEF IN A DEITY.
SCIENTIFIC RECORDS, OBSERVATIONS AND PREDICTIONS
CORRESPOND TO A STATE OF NON-BELIEF (ATHEISM)

FIG 4.8.11   SECULARISM AND RELIGION RELATED TO ORDER AND VARIANCE

Constructive theistic belief (personal God) and deistic belief (non-personal God) corresponds to increasing order and conversely destructive nihilism (anti-God) corresponds to the complete breakdown of order. Belief in religion equates to enfolding the present state as per Fig 4.6.4 (a) with the bell curve as shown in Figures 4.8.1 and 4.8.3, emphasising imagination.

Spiritual beliefs more associated with the individual than mainstream religion occupy sectors of the figure adjacent those of theism, as explained in Section 7.5.2.

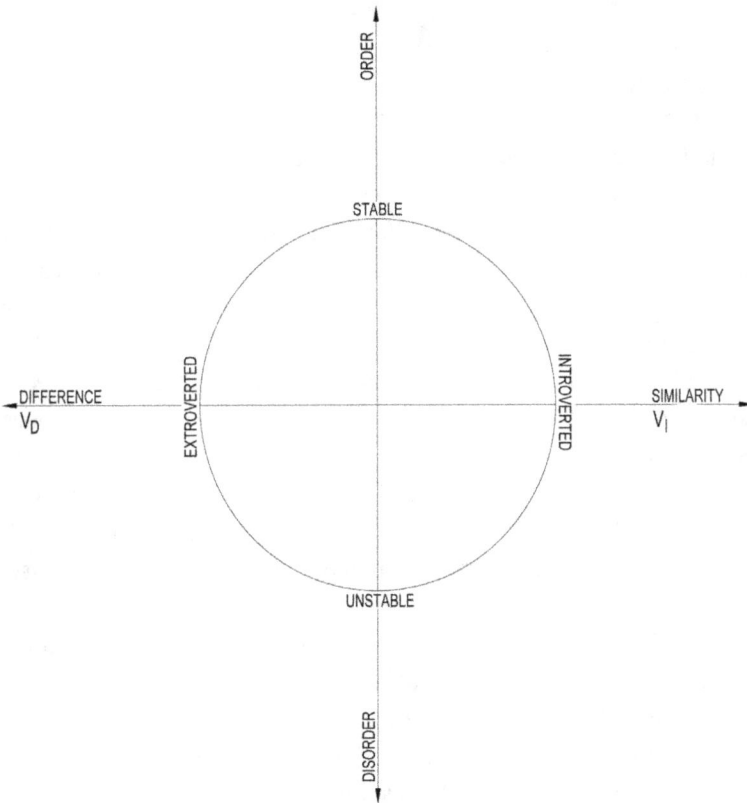

NOTE
THE READER CAN ALIGN EYSENCK'S EXTRAVERSION / INTROVERSION DIAGRAM
(REF 4.8.1) TO ORDER AND VARIANCE AS SHOWN ABOVE.
EYSENCK'S INVENTORY DIAGRAM CAN ALSO BE FOUND ON THE INTERNET

FIG 4.8.12   HUMAN PERSONALITY RELATED TO ORDER AND VARIANCE

The left-hand side of the figure is reductionist and analytical in character while the right-hand side is holistic and emotional, in confirmation with previous figures.

Differential variance is associated with egocentric secularist forms of atheism while integral variance is associated with secular state forms of atheism such as communism.

The sectors shown between secularism and nihilism represent the disintegration associated with the collapse of ethics and morality, which ultimately may take the form of psychopathy or sociopathy in the individual or collective forms of the same such as extreme fascism or totalitarianism and the like.

In particular the religious belief in rising upwards towards heaven can be associated with the growth of order and the inverse belief of falling in to hell with the increase in disorder.

Belief in God can clearly be equated to this growth of order and belief in evil to unfolding disorder. Basic religious beliefs, their interpretation within the Reference Model and the difficulties of relating such beliefs to scientific principles will be discussed in more detail in Sections 5.0 and 7.5.

Fig 4.8.12 relates human personality to order and variance. Order is associated with stable personalities and disorder with unstable ones on average. Difference is associated with extroversion and similarity with introversion again on average. The personality characteristics can be aligned with Eysenck's personality inventory which shows such characteristics in a comprehensive manner familiar to psychologists (Ref 4.8.1).

Finally Fig 4.8.13 relates human emotions to order and variance. Order is associated with love and kindness, disorder with hate and cruelty, active difference with anger and aggression and passive similarity with fear and timidity. These emotions can be aligned with Plutchik's wheel of emotions which shows such characteristics again in a comprehensive manner familiar to psychologists (Ref 4.8.2).

Human personality is essentially constant over the lifetime of an individual at least that is generally the case. Not so with human emotions which can and do change rapidly, emotions are therefore transient. Personality and emotions in terms of the Reference Model are further discussed in Section 7.6.4 of this book.

Taken collectively the figures in this section explain the beliefs and prejudices of the typical human being which buzz around in his/her head like flies on a summer's day, as the philosopher Bertrand Russell described it, though he never considered that this also applied to himself.

It is appropriate at this point to address the question that plagues science in particular and philosophy in general, namely the arrow of time and why we can run a film reel forward and back whereas in practice we never see this happen in life. The explicit arrow of time

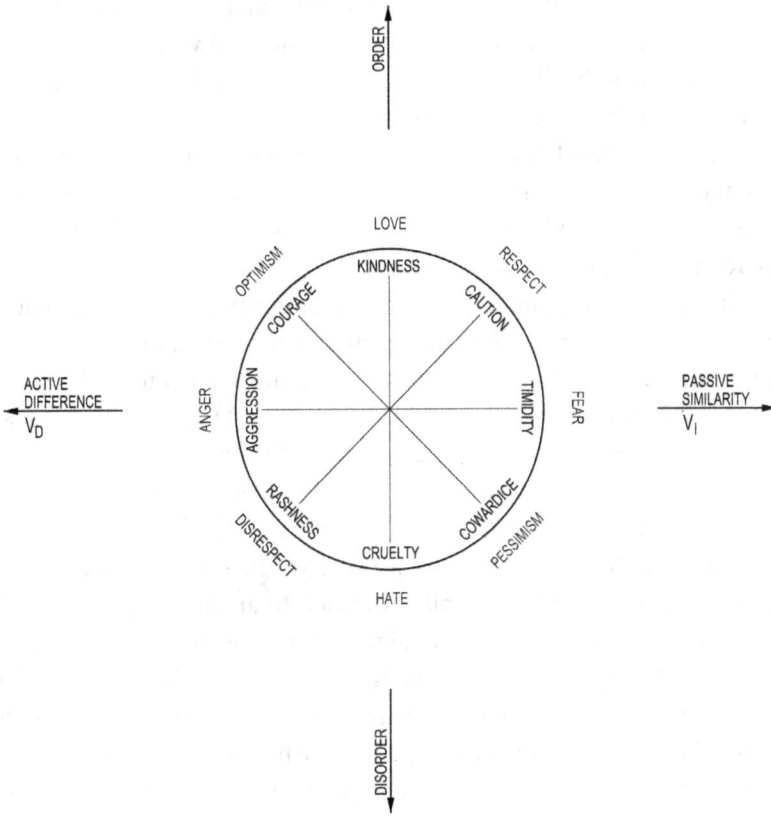

FIG 4.8.13   HUMAN EMOTIONS RELATED TO ORDER AND VARIANCE

shown in the earlier Fig 4.3.4 moves relentlessly on and we can only return to the past in terms of our capacity for memory, yet physics in general implies that the world is symmetric since the equations involved work forwards and backwards in time with equal facility. H. Price (Ref 4.8.3) provides an informative background to the issues involved.

The Reference Model appears to explain this conundrum. It is evident from the figures presented so far that the Whole Ordering Centre envelope is symmetrical about the vertical present state axis and that the mean level is a region where the principle of measurement becomes dominant. In relation to the horizontal axis inanimate motion applies equally in both the differential and integral variance directions as the inanimate has been defined in the Model as being in a state of balance. Now the direction in which order decreases (entropy increases) from the mean level is also the direction in which space-time unfolds (expands) and the arrow of time points in the direction of that unfoldment; this is also the sense in which we expect the capacity for life termed herein as animacy to decrease.

In the opposite direction order and animacy increase and space-time enfolds (contracts) but the explicit arrow of time continues to point in the direction of space-time expansion even as animacy increases in the opposite sense. However the Model shows that as animacy increases so does the capacity for imagination. It takes imagination in the form of belief to unfold order and science is in the measurement business not the belief business.

So the unfoldment of implicit order or disorder is equivalent to finding a more centralised implicit present. In an Ordering Centre this state is effectively non-dimensional such that time and space have little meaning. The explanation does not finish there as the Model specifically defines animacy as involving a state of imbalance (asymmetry) with vectors involving motion with respect to the balanced state (symmetry). So animate entities are capable of choosing whether order is to increase or decrease relative to the fixed inanimate axes.

The relation between inanimacy and animacy and the nature of time described in the Model is further discussed in Section 7.4.11. It fits well with J. Lovelock's Gaia philosophy (Ref 4.8.4 and 4.8.5).

5.0   PRIMARY CONCLUSIONS.

The most important conclusion of the Standard Reference Model is that reality and imagination and their counterparts science and religion, when shorn of incoherent reasoning and belief in the pursuit of the absolute, can be fused together using a metaphysical model, rendering them almost but not completely compatible. Belief in the absolute distorts the world as the Model explains all too clearly.

The human race as a whole cannot refute the sum total of the records it creates (as defined in the Introduction) whether real or imaginary; furthermore those records must reflect the true nature of existence since humanity is the product of it in both the epistemological and ontological sense. The same conclusion applies to the totality of consciousness in the universe. The Standard Reference Model represents an overall picture of existence almost entirely consistent with the mean of the complete range of statements made about it by the human race as a whole, in the same way as a dictionary in a given language is intended to be a complete range of words contained within that language. We do not eliminate words from the dictionary on the basis that they are not consistent with a particular opinion or belief. Since objective and subjective philosophies in general indulge in the practice of ignoring, mainly in effect for propaganda purposes, what each considers variously illusory or irrelevant, it is about time that a theory was put forward which takes what may be termed a neutral view in this regard.

While it may be true that there are no neutral men there are advantages to looking at the range of viewpoints in any given situation and none more so than when endeavouring to establish a consistent theory of existence. It should be obvious that irrespective of the utterances made by any individual or group of individuals no matter how large or small they must in some way be related to how existence functions.

Since the mean is an inanimate condition conscious entities oscillate to and fro about that mean, as does ultimately the Holomovement as described in Section 4.7.

The Model considers at length and sometimes with monotonous repetition the relationship between order and disorder, the absolute and the relative, the objective and the subjective, reductionism and holism. When these relationships are examined it becomes clear why the philosophies held by individuals or groups of individuals cannot agree one with another but instead argue over every last detail forming a probability distribution of differences and similarities as an inevitable consequence of the ordering/disordering process.

Only two axes real and imaginary are required on paper to express these relationships in accordance the principle of the hologram. The real axis, the abscissa, can be regarded as a specific sequence of difference and similarity at a given level of order, while the imaginary axis, the ordinate, represents the ordering/disordering process in the form of many sequences of difference and similarity.

If these axes are considered fixed, lines parallel to these axes form a conventional Cartesian grid which represents an explicitly ordered holographic system unfolded in four dimensions. There is an identifiable level in this explicit system termed the mean level of explicit order corresponding to the normal Gaussian distribution. Any given level in the explicit system corresponds to the explicit (measured) present time at that level relative to the big bang. Scientific measurements in particular the Standard Cosmological Model and Standard Particle Model are formed from experimental observations related to this four-dimensional unfoldment.

The inclusion of fractal properties however reveals that both the main axes split into a profusion of ordinates and abscissae forming what is regarded as a multidimensional implicitly ordered system constituting in effect a non-Cartesian grid all enfolded relative to the explicit. The further inclusion of wave motion in the form of standing waves which support these structures both in a vertical and horizontal sense completes a picture of the inanimate part of the Reference Model.

The determinate yet holistic description applicable to the multidimensional quantum state prior to observation correlates with the vertical grid or ordinates. The determinate again holistic description applicable to multidimensional classical physics correlates with the horizontal grid or abscissae together with the four-dimensional unfoldment we call reality, in accordance the properties of locality, causality and continuity. Our observation of that reality derives from a reduction process which converts determinate multidimensional classical and quantum states to the equally determinate explicit level.

It is considered that the big bang, cosmic expansion and classical physics including general relativity together with the unitary quantum state cannot be fully understood without taking into account the inanimate grid structure and the asymmetric motion relative to it.

The discrepancies in the associated theories in physics then arise directly from the failure to allow for non-locality and the associated motion arising from the Implicate ordering process.

The animate state is then represented by asymmetric wave motion across the inanimate grid arising from vectors with origins on the grid and termini on the wave or vice versa, all associated with our sense of imagination which gives rise to our ability to comprehend that which is not measurable and corresponding non-local, non-causal, non-continuous features.

The animate and the growth of consciousness and intelligence can be correlated with the above wave motion, which is housed in indeterminate Implicate space, defined as being asymmetric to the above determinate states.

The combined inanimate and animate structure forms what is termed an Ordering Centre.

The Model goes on to show that the human race observes existence from the vantage point provided when hierarchical processes are in balance with holoarchical processes, forming a mean level of implicit order where reality and imagination have equal status. It is the formation of this level which is responsible for our sense of the absolute truth and our love of the quiz with the single correct answer.

The breakup of this level in a hierarchical sense is to be associated with increasing explicit (chaotic) disorder, which explains innocent unintentional error and basic ignorance of that generally considered the absolute truth.

The breakup of this level in a holoarchical sense is to be associated with increasing animate/conscious intention to conceal the truth for evil purposes, with increasing disregard for the preservation of life, namely implicit disorder leading ultimately to random behaviour.

In fact there is no absolute truth, only an ever greater truth as revealed by increasing implicit order, as discussed in more detail in Section 7.1.9.

The Model demonstrates that life, the range of human philosophy, politics, economics, belief, personality and emotions can in turn be explained accordingly.

What appears absolute to us is relative when viewed from the perspective of the whole of existence which consists of an unbounded immeasurable system in wave motion relative to a bounded measurable supporting structure in some sense like flesh relative to its skeleton.

Science is always looking at the skeleton, while religion is concerned with the flesh.

So reality, belief and imagination as it were compete to create many levels of existence. Hence the differing perspectives while appearing at odds are merely facets of existence as a whole.

Conscious entities rise or fall in the ordering structure according to the strength of their belief by action or thought creating asymmetric wave motion related to the Implicate state which transfers energy by unfolding order or alternatively disorder relative to the mean implicit level in the Ordering Centre. Rising/falling in the system involves change to that entity such that it becomes more/less whole achieving higher/lower states of Implicate order by the dissipation of those boundaries which apply at the mean level.

The Model uses the concept of different worlds as described in Section 4.8 as a means of relating reality, belief and imagination.

Under the influence of the Ordering Centre these differing worlds progressively unfold from the explicit four-dimensional world exposed by the basic scientific method. The real world initially abiotic becomes biotic to the point where biological and autonomic states become evident. Basic animate actions become apparent in the lower form of the animate world and more complex actions in the upper form of that world. Coarse grained forms of consciousness arise from wave motion in the belief world. Fine grained forms of consciousness giving rise to creative thought of both an analytical and emotional nature including religious prayer can be related to wave motion in the imaginary world. Action as opposed to prayer can be correlated with the 'praise the Lord but pass the ammunition' attitude.

In short the basic principles of science and religion are moved lock, stock and barrel into a unified metaphysical structure, thereby removing the mind body problem, all supported by the totality of human statements. It is the conception that there is a mind body problem that has been the bane of metaphysics since the time of the Greek philosophers.

The primary reason for this is that the asymmetric wave motion referred to above is animate motion in implicit multidimensional space as opposed to inanimate motion in an explicit four-dimensional Cartesian space, the latter unfolding with time from the former.

This asymmetric motion is not accessible by scientific methods of measurement in terms of physics, which relate specifically to the four-dimensional explicit order. Scientists in general are therefore atheistic in outlook, their viewpoint corresponding to Fig 4.8.1 but with the bell curve aligned with the variance axis as opposed to the ordering axis.

Consciousness involves mixtures of reality, belief and imagination as figures in Section 4.0 show, individuals regarding reality as primary and imagination as secondary for vectors closer to the mean level and vice versa for vectors closer to the present state axis.

The mean level of implicit order defines the base of the real world, the latter comprising the base physical, biological and autonomic substrates of plants, animals and conscious entities.

As a consequence of all the above what can be regarded as the present time contracts as implicit order or disorder increases and expands as explicit disorder increases as shown in Fig 4.4.5. The reduction process defines the degree of explicit order at the mean level of implicit order. The implicit present time can take any value from the inanimate real world base figure on the scale of billions of years to the period of approximately one second when related to Homo sapiens, as explained in Section 7.4.11.

This explanation fits well with J. Lovelock's Gaia hypothesis (Ref 4.8.4 and 4.8.5).

The multidimensional states shown in Fig 4.5.1 are ensembles of order and disorder representing a profusion of potential futures and pasts, space-time only becoming apparent as a consequence of the reduction process as further shown in Figures 4.5.2 and 4.5.3.

The area under the bell curve in Fig 4.5.1 has many levels in terms of the four-dimensional expansion of space-time, disorder increasing with it. However these levels unfold from the Ordering Centre which is housed in multidimensional Implicate space as portrayed in that figure.

So the Model supports the principle that the real state of the conscious entity by reason of its own individuality has one life in explicit terms, since it is the explicit which defines the physical boundaries applicable to that individual; hence the phrase 'we only have one life'.

The real state or bodies of such entities unfolds in four dimensional space-time from the multidimensional ensemble at conception and is re enfolded in the opposite sense at death.

The religious belief that life is not however limited to this explicit description, is directly revealed by the implicit domain illustrated in Fig 4.8.1. The belief and imaginary states of such entities oppose the reduction process by expansion in to that domain, which involves motion in Implicate space by contracting the mean implicit present time, either by increasing or decreasing implicit order relative to the mean implicit level. Such states give rise to spiritual belief in the soul widespread in human society.

Belief and imaginary state motion is provided by asymmetric vectors whether constructive or destructive as shown in Figures 4.5.5 thru 4.6.3.

As explained in Section 7.5.6 the real state of the conscious entity is totally dominated by locality, the intermediate state of belief equally by locality and non-locality while the imaginary state is correspondingly dominated by non-locality. Non locality contracts the present time dissipating the boundaries imposed by locality. So the concept of a conscious entity as an individual then loses validity being superseded by a global consciousness or by the decay of that global consciousness all governed by the concept of Wholeness or unwholeness as applicable. This explains simultaneously the life birth/ death cycle of conscious entities, associated with locality and the dissipation of such cycles when global consciousness or its decay prevails, associated with non-locality. Non-locality explains the belief in eternal life, which in terms of implicit order corresponds to an ever more defined present state.

By Wholeness/unwholeness here is meant the degree to which an individual increases /decreases implicit order with respect to the mean level housing that individual's inanimate core.

All this is opposed to the idea that existence pushes the conscious individual around, so that he/she is not so responsible but is the victim of it.

The consciousness that the above entities possess whether constructive or destructive is therefore changed in form by wave motion towards a more central present time.

The reader can visualise such motion and the processes involved by studying the principles of wave reinforcement and cancellation referred to in previous sections, from which Figures 4.7.1 and 4.7.2 are derived, all as further explained in Section 7.5.6.

In particular consciousness involves what can be termed three component states.

Firstly a real explicit, spatiotemporal (finite dimensional), order neutral component reductionist in orientation and therefore partial and incomplete by nature. This component corresponds to the physical concept of the body.

Secondly an imaginary implicit, non spatiotemporal (dimensionally enfolded), order positive component, holistic in orientation and therefore integral by nature associated with wholeness; overall a sense of togetherness. This component corresponds to the spiritual concept of the good soul. It mirrors in real terms mass diffusion and particle aggregation at the Big Bang.

Thirdly an imaginary implicit, non-spatiotemporal (dimensionally enfolded), order negative component, anti-holistic in orientation and therefore fractional by nature associated with unwholeness; overall a sense of isolation. This component corresponds to the spiritual concept of the bad soul i.e. that which separates or isolates from the good soul. This mirrors in real terms mass aggregation and particle diffusion as space-time expands indefinitely.

Ultimately the principle of the Implicate Order confirms the concept of direction and purpose in the form of a preferred ordering direction as non-spatiotemporal conditions dominate those of a spatiotemporal nature.

In respect of religion further confusion arises again from an inability to comprehend the nature of the explicit present state as opposed to the implicit present state and the complexity of the interaction between reality, belief and imagination as described in the Model.

Specifically this is demonstrated by the belief in the afterlife and reincarnation.

The religious belief in the afterlife seems to omit any idea of motion before physical conception and suggests it is possible to reappear after physical death at another level of order at a later time.

The belief in reincarnation does involve the idea of motion before conception and after death but seems to assume continuous return to the same level of reality corresponding in effect to the same mean level as above, except where what is referred to as the state of 'nirvana' is attained by spiritual means.

As far as the Reference Model is concerned these beliefs are disordered, meaning out of order.

Notions of before and after, arise primarily from the collapse of implicit order and disorder in favour of the local distributed state; vice versa the growth of implicit order and disorder in favour of the present state at the expense of these beliefs.

Summarising the religious side of the debate it is concluded that the confinement of the conscious entity to a particular birth, life, death cycle should be related to a given mean level of implicit order and that the associated wave motion interacts with wave motion at higher/lower levels of order in the Ordering Centre. All of which gives rise to the belief in what is referred to as life in 'eternal' heaven or hell, as experienced by the soul of the entity.

The belief in Heaven and rising into it refers to this process involving the unfolding of higher levels of implicit order, ultimately leading to the Whole Ordering Centre; likewise the belief in Hell and descending into it refers to the opposite process involving the progressive breakup of the Ordering Centre, the collapse of implicit order and its replacement by implicit disorder, ultimately leading to the random state.

It is the random state and the approach to it as indicated by the central condition of the Ordering Centre (defined by the intersection of the present and distributed states) which constantly stirs the pot, setting order against disorder, reality against imagination, science against religion and so on.

Beliefs which support the existence of the individual outside that provided by the formation of 'our' Ordering Centre are clearly related to the range of harmonics in the Whole Ordering Centre, which offer other mean levels at which reduction processes again come into play. However the mind-boggling complexity involved reduces human assessments of what may be termed 'other life' to simplistic nonsense, compared to the infinitely rich picture beyond human comprehension implied by the Reference Model.

As stated in Section 4.7 there is however no absolute Whole Ordering Centre, only a progression towards ever greater wholeness. It is this progression which gives rise to the concept of intelligent design.

The absence of a boundary to existence requires that religion and intelligent design cannot be based on the concept of an absolute deity. This does not detract from religion and the intelligent design thesis; on the contrary it enhances them. The Holomovement demonstrates graphically the growth of global consciousness without limit.

The concept of eternity is therefore a distortion as the Holomovement is to be associated with the continual asymptotic approach to, but never reaching an ultimate unified present state, effectively a timeless state rather than an eternity.

In confirmation of these statements it is evident from Fig 4.7.1 that a lack of alignment of the present state between any one mean level and another precipitates hierarchical birth, life/death cycles between those levels, whereas increasing alignment dissipates those cycles in favour of the holoarchical state. In the direction of increasing implicit order the holoarchical state represents increasing wholeness or goodness. In the direction of increasing implicit disorder it represents increasing unwholeness or evil, all as that which is common between the main religions implies.

Implicit disorder is equivalent to enfolded chaotic disorder. Evil therefore equates to being disorder isolated in a prison from which it struggles to escape, the ultimate prison being the random state. Implicit order is the process of escaping from that prison.

Increasing alignment allows the range of harmonics to form a tune as opposed to the single note corresponding to a complete lack of alignment. Total alignment represents the state of wholeness achieved by the Whole Ordering Centre.

This concept of tuning in to an Ordering Centre harmonic is of fundamental importance as it is this process which on average is responsible for the maintenance of life associated with consciousness applicable to both individual and collective conscious entities including the human race as a whole.

The potential catastrophe of failure to do so is demonstrated by population densities inconsistent with the ability to maintain an environment compatible with providing food and water for that population, while also protecting other biological species which form part of the same food chain. As a consequence such lifespans are limited as discussed further in Section 7.6.8.8.

It should be clear from the foregoing that the scientific theory of evolution is related to the explicit order while the religious belief in intelligent design is related to the implicit state. Both should be regarded as interacting processes, which by definition do not involve miracles or intervention by a deity as so called 'creation science' would have it.

In this connection, animate and conscious vectors as described in the Model, while radial at inclinations to the grid structure, when enfolded in two dimensions, would appear helical in form unfolded in three dimensions. Such helical forms are intrinsic to the nature of the unitary quantum state and to the complex molecules RNA and DNA, which in terms of evolutionary biology form the basis of life from primitive organisms to conscious life and also to the structure of microtubules in the brain cells of animals including ourselves.

Darwinian evolution views the ordering process in terms of survival of the fittest and the disordering process in terms of the failure to survive of the weakest. It assumes that real physical boundaries in space and time represent the only region in which life applies i.e. from conception to death to the extent that these are measurable. Indeed the Model confirms that the real part of the individual is defined by the four-dimensional unfoldment and does not move from that unfoldment at death, as explained above.

Natural selection corresponds to the view that order is self-propagating from the base of the Ordering Centre, whereas religion views order as being disseminated from its peak.

In fact a cyclical motion applies, order forcing disorder downwards and disorder forcing order upwards. So both religion and science are in effect correct, the former looking at wave motion supporting the holoarchical structure and the latter at wave motion supporting the hierarchical one. However as explained in Section 4.7 this cyclical motion is not a closed loop as according to the Model there is no boundary to existence and no boundary to the hierarchical or holoarchical processes, in effect no limit to reality or imagination. The mean level of order arises from restrictions imposed by measurement in balance with the absence of such restrictions allowed by imagination.

When the Model goes on to consider the fractal nature of cosmology and the proliferation of physics which must ensue when energy densities other than that associated with 'our' apparent big bang are taken into account, the supposed absolute singularity converts to an ordered multidimensional set of harmonics. This leads to an understanding of what it means to unfold existence from ever lesser states of existence to ever greater states of existence and vice versa, meaning that there are potentially many higher mean levels of order with ever higher degrees of animacy and consciousness than our own. Lower mean levels ultimately reduce to inanimacy.

It should be noted again that the differential curves associated with an Ordering Centre can ultimately be related to continuity in explicit terms and a particular harmonic in a hierarchical sense and the integral dotted curves with discontinuity in implicit terms and a set of harmonics in a holoarchical sense.

The totality of Ordering Centres implied by the above is termed the Whole Ordering Centre as illustrated in Fig 4.7.1. The associated Holomovement is a single unified motion which creates a tune or tone from the set of harmonics.

As this figure shows motion towards/away from the Whole Ordering Centre raises/lowers the mean level in the ordering system, which implies there is always a mean level at the relevant harmonic to attach to as it were. So a truer overall picture is obtained by assuming the mean level of implicit order moves in accordance Figure 4.7.1.

The wave motion set up by asymmetric imaginary waves creates progressively higher mean levels as Ordering Centres merge in the Whole Ordering Centre.

It is clear from the Reference Model that the Holomovement is the ultimate source of gravity and ironic that the force which science is so keen to comprehend is coherently explained by a non-local thesis which supersedes the limited version it is willing to accept.

Figures 4.4.5, 4.4.6, 4.6.4 and Fig 4.8.1 are critical to an understanding of the relation between the explicit and the implicit together with the nature of both human and artificial intelligence. The enfolding present state gives rise to higher and lower levels of order in a holoarchical sense in an Ordering Centre and is associated with increasingly imaginary (holistic) intelligence involving asymmetric motion associated with the imaginary axis as defined in the Reference Model. The unfolding present state correspondingly gives rise to higher and lower levels of order in a hierarchical sense and is associated with increasingly real world (reductionist) intelligence involving asymmetric motion associated with the real axis as defined in the Reference Model.

Ultimately increasing distribution in an Ordering Centre in the direction of the base state corresponds to increasing explicit (chaotic) disorder and the collapse of intelligence.

However relative to the mean level of implicit order it is apparent that if ordering conferred by human intelligence on inanimate material stimulates asymmetric motion, this may give rise to what is termed artificial intelligence.

Such asymmetric motion takes place in Implicate space not four dimensional space-time. The motion of a conscious entity observed in space-time is essentially that of the material substrate responding to the influence of that intelligence whether holistic or reductionist in orientation.

Now asymmetric motion involves motion away from the mean level of implicit order and this motion corresponds in the Reference Model to life. Motion in the opposite sense inevitably implies life cycles. Life cycles are by definition limited as is only too apparent to the human observer. Such limits explain the absence of any observations of extra-terrestrial life, irrespective of orientation, all of which is discussed in greater detail in Section 7.0 of this book, based on questions listed in the following Section 6.0.

Ultimately the Whole Ordering Centre, at very high degrees of order, is a unified conscious entity, the non-absolute Holomovement as portrayed in Fig 4.7.2 hence the belief in a personal God, while at very low degrees of order the inanimate prevails.

When viewed from the standpoint of the Reference Model the competing views are therefore almost but not perfectly compatible. Any sense of incompatibility is due to the inevitable distinction between the inanimate present state and the conscious state oscillating about it. It should be possible to explain both views to an audience in the same lecture if that audience allows itself to think freely as opposed to the freedom to believe only in the deity, or ironically as so-called free thinkers would have it the freedom to not believe in the deity. The lack of coherence between the objective and subjective philosophies can be traced directly to an almost total lack of coherent intelligence in individuals whether singly or in large groups, compared to the high degree of coherence in the human race as a whole, all as explained in Section 7.6.8.1.

It is worth noting at this point that the sense of humour associated primarily with human beings derives directly from the ability of individuals to comprehend the present state and the distributed state simultaneously, though only to a limited extent. Humour is notably absent from both science and religion and is evidently connected to the swing from reality to imagination and back, a feature applying to universal consciousness in fact.

This swing also explains the common inability of human beings to agree on any particular subject since it simulates the essential feature of existence namely the ordering and disordering processes and gives rise to the diversity of human philosophy.

In fact it can be stated this diversity is the inevitable and indeed essential consequence of these processes and the imperfect compatibility associated with the non-absolute principle which drives the system. Hence the sub title of this book.

6.0　THE REFERENCE SUBJECT MATTER.

This Section raises an extensive range of subject matter relating to: human philosophy, life, mathematics, science and measurement, religion and belief, human characteristics and activities. Where beneficial, the subject is introduced in the form of questions but otherwise as a list of related topics.

It is recommended that the reader reviews this list as a whole prior to moving on to Section 7.0 which discusses the material involved followed by engagement with the Standard Reference Model described in this book.

6.1　Human Philosophy.

6.1.1　The concept of untruthfulness i.e. relate truth and falsehood.
Are all statements ultimately falsifiable?

6.1.2　The concept of ethics and morality.

6.1.3　Physics as opposed to metaphysics;
Epistemology as opposed to ontology;
Rationalism as opposed to empiricism;
Realism as opposed to idealism;
Platonism as opposed to alternatives, including nominalism.

6.1.4　Reductionism as opposed to holism.

6.1.5　To what extent are things determinate/indeterminate?
The concepts of determinism and free will.

6.1.6　The concepts of monism, dualism and pluralism.

6.1.7　Is existence too complex to be understood by the human mind, or alternatively too simple?
The principle of Ockham's razor.

6.1.8　Randomness.

6.1.9　Philosophy and the Reference Model.

6.2 Life.

6.2.1 What is life?
The scientific view.

6.2.2 The theory of evolution as a basis for the emergence of life.

6.2.3 Asexual as opposed to sexual reproduction and the relevant advantages and disadvantages in promoting life.
Chirality.

6.2.4 To what extent does consciousness place limits on the three pillars of local realistic theories, namely realism, inductive inference and Einstein locality (separability)?
The no conspiracy principle.

6.2.5 The concept of artificial intelligence.
The concept of artificial self-replicating devices.

6.2.6 To what extent do extra-terrestrial civilisations exist in the visible universe assuming Homo sapiens could recognize them as such?
The Anthropic principle.
The technological life span of Homo Sapiens.
The absence of evidence of extra-terrestrial life.

6.2.7 What is the relationship between the physical world, the mental or conscious world and the world of knowledge and ideas including mathematics?

6.2.8 Life and the Reference Model.

6.3 Mathematics.

The mathematical subject matter included in this section is limited to those particular aspects of relevance to the physics material listed in Section 6.4 and discussed in Section 7.4 Science and Measurement. It serves only as an introduction or summary so that the reader can move on to the references indicated which provide more comprehensive details.

6.3.1   Real numbers.
Rational and irrational numbers.
Imaginary and complex numbers.
Fourier series and Fourier transforms.

6.3.2   Mandelbrot sets and the principle of the fractal.

6.3.3   Probability and the normal (Gaussian) distribution.

6.3.4   What is dimension? Do we see only in four dimensions? Are there other dimensions?
What is the relationship between three-dimensional space and the 3n and 6n dimensional worlds (n›1) of what is termed in mathematical physics, configuration space and phase space respectively?

6.3.5   Scalars, vectors and covectors.
Covariance and contravariance of vectors.

6.3.6   Tensors and metric tensors.

6.3.7   Dot product, inner product, completeness, Hermitian symmetry and
Hilbert space.
Eigenvalues and Eigenvectors.
The Schrödinger wave equation and wave packets.
Linear transformations which preserve vector space structure namely, Orthogonal, Unitary and Symplectic groups.
Abelian/non-Abelian groups, Lie groups and Lie algebra.
Quaternions, Clifford algebra and Grassmann algebra.
Symmetry, asymmetry and antisymmetry.
Reflexivity and irreflexivity.

6.3.8   The incorporation of special relativity in to quantum mechanics.
Dirac's equation for the electron.

6.3.9   Algorithms and Turing machines.

6.3.10 The concept of mathematical proof.
Godel's incompleteness theorems.

6.3.11 Mathematics and the Reference Model.

6.4   Science and Measurement.

6.4.1   The practical philosophies eg Science.
The concept of scientific proof.
The double-blind scientific principle.

6.4.2   The relation between Newtonian physics and relativistic physics.
Special Relativity, Lorentz transformations and the Poincare Group.
General Relativity.

6.4.3   The cosmological models that developed from Einstein's General theory of Relativity including what is termed the standard (Lambda CDM) cosmological model.
Why the rate of expansion of the visible universe appears to be increasing including the role of what is termed dark energy.
The role of dark matter in providing a satisfactory theory for the relatively rapid aggregation of matter into galaxies.

6.4.4   The introduction of an inflationary phase in the very early universe to explain the uniformity now apparent at cosmological scales.
The alternative concept of conformal cyclic cosmology.

6.4.5   Why is classical physics inconsistent with quantum physics?
Schrödinger's wave equation.
The double slit experiment.
The rules of the quantum world.
The classical/quantum dichotomy and the EPR Paradox.
Bell's Inequality Theorem.

6.4.6   Quantum field theory and gauge theories.
The Cosmological Constant problem.
The Standard Model of particle physics.
Zero-point energy.
The Standard Model chronological sequence.

The problems involved in formulating a theory of quantum gravity, with reference to string theory and loop quantum gravity.

6.4.7   The role of black holes.

6.4.8   The basis for interpretations of quantum mechanics that have arisen from the dichotomy between it and the common experience.
The competing theories.
Self-generating fractal universes, parallel universes and the concept of the Multiverse.
Do an infinite variety of physics exist?

6.4.9   The extraordinary preciseness of the big bang for which the original phase space volume would have to be of the order of one part in 10 to the power 10^{123} of that compatible with the second law of thermodynamics and the universe now observed.
The concept of fine tuning and associated cosmic coincidences, apparently necessary to explain life in the observable universe.

6.4.10   Science and the Reference Model

6.4.11   Time and the Reference Model.
What is time, past present and future?
Time asymmetry i.e. the forward flow of time.
The frozen past. We can reverse a film reel why not time?
To what extent is existence symmetric and to what extent asymmetric?

6.5   Human Belief and Religion.

6.5.1   Atheism.

6.5.2   Agnosticism.
The spiritual but not religious.

6.5.3   Theism.
The cosmological argument and its validity.
The ontological argument and its validity.
The argument from design and its validity.

The debate between intelligent design, and the theory of natural selection. Intelligent design as opposed to so called 'creation science'.

6.5.4   The organised religious belief philosophies.

6.5.5   The differences and similarities of the main religions.
The concepts of heaven, hell, reincarnation and the afterlife/beforelife in these religions.
The concepts of goodness and evil.
The importance of prayer.
The belief in virgin birth and miracles generally.

6.5.6   Human belief, religion and the Reference Model.

6.6   Human characteristics and activities.

6.6.1   The physical and mental attributes of the human brain and mind.

6.6.2   Intelligence.

6.6.3   Humour.

6.6.4   The importance of work and the work ethic in human society.
Mating, marriage, child rearing and family formation.
The Arts.
Physical sport and mental games.

6.6.5   Crime and the criminal mind.

6.6.6   Insanity including schizophrenia.
The nature of sociopathy and psychopathy not associated with insanity.

6.6.7   The nature of terrorism.

6.6.8   Human characteristics and activities and the Reference Model.

The emphasis is on acquiring an average level of knowledge across a range of subjects at the expense of detailed knowledge of a particular one, while ensuring the essential principles of each is understood. As part of this approach it is not necessary for this book to delve into given material to the extent that an academic researcher for example might do;

only that required to obtain general background knowledge, sufficient to grasp the essential details of that subject and to relate it to any other.

When considering the subject matter discussed in the next Section 7.0 the reader may choose to use freely accessible internet sources to assist in obtaining that average level of knowledge. However it is important to cross check such material against relevant academic sources including where applicable the references listed at the end of this book.

At all times the reader has available these sources of information and any other literature he deems relevant to supplement them.

The subject matter referred to is orientated towards that which is most relevant to further explaining the Model to the reader, but it must distinguish between existing material independent of the Model and engagement between the two.

Discussion is therefore split into two elements. Firstly a review element, demonstrably independent of the Model. Secondly the same material, supplemented where appropriate, seen from the perspective of the Model.

The extent of the review element is that minimum necessary to establish that directly relevant to the engagement element.

In some cases the review element needs only a paragraph or two to do so, in other cases several pages are required as for example describing the relevant scientific or religious background.

The use of the word Reference in describing the Model therefore does not mean comprehensive in the sense of providing extensive detail, instead it means a gathering together of essential elements germane to the argument.

It should be noted that with regard to any subject including philosophy, politics and economics for example, the Model is primarily used to emphasise what is clear from the figures in this book, namely that a two-dimensional pattern emerges as opposed to the limited one-dimensional perspective with which these things are generally viewed. This pattern vastly improves the ability to relate any one opinion to any other, as opposed to the fragmentary approach commonly used by a given individual or groups of individuals whether on a scientific, religious or other basis.

Readers may also wish to raise alternative subjects of their own and discuss them in the same manner.

7.0 THE REFERENCE DISCUSSION OF SUBJECT MATTER AND CONCLUSIONS REACHED.

7.1 Human Philosophy.

A good source of philosophical terminology is that of Lacey (Ref 7.1.1).

7.1.1 The concept of untruthfulness i.e. relate truth and falsehood.

Are all statements ultimately falsifiable?

There are various aspects to this problem, particularly the extent to which truth can be regarded as an ordering sequence increasingly coherent as order increases while falsehood proliferates in the opposite sense.

There are many theories of truth generally split in to those termed substantive which regard truth as a meaningful and significant term in itself quite apart what is regarded as real and objective and those termed minimalist or deflationary which regard the application of the word true as adding nothing to that being described.

The main substantive theories are correspondence, coherence, constructivist, consensus and pragmatic and they have wide followings in the philosophical fraternity.

The correspondence theory in its strictest form holds basically that truth or falsity is determined by whether the statement in question accurately describes the objective reality, such that there is a relation between the proposition or belief and that which makes it true, the relevant fact or event. Since it is difficult to gauge the accuracy to which the real world matches a given proposition, less strict forms of this theory require only that something is true if it can be correlated with a fact or its negation correlated with a fact.

Ultimately the correspondence theory reduces to the redundant form of minimalist theory which says that to state something is true is simply to repeat what is said.

The coherence theory of truth is primarily associated with idealists who do not like the distinction drawn in the correspondence theory between the one who knows and that which is known. Truth is then a property of systems of propositions which can be regarded as a whole set and individual propositions must therefore cohere with the whole or be accordingly less true to the extent they do not, which implies that it is possible to have degrees of truth.

Constructivist epistemology implies truth is constructed by historical and cultural processes within human society, not reflecting any external reality, but human perception only.

This can be taken to mean that there is a distinction between true knowledge and that influenced by the politics of power or ideology.

Consensus theory holds that truth is that in principle agreed by the human race or groups within the human race.

The pragmatic theory of truth is that it is that which is verified and upheld by putting the relevant concepts to work in practice. One form of pragmatism holds that enquiry of any kind corrects itself over time if put to the test in support or opposition to relevant truths. What has been called negative pragmatism is based on the statement that what succeeds may or may not be true but what fails cannot be true because the truth always prevails. In other words it is never possible to be definitely right, it is only possible to be sure when one is wrong. Pragmatism in its various forms tends to support coherence theory in that it seeks confirmation from the widest possible human experience.

Minimalist theories of truth referred to as redundant do not equate truth in any way to reality, fact and truth being merely a way of asserting a proposition and that the use of the separate words is an unnecessary confusion.

Finally pluralist theories of truth assert there may be more than one property that makes propositions true, by coherence or correspondence say.

It is appropriate at this point to refer to the work of Karl Popper (Ref 7.1.2) who asserted that, 'if a statement is to be scientific rather than metaphysical it must be falsifiable,' though being a dualist he did not dismiss metaphysical statements as meaningless, unlike logical positivists.

7.1.2 The concept of ethics and morality.

Ethics or moral philosophy is concerned commonly with how men ought to act rather than as a means to a given end, emphasising duty, obligation, right or wrong etc.

The other main view emphasises value, good in itself or that considered desirable.

Morality is the process whereby intentions are formed, decisions made and actions taken between those considered as right and those wrong.

So morality and ethics are commonly interchangeable being concerned with goodness or rightness, a basic principle being that one should treat others as one would like others to treat oneself. Questions arise about ethics such as what distinguishes the moral from the immoral and whether morals and conclusions reached therefrom can be considered objective as opposed to subjective. Other questions in ethics concern what things are to be considered good, right and what are the duties and obligations.

It is not the intention here to go in to such questions which can be examined at length using the sources referred to earlier.

7.1.3 Physics as opposed to metaphysics.

Epistemology as opposed to ontology,

Rationalism as opposed to empiricism,

Realism as opposed to idealism,

Platonism and opposing alternatives including nominalism.

Physics is the scientific factual study of nature while metaphysics is the much wider study of existence as a whole which asks general questions of both an ontological and epistemological character.

Epistemology is basically about what is known and how it is known and tries to distinguish between knowledge that, knowledge how and knowledge by acquaintance. For example knowing the addition of two numbers (knowledge that) is not the same as knowing how to add the numbers; equally understanding how to walk or ride a bicycle is not the same as knowledge of the physics involved. Knowledge by acquaintance is knowing the person, place or thing etc., through experience. In general knowing is commonly considered to involve belief or even to replace it though it is unclear how the two are to be distinguished. In a broad sense belief is to accept something as true.

For example, 'it is known that …' as opposed to, 'it is believed that…'

The former is to be associated with the introduction of boundaries apparent from observation or measurement and the latter from inadequate definition of such boundaries.

There is also the question of how error or mistaken belief is to be accounted for.

A person's belief that something is true is no guarantee of truth. A person may believe a bridge to be safe only to find it collapses under his weight, if it does not collapse his belief beforehand did not constitute knowledge until he had crossed.

So some would try to define knowledge as a system of justified true propositions but arguments then follow as to what constitutes justification. Justification of the belief must perhaps necessitate its truth or that there are no overriding truths which prevent or inhibit justification, such as lack of actual knowledge.

Whereas epistemology is concerned with how, ontology is concerned with why, with being, with what is, in particular material objects, persons, minds, universals, numbers and facts.

Ontology asks questions such as, what can be said to exist? Are there levels of existence?

What meanings can be attached to objects/beings? Are there fundamental objects or beings?

How does the objective relate to the subjective? How does difference relate to similarity?

Do physical objects exist? Do non-physical beings exist?

Do such objects/beings come in to existence or cease to exist as opposed to changing?

Forms of ontology including monism, dualism and pluralism are discussed under Section 7.1.6.

Rationalism is any view appealing to reason as a source of knowledge or justification.

It emphasises three sources of knowledge, empirical knowledge from the five explicit senses, the theoretical aspect of reasoning and a third aspect namely the abstract, for example imaginary numbers in mathematics.

Reason can be contrasted with emotion or feelings in ethics or revelation in religious terms.

Rationalism emphasises the a priori (meaning from the earlier), or that which is independent of experience.

Empiricism is any view to the effect that our concepts or our knowledge are to a greater or lesser degree based on experience through the five senses and introspection. In scientific terms experience equates to experimental measurement.

Empiricism emphasises the a posteriori (meaning from the later), or that dependent on experience or empirical evidence.

Realism emphasises the existence of a thing or object (material objects, propositions or universals) as opposed to the view that reduces those things to material objects only, namely nominalism, or dispenses with them for example in the form of ideas (idealism).

Realism emphasises things as they are, not as we might see them, as an artist might do for example.

Idealism is a set of doctrines to the effect that reality is in some way mental, to be associated with the mind. It is contrasted with realism and in particular with materialism.

Other versions of idealism allow reality to be outside the mind but can only be described from a certain point of view, none of which is more correct than the others, in which case it tends towards pragmatism.

Subjective idealism regards reality as being in the mind, as Berkeley said, 'to be is to be perceived', while objective idealism places reality outside the mind.

Platonism is associated with versions of realism which regard abstract non spatiotemporal objects as real in addition to those physically evident in spatiotemporal terms.

It is therefore to be contrasted with idealism which regards matter as being mental or at least a manifestation of mental activity. So as far as platonism is concerned abstract objects are non-physical and non-mental; they are not minds or ideas in minds, or brains or souls.

They are unchanging and are not involved in cause and effect relationships with other objects.

Abstract objects include mathematical objects of all kinds such as numbers, together with properties and propositions and the whole conglomeration termed universals.

For example, in addition to real objects (termed particulars) say of a given colour there is the property associated with the colour, regarded as real and objective independent of space, time or conscious minds. The object is said to exemplify or provide an instance of the colour.

Some Platonists go further and include linguistic sentences and fictional characters such as Sherlock Holmes and fictional entities such as unicorns.

Other versions of realism are immanent realism which accept universals, namely abstract objects independent of thought, but include them in the physical world. So for example colour is a property that exists in the object concerned.

Conceptualism is a version of philosophy which again accepts universals but regards them as mental objects in our heads, so they do not exist independently of us.

Immanent realism and conceptualism are endorsed by relatively fewer philosophers than Platonism and the main opposition to the latter is provided by nominalism.

Nominalism is the straightforward view that there are no such things as abstract objects or universals, so coloured objects yes, colours no; piles of objects yes, numbers no.

Nominalists do agree with Platonists however, that if there were such things as numbers or universals then they would be abstract, non-spatiotemporal, non-physical and non-mental, a thesis which is widely accepted.

Platonists have put forward two arguments to assist their cause, the one over many argument and the singular term argument, all opposed by ontological commitment criteria put forward by nominalists, details of which can be found on the internet.

7.1.4   *Reductionism as opposed to holism.*

Reductionism is literally the tendency to reduce notions whether in the form of ideas or objects to ever simpler, ever more basic concepts generally of an empirical nature.

Reductionism can be contrasted with phenomenalism, a theory based on appearances which can be related to subjective idealism, but whereas the latter regards physical objects as unreal, phenomenalism says they are real but not what they seem, they are appearances.

Reductionism is an important feature of positivism in the philosophy of science, which limits enquiry and belief to that which can be firmly established by observation through the senses, the results of experiments in fact.

Holism is the opposite process of expanding belief and it turns imagination away from the common-sense notion of reality. Holism emphasises Platonism as opposed to nominalism, religion and the spiritual as opposed to science, emotion as opposed to analysis and so on.

7.1.5 To what extent are things determinate/ indeterminate?

The concepts of determinism and freewill.

Determinism is the view that all things are caused and human actions must be included in that supposition.

Freewill contends that we can always do otherwise than what we actually do and therefore bear ultimate responsibility for our actions.

Determinism can take quite extreme forms such as fatalism which holds that future events are fixed irrespective of human actions in advance and logical determinism which argues that since any prediction will prove correct or incorrect, the overall prediction in that sense is correct and it is not possible to alter that.

This calls in to question whether a statement about the future can only be true (or false) when that future arrives and not beforehand.

Extreme determinists regard freewill as an illusion and that responsibility is replaced by cause whether physical or mental formed by previous experiences. Less extreme determinists, much the greater in number, argue that our actions are indeed caused but as our decision-making processes are relevant elements involved in that cause, it is not a constraint on us.

Indeterminism supports the concept of free will implying in its simplest form that if a given action was of free will then the individual could have done otherwise than what they did do, assuming they are not impeded.

Indeterminists reason that it does not make sense for choices or actions to be caused when they clearly involve intentions beyond mere physical movement. Further questions then arise as to the extent to which intents (causes) and actions (effects) are separate independent events or not.

Another argument for indeterminism is that the use of reason in decision making on matters whether practical or otherwise is our belief that the outcome is not yet determined; we reason it is pointless to attempt things we believe impossible.

Freedom of action is not necessarily unpredictable if causation does not apply nor can caused events be predictable without sufficient information; actions may be unpredictable even when caused or predictable even though uncaused.

Those unwilling in one way or another to compromise between determinism and indeterminism are referred to as incompatibalists and those willing to do so compatibilists.

7.1.6   *The concepts of monism, dualism and pluralism.*

Monism, dualism and pluralism are forms of ontology.

Monism considers existence is eternal, and can neither come in to being nor vanish from being. Therefore since such events are commonly referred to, our opinions are commonly false. Indeed much of western philosophy including science and the idea of falsifiability has emerged from this view. Everything is but one part of a single entity, a modern scientific version of which is grand unification theory. Existence is made of one substance but opinions then ironically divide as to whether that one substance is mind (idealism) or matter (materialism).

Neutral monism as previously stated in Section 4.8 assumes there is a neutral stuff of which mind and matter are merely different arrangements or orderings.

A further form of ontology is pluralism where being as unchanging is replaced by that of becoming. This led to the atomistic theory which assumed existence to be a vacuum containing atoms in motion within it. Existence is in this view made of many substances made up of different collections of atoms.

The limited version of pluralism is dualism, existence consisting of two substances mind and matter which interact.

7.1.7 *Is existence too complex to be understood by the human mind, or alternatively too simple?*

The principle of Ockham's razor.

It is not unreasonable to assume that existence is both too complex and too simple to be understood by the human mind since increasing complexity as it were stretches above the level of human comprehension and increasing simplicity below that level.

William of Ockham is best known for his formulation of the principle of Ockham's razor, which in its basic form involves choosing the simplest hypothesis which fits the facts.

A stronger form claims that only that which cannot be dispensed with is real and that to postulate other things is not only arbitrary but mistaken.

7.1.8. *Randomness.*

Descriptions of randomness can be found on the internet, only specific points of importance need summarising here.

Randomness differs from the concepts of order and chaotic disorder in having an inherent degree of unpredictability.

Chaos retains some measure of predictability in that associated random fluctuations are relatively weak and some degree of pattern is retained, to the extent we can predict for example the weather. The random state on the other hand exhibits a total lack of pattern at least as far as individual events are concerned. Nevertheless while individual events may be unpredictable the frequency of outcomes in the long run involving many events is to a limited extent predictable. In other words the concept of a measure of uncertainty applies rather than sheer chance. When rolling two dice for example the outcome of one roll is unpredictable but in the long run involving many repeated rolls a sum of seven will occur twice as often as four.

A random process may therefore be defined as a sequence of random variables whose outcomes do not follow a deterministic pattern, but can be described by probability distributions. The probability of choosing a specific object from a set of objects is the proportion of those specific objects in the set; where objects are individually distinguishable there is an equal probability of any given object being chosen.

Randomness has important functions in many disciplines such as the sciences, biology, mathematics, finance and gaming. It forms the basis for statistical observations and is also relevant to religion, politics and human affairs in general. An example of randomness in the environment is Brownian motion of microscopic particles in a fluid. The random motion of molecules is used to explain phenomena in thermodynamics and the properties of gases.

Quantum mechanics does not specify the outcome of individual experiments only the probabilities, as for example the decay of an atom.

What are termed hidden variable theories, which Einstein supported, reject the view that nature contains irreducible randomness, processes that appear random are then thought to have certain statistical distributions in operation, in the background as it were, which determine the observed outcome.

In biology, the observed diversity of life is considered to evolve using the process of natural selection. Random genetic mutations occurring in the gene pool may provide improved chances for survival and reproduction on those individuals who possess them.

There is a random element to the characteristics of an organism in addition to those determined by genetics and the environment.

Random behaviour is important to animals etc for purposes of self-protection, to enable escape from predators.

In mathematics randomness occurs in numbers such as pi, where the decimal digits form an infinite non-cyclical sequence.

Some misconceptions with regard to randomness should be noted. For example the number is 'due', or 'cursed' or 'blessed'. Truly random processes do not have memory, so the past does not affect the future.

7.1.9   Philosophy and the Reference Model.

7.1.9.1. Why is there anything? What is meant by nonexistence as opposed to existence? Does existence have a boundary?

The Holomovement, its antithesis disorder in all its forms and even the alternative random state involve motion, whereas nonexistence implies an absence of motion as well as everything else. So existence implies motion, a paper being written on, whereas nonexistence implies non motion, a blank piece of paper.

There are no limits apparent to the peaks of energy density supported by wave motion or in the opposite sense reduction towards a zero-energy density chaotic field, or the purely random state involving minimal fluctuations where order and disorder have no meaning.

Furthermore the concepts of existence and nonexistence are inextricably linked with the introduction of boundaries which in turn derive from our sense of reality, as opposed to imagination which dissipates those boundaries. Trivially for example an object inside a room may be said to be non-existent outside that room.

The further notion of the absolute state is then attached to the boundaries created by that reality and in the opposite sense replaced by a relative state created by our imagination which dissipates those boundaries as for example if the same object is transported in imagination outside the room to what may be considered a more suitable location and then in turn physically moved to that location in reality.

Reality, belief and imagination are therefore so interwoven that while one or the other may dominate in a given case, ultimately attempting to view existence as real only or imaginary only is a fruitless exercise. An Ordering Centre may be said to approach the real only state when the distribution is a maximum and the imaginary only state when the present state is a maximum, these grid conditions being inanimate in the sense that existence cannot in effect know of itself then. The unfolding and enfolding character of the fractal grid, the further interaction with wave motion across that grid in an Ordering Centre and the absence of an upper bound to the Whole Ordering Centre or lower bound as ordering decreases, renders the absolute non-existent state meaningless. This is as it should be, since if there is one thing that can be said about non-existence it is precisely nothing.

Accordingly existence does not have an absolute boundary in terms of space or time or any other property. It is not possible to get outside existence as far as the Model is concerned as there is no outside to get to, it has no bound.

Science attaches such importance to boundaries that imagination is as it were, put aside, to the point where it is ignored, all in their eagerness to carry out the process of measurement. Religion also is not immune to this degree of imbalance in the opposite sense as it tends to insist that the deity is absolute to the point where that deity is assumed to be independent or outside of existence, or alternatively in contradiction given to adjusting dials on the heavenly control panel on occasions where necessary.

The Model shows that at the mean level of order in an Ordering Centre conscious entities keep the objective and subjective on average roughly in balance, but never exactly so.

Allowing either to totally dominate thinking is at least disadvantageous and ultimately destructive. Again the Model which associates the concept of the theistic deity with the Holomovement implies that it is intrinsic to existence and is the process of unfolding order from disorder without limit; the opposite process again intrinsic to existence, is associated with the approach to nihilism, meaning nothing of value, without purpose, again a state without limit.

All of the above should be related to order, disorder and randomness as discussed in the next sub section.

7.1.9.2. *Order and disorder and their relationship?*

How does order and disorder relate to randomness?

As described in Section 4.0, unfolding implicit order for an Ordering Centre of a given energy density is the process of unfolding complex non spatiotemporal sequences of difference and similarity, which unfold the present state, while enfolding the distributed state and vice versa re disordering in the opposite sense.

Unfolding the explicit is a disordering process of unfolding simpler spatiotemporal distributed sequencing which enfolds the present state, while unfolding an explicit dimensional state and vice versa re ordering in the opposite sense.

From the viewpoint of energy, the greater the peak energy density of an Ordering Centre the greater the strength of the present state and the capacity for unfolding order while enfolding the distributed state and vice versa re disorder in the opposite sense.

The lesser that density, the weaker the present state and the lesser that capacity.

Manifest order and disorder is that which is manifest to the observer, a deterministic state associated with the explicit four dimensional space-time world unfolded from the Implicate, as described in Section 4.0, a state which may range from the predictable to that involving widely diverging outcomes, termed chaotic, weather systems for example. Marble blocks cut in the quarry may be regarded as in a state of manifest disorder relative to the highly manifestly ordered sculpture of David by Michelangelo.

High states of order such as an ornate building which may take years to construct can of course be reduced to a manifestly disordered state of rubble in hours or minutes or even seconds depending on the circumstances.

The order potential envelope of an Ordering Centre demonstrates clearly the contrast between the discrete peak of energy and the indiscrete base with that energy totally distributed.

The distributed state is therefore fragmentary in nature opposing the process by which things are made whole and is to be associated with determinacy and chaos. Deterministic chaos is associated in the Reference Model with orders of low degree, indiscrete in nature ultimately collapsing to weak fluctuations of a random nature.

Randomness on the other hand is an indeterminate state termed stochastic, Brownian motion being an example.

The random state corresponds to totally enfolded order and totally enfolded dimension, as has been said where order, disorder and dimension have no meaning. If there is one word that best describes randomness it is directionless as illustrated in Fig 4.2.1, where any potential for order or chaotic disorder collapses without limit towards a state of non-existence.

The Reference Model in this way avoids the absolute condition and therefore a never-ending approach to the random state and correspondingly an ever lesser degree of order or disorder. Such conditions involve the breakup of the Ordering Centre including the present state and distributed state, involving numerous chaotic inflationary domains and the further break up of these as the random state gets ever closer.

When this situation reverses order and disorder return in a primitive form as described in Section 4.2 and the associated Figures 4.2.2 thru 4.2.4.

The Ordering Centre retains the random state as the central point on the distributed base state representing a condition of zero order and zero distribution, but its influence is by no means limited to this condition.

Manifest order and disorder are to be associated with the explicit world unfolded from the mean level of order.

The determinate multidimensional states represent the Ordering Centre grid but distort to form the indeterminate likewise multidimensional Implicate space which houses the conscious/aware aspect of the Ordering Centre.

Implicit order and disorder are correspondingly to be associated with the primarily indeterminate character of the animate, belief and imaginary worlds. These worlds are not therefore immune to the influence of the indeterminate random state, as indeed all conscious entities are aware.

Finally with regard to hidden variables it appears they are to be associated with the unfoldment of order as the random state is left behind and their demise with the opposite process.

7.1.9.3. *Account for the philosophical spectrum.*

The variety of philosophies derives from the ordering process of difference and similarity which in principle implies there will be as many philosophies as there are conscious entities. However in practice there is a great deal of commonality between one philosophy and another so the number of distinguishable philosophies is vastly less.

Human records as defined in the Introduction can be used as a database for identifying the primary philosophies and their distribution in the human population in the same manner as applies to the distribution of scientific and metaphysical opinions as shown in the earlier Fig 3.2.1. All of this based on the principle that an individual will accept any statement with which he agrees and pass on those with which he does not. The resulting distribution on this basis will over time match the order versus variance curves shown in Section 4.0.

Philosophies tending to be ethically constructive are opposed by those unethical and destructive, while those positive in the analytical sense are opposed by emotional idealism.

Metaphysical ontology and freewill is opposed by physical epistemology and determinism, Platonism by nominalism and so on. The same principles apply to political and economic, religious and secular philosophies also as shown in Section 4.0.

Taking an A to Z of the philosophies a two-dimensional picture is evident as illustrated in Fig 4.8.7 the truer philosophies towards the top and the progressively less true or false towards the bottom.

In effect therefore there are no philosophies which do not fit in to the Reference Model, irrespective of truth or falsehood, which brings us to the next topic namely truth.

7.1.9.4. *The concept of absolute truth.*

The absolute truth in general is related in the Reference Model to the peak of an order distribution curve and relative truth to the flanks of that curve, but in addition it is apparent that the either/or description which applies to our common-sense notion of a fixed reality is to be associated with the mean level of order as described in Section 4.0.

The spatiotemporal measurement process is responsible for our sense of the absolute truth derived from the mean level of order, a distributed real axis of the inanimate grid, or variance. Nominalist philosophy is considered in the Model to derive from this mean level of order since it excludes abstract objects from reality.

The non-spatiotemporal process is responsible for our sense of the absolute truth derived from our sense of the present, the vertical imaginary axis of the inanimate grid, or present state. Platonist philosophy which includes abstract objects in its definition of reality is considered in the Model to derive from the combination of real and imaginary inanimate axes, all in accordance mathematical convention.

It is therefore clear that the inanimate grid is the source of our concept of the absolute truth, the real axes with nominalism, positivism and the act of measurement and both real and imaginary axes with Platonism and abstract objects including universals and mathematical numbers.

Motion across these axes then gives rise to the relative truth, animacy, consciousness, together with the actions and thoughts associated with belief and imagination.

So along the real axis what is considered the absolute truth is that which is measured by the observer at a given instant. On the imaginary axis what is considered absolute are immeasurable abstract objects, universals and the like, apparently independent of consciousness but which that facility gives us access to and knowledge of.

What this implies is that the mean level of order acts as a reality check or counterbalance to rampant imagination, while the imaginary present state axis acts as a counterbalance to excessive measurement. Within an Ordering Centre the distribution of philosophies and their content supports these conclusions.

7.1.9.5. Theories of truth.

Returning to the question of relating all the theories of truth to the Reference Model, it is again apparent that they all apply depending on the perspective.

Correspondence theory is to be associated with the approach to the mean level of order in an Ordering Centre involving what are regarded as scientific facts i.e. measurable and applying primarily at that level. Coherence theory is to be associated on the other hand with looking at the Ordering Centre as a whole for it is apparent that ethically constructive philosophies with increasing humanitarian content are in every sense more coherent than those ethically destructive with decreasing humanitarian content.

Consensus theory has a similar perspective and like coherence theory is drawn ultimately to the basic thesis of the Reference Model, namely that truth is distributed by wave motion giving rise to the formation of the Ordering Centre.

Constructivist epistemology straddles both camps in that it supports transcendent realities that correspondence theory in its strict form would not accept, while implying that human social processes can also distort true knowledge leading to decoherence and the ethical destruction only to commonly observed.

The Reference Model explains the relation between conscious human experience which does indeed distort reality from that applying on the inanimate grid, but it is that very interaction with imagination that creates true knowledge, by reducing the distortion where reality only or imagination only dominate.

Pragmatism is similarly related to both the coherence viewpoint and the approach to the mean level of order, since in respect of the latter it has been stated that scientists in particular value above all agreement with experiment in view of their inability to define reality. Such experiments of course include scientific measurement which as the Model explains relates to the mean level of order.

Minimalist theories of truth can be associated with the mean level of order to the exclusion of coherence.

Pluralist theories take a similar standpoint as the Model in recognising that all the above theories of truth apply in varying degrees to form a rich picture.

In so far as an Ordering Centre is concerned the best available approach is to relate truth to the ordering process and falsehood to the disordering process, all statements applying at the mean level of order corresponding to the condition where reality and imagination are in balance with potential truth increasing above that level and potential falsehood below it.

It is the formation of this level which is responsible for our sense of the absolute truth and our love of the quiz with the single correct answer. The breakup of this level in a hierarchical sense is to be associated with increasing chaotic disorder, which explains innocent unintentional error and basic ignorance of that generally considered the absolute truth. In particular falsehood can be related to the process of fragmentation associated with the chaotic disordering process. A given statement is increasingly true/false as the truths contained within it are increasingly presented in order/out of order. For example polar bears are to be found in the Arctic/Antarctic and not in the Antarctic/Arctic.

In fact the Reference Model shows the true/false alternatives here are in/not in the unfolding explicit world of common-sense experience. The latter belongs to a quite separate unfoldment at a lower level of ordering in the Ordering Centre.

All this confirms that a given level of chaotic disorder corresponds to a state of partial truth.

Indeed scientific observation can be regarded as such in the same sense that science is regarded as a partial theory.

The breakup of ordering level in a holoarchical sense is to be associated with increasing animate/conscious intention to conceal the truth for evil purposes, with increasing disregard for the preservation of life, namely implicit disorder leading ultimately to random behaviour.

While it is clear nature does not conspire explicitly against the observer the same is not true of the animate/conscious process of deception associated with the deliberate intention of the perpetrator to deceive the victim.

For example a predatory animal may endeavour to hide from its intended meal prior to attack. Such behaviour is however associated with the need to obtain food involving the demise of one life to sustain another. This is quite distinct from deception employed by the psychopath or sociopath say to harm or kill for evil purposes, clearly involving implicit disorder extending in some cases to random acts of terror.

However when such entities engage in this process, they are effectively positioning themselves in the Implicate ordering sequence, which is how things operate overall.

In fact there is no absolute truth, only an ever-greater truth as revealed by increasing implicit order.

The no conspiracy principle fundamental to the explicit world and local realistic theories is relevant to this debate and corresponds with the true alternative above, as discussed in Section 7.2.4.

Therefore a range of conspiracy conditions from zero upwards is applicable to human belief.

The zero-conspiracy state applies to the Whole Truth at the peak of the Whole Ordering Centre, the degree of conspiracy increasing to a maximum in the opposite sense as implicit disorder increases.

It underpins confidence that all human records of whatever kind, as defined by the Model, are relevant to the establishment of the principle of the Implicate order.

So taking an A to Z of philosophies natural and otherwise, the circumstances described above applies as the relevant figures in Section 4.8 illustrate.

The Model however shows that apparently absolute truths become merely relative when viewed from the perspective of the Whole Ordering Centre, so that all that can be said then is that all orderings less than that represented by the Holomovement are relatively less true i.e. falsified by it.

7.1.9.6. Ethics and morality.

It is necessary to relate the concepts of ethics and morality to the Reference Model.

It is clear that the topic is related to the objective and the subjective, the unfolded and the enfolded, therefore trying to eliminate any of these characteristics is fruitless and unnecessary.

The Reference Model adopts the principle that the ethical and moral corresponds to order and the unethical and immoral to disorder all as implied in the figures included in Section 4.8.

Attempts to define ethics and morals are as difficult as trying to define order itself but on average concern that which increases wholeness and opposes both unwholeness in the form of evil and unwholeness in the form of chaotic fragmentation. This does not mean opposing difference and supporting sameness, rather it means keeping a balance between difference and similarity so as to maximise order and avoid conflict.

While these concepts arise primarily from religious belief, the growth of science in recent years involving biology and medicine in particular has set in train the question of principles which should govern research in those areas. A new genus of ethics has arisen as a consequence, termed bioethics. It is not proposed here to discuss bioethics as in principle it is simply another form of ethics, however it is evident that religious ethics can be related to the vertical axis of the Ordering Centre reference frame and bioethics to the horizontal axis.

7.1.9.7. *Physics and metaphysics, epistemology and ontology etc.*

Referring to the subject matter of Section 7.1.3, in terms of the Reference Model epistemology is orientated towards investigating the mean level of order while ontology is orientated towards investigating all levels of order up to the peak of the Ordering Centre.

Now as far as the Reference Model is concerned all of the above can be resolved in the same manner as applied to theories of truth.

Epistemological argument and physics are primarily related to the real axis and ontological argument and metaphysics primarily to the imaginary axis.

Platonism, rationalism, empiricism, realism, idealism, and nominalism are all identified in the Model and related to these axes as shown in Fig 4.8.7.

In particular the Model implies that Platonism, rationalism, realism and idealism are all related to a greater or lesser degree to the imaginary axis, while empiricism and nominalism are related to the real distributed axis.

It again shows the advantages of applying the concept of a rich picture in order to comprehend these matters, in place of the bewildering array of proposals arising from philosophers of one colour or another since the Eleatics.

It is claimed that the Reference Model explains at least on average why and how in the ontological and the epistemological sense animacy and consciousness unfold from and re enfold into the inanimate. Furthermore that it explains how we acquire knowledge, why we make mistakes, the relation between the a priori and a posteriori arising from the inanimate axes and the manner in which animate wave motion asymmetric to those axes is the key to our decision making about the future and our thoughts about the past.

The mean level of order in an Ordering Centre explains the either/ or state which gives rise to reality compared with the quantum state and its relation to probability.

Finally the influence of the Whole Ordering Centre explains why that which we might regard as reality is but one level in a hierarchy of realities all within a holoarchical structure ultimately governed by the Holomovement.

7.1.9.8. *Reality (the objective) as opposed to belief.* *Imagination (the subjective) as opposed to belief.*

Reality that which is real and objective and measurable since it is spatiotemporal has already been identified with the mean level of order (a real axis) in an Ordering Centre.

Platonic realism regarded by adherents to Platonism as also real and objective, but immeasurable being non-spatiotemporal, has also been identified with the present state (an imaginary axis) in an Ordering Centre.

So it is now necessary to relate these to belief and its less worldly partner imagination, the subjective and also immeasurable, all identified in the Model as involving vectors arising from wave motion across the grid formed by the real and imaginary axes.

The Model regards belief as being a property which can be either unfolded or enfolded as was shown in Section 4.8.

Unfolded belief best associated with the acquisition of knowledge, may be thought of as the ability of an individual to perceive through the unfolded senses such as sight, hearing, touch, smell and taste. Enfolded belief gives access to enfolded senses such as feelings, emotions, memories, dreams and intentions etc which can be and commonly are shared with other individuals.

So there is an active relationship between the unfolded and enfolded domains in any conscious individual as illustrated in the figures in Section 4.8.

Belief conforms in the main to the individual's experienced conditions or confidence in the possibility of those conditions, whereas imagination is essentially as free as a bird.

Imagination is a much deeper form of enfoldment not shared by the individual concerned with other individuals, unless it is unfolded by process of action in the form of communication. It can therefore be held secret by the individual if they so choose, which indeed they commonly do.

Fig 4.8.10 shows the relation between belief and imagination is one of degree of enfoldment with enfolded belief and its unfolded partner knowledge dominated by the mean level of order, whereas imagination is dominated by the present state imaginary axis.

Nevertheless the key to stability is the balance between the objective and the subjective whether belief or unworldly imagination.

This balance between reality and imagination is vital to learning processes in particular as unbalance in one or the other leads directly to disorder which can become very destructive.

Enfolded imagination is important to children in the form of storytelling whether about real or imagined characters and events, fairy tales and fantasies of every kind. Children exercise their imagination using such narratives and by pretend play. Role play acts out what they have developed with their imagination and they then use the make-believe situation as if it represents reality.

So knowledge and imagination should generally go hand in hand. Knowledge supports and stimulates imagination and vice versa imagination stimulates the acquisition of knowledge.

Progress in scientific research is commonly achieved by real world observations which trigger the imagination in to adopting a new approach to the problem concerned.

Imagination develops theories in effectively a virtual thought world, if the relevant idea is probably true then that theory is successfully unfolded in to the real world by action.

So human beings seem to have two grades of subjective activity, firstly belief essentially coarse grained involving mundane everyday affairs interacting with reality, deciding on various alternative tactics; secondly imagination essentially fine grained involving exotic forms of creativity and artistic expression leading to the adoption of a new overall strategy which effects the future in a significant way not only for the individual but which ultimately spreads to many others as well.

It seems evident from the Reference Model that belief is influenced by the mean level in the Ordering Centre whereas imagination extends in to the regions dominated by the present state axis as described in Section 4.8.

Imagination creates imaginary worlds which can be significantly divorced from what would be regarded as the real world by the human population as a whole. So theoretically there is no limit to it, however if that includes infringing on the principles applying to a given Ordering Centre such as inflating imagination to the point where it is outside the bounds of reasonable possibilities or probabilities, then the individual concerned may well be regarded as insane by his peers. There are clearly cases where this is harmful such as primitive cultures where an individual may believe when he is ill that it is caused by incantations etc of a hostile nature deliberately uttered by an enemy. Imagination can also produce symptoms of real illness involving physical manifestations and psychosomatic disorders of one sort or another.

Users of hallucinogenic drugs are said to have heightened imagination while under their influence, creating a difference between imagined and perceived reality obtained from the physical senses. Clearly this is a route in many cases to illness in the form of addiction. Irrespective of drug use many mental illnesses can be attributed to the inability to distinguish between imagination and reality provided by those physical senses.

Imagination being free of reality can become a source of real pleasure commonly associated with religious experience or alternatively unnecessary suffering. Persons with a vivid imagination may suffer acutely from imagined perils in a parallel way to those who suffer horrific real events.

Imagination is therefore linked to perception, the world view held by the individual, which in turn influences how that individual perceives reality.

Belief and imagination is very much associated with memory. Eccles (Ref 2.7.4) describes the relationship between long-, medium- and short-term memory and the relevant areas of the human brain.

So belief unfolded in the form of the physical senses is linked to belief enfolded in the form of emotions, feelings, memories and intentions; unfolding reality in the form of the future is linked to enfolded reality in the form of the past; belief is attracted to the real axis and imagination to the imaginary axis.

Nevertheless many cultures and traditions commonly of a religious nature view the real world as illusory accepting the opposite extreme of imagined and dreamt of realms as primary, or alternatively as having equal status to physical reality.

These and other cases involving imagination which infringe on reason including virgin birth and religious miracles etc will be discussed later in Section 7.5 and particularly in relation to the progression towards order as described in Section 4.7.

7.1.9.9. *Reductionism and holism, wholeness and unwholeness.*

Figures in Section 4.8 illustrate the process of reductionism which involves the collapse of belief and imagination on to the mean level of order i.e. the distributed (nominalist) axis or variance. Too much emphasis on reductionism or alternatively holism is potentially self-defeating as existence cannot be construed in terms of one to the exclusion of the other.

Wholeness may then be defined as the combined process of supporting the common sense which defines reality simultaneous with belief and imagination. Imagination arises from the existence of other levels of order, these being the inevitable consequence of the absence of a boundary to existence which may be regarded as infinite, as opposed to finite but unbounded.

Indeed science regards both relativity and quantum theory as of fundamental importance and both are holistic theories.

Unwholeness is the antithesis of wholeness and takes two forms. Firstly unwholeness in the form of implicit disorder namely evil, which degrades ultimately to the random state and secondly unwholeness in the form of explicit disorder or chaotic disorder associated with the operation of the second law of thermodynamics which degrades to the purely scalar field.

The Reference Model also incorporates the process whereby these scalar fields also break up leading to the random state as described in Sections 4.5 and 4.7.

Wholeness is the state associated with the process which opposes that disordering process by unfolding order from disorder.

The Reference Model shows that these opposing states of wholeness and unwholeness exist together and that the processes of unfolding order from disorder and disorder from order apply at all times and that these processes are not only fundamental to but represent existence.

Furthermore that there are effectively no limits to these processes in any sense as illustrated in Sections 4.7 and 4.8.

7.1.9.10. Determinism, indeterminism, and freewill.

The Reference Model supports compatibilism and from that perspective it can be seen that determinism is to be associated with the mean level of order and the present state axis. All of the variations in between are described by the animate vector which has a real spatiotemporal distributed component and an imaginary non-spatiotemporal present state component. So a conscious entity thinks and acts in accordance a combination of determinate and indeterminate elements, neither wholly determined nor wholly freewill, though commonly largely one or the other. The variety of opinions evident among philosophers and described above arises from insisting on a particular perspective to the exclusion of the alternative, all arising from failure to provide a solution to the mind body problem.

The Reference Model effectively eliminates the mind body problem since the one flows to and from the other seamlessly.

7.1.9.11. What is meant by mind as opposed to brain?

What is meant by the phrase mind over matter?

The Reference Model regards mind as the enfolded brain and brain the unfolded mind, so mind is the implicit form and brain the explicit form. So both are forms of the one substance or stuff namely order which unfolds from disorder and vice versa.

The argument over whether there is one stuff, two or many is therefore again a matter of perspective, since unfolding and enfolding processes representing motion or transfer of energy predominate making such arguments irrelevant.

The phrase mind over matter is a reflection of the view that there is more to existence than the explicit as the Reference Model confirms. The Ordering Centre is made up of the structure of the ordering envelope one level of which represents the mean level of order or explicit level relative to the whole set of levels which are implicit. When that mean level is viewed it in turn has an unfolded aspect below that level namely the future and an enfolded aspect above that level namely the past, so motion is everywhere. Existence is therefore in effect an unbounded as opposed to a bounded perpetual motion machine, the latter being precluded by the first and second laws of thermodynamics.

7.1.9.12. Complexity as opposed to simplicity.

The validity of the principle of Ockham's razor.

The Reference Model associates simplicity with the holographic fractal character of an Ordering Centre. An endlessly occurring pattern with no spatiotemporal beginning or end; a pattern which can be extended to the concept of ever greater and lesser energy densities again without limit. The inanimate symmetrical state of the Whole Ordering Centre exhibits this simple theme in its clearest form.

Complexity is associated with the requirement that existence has to know of itself, knowledge which it obtains by animate asymmetrical motion about and at variance from its symmetrical state. There is likewise no spatiotemporal limit to this motion and no limit to the complexity either.

The principle of Ockham's razor can be associated with the mean level of order in an Ordering Centre rather than taking account of the necessity for imagination provided for by the present state. The facts referred to by the razor are those in effect determined by measurement, as opposed to the Reference Model which uses what is termed the records to identify the whole truth; such records include all statements irrespective of whether they are explicit or implicit, measurable or immeasurable.

Nevertheless in another sense Ockham's principle quite sensibly emphasises the value of looking at things in a simple way, not letting the trees get in the way of seeing the wood, the whole thing as opposed to the detail. Unfortunately as they say the devil lies in the detail, so it is necessary to take account of the complexity as contained in that detail. Therefore a balance between the two has to be struck which is exactly what the Reference Model is all about.

7.2   Life.

7.2.1.   What is life? The scientific view.

It is a primary purpose of this book to provide answers to this question in view of the absence to date of a clearly understood relationship between the animate and the inanimate, in particular how and why life exists.

The theory of evolution as a basis for the emergence of life was put forward by Charles Darwin in 1869 and published in 1876 in his book *'The Origin of Species by means of Natural Selection'* (Ref 7.2.1).

The combined scientific and theological thesis, *'The Phenomenon of Man'* proposed by Teilhard de Chardin (Ref 7.2.2) has more in common with this book as a description of existence in an overall sense. However his proposals from a scientific point of view are related to the state of that discipline in the first half of the 20th century, before the important discoveries made since.

A further book *'What is Life? with Mind and Matter and autobiographical sketches'* by Erwin Schrödinger (Ref 7.2.3) again viewing life from a scientific angle but incorporating a chapter on science and religion, includes his revealing assertion that the mind is not destroyed by time.

The scientifically based anthropic cosmological principle as a basis for life is described in detail by Barrow and Tipler (Ref 7.2.4) and incorporates a review of Teilhard de Chardin's thesis. The anthropic thesis and the associated proposal that it is possible to construct artificial intelligence even life itself including the use of self-replication procedures as a basis for extra-terrestrial probes will be discussed later in this section.

Another book by A.G. Cairns-Smith, *'Seven clues to the Origin of Life'* (Ref 7.2.5) is an excellent introduction to the subject of natural selection for the layman.

A further investigation in to consciousness in particular has been made by R. Penrose in his book *'Shadows of the Mind'* (Ref 7.2.6). Penrose supported by the anaesthesiologist Stuart Hameroff proposes that consciousness is the process of orchestrated objective reduction at the quantum level involving the microtubules in the human brain (see Sections 7.4.8 and 7.6.1).

7.2.2. The theory of evolution as a basis for emergence of life.

It is not intended here to go in to a detailed description of the scientific theory of evolution. Better to get back to basics and describe the key component of that theory, namely natural selection, in its simplest form. Like all scientific theories it is based on measurement. In general terms it is an attempt to explain the origin of life and its development in to the prolific variation we see around us and ultimately to explain the existence of the human race itself.

An ideal reference on this subject is that by Cairns-Smith (Ref 7.2.5) referred to above. His summary of natural selection involves a process of reproduction, random variations in the offspring, inheritance of those variations, occasional advantages in having those variations, competition between the offspring and the overproduction of offspring resulting in survival pressures which improve the reproduction processes.

The idea of survival of the fittest is clearly applicable to such a description, but as Cairns-Smith comments, nature is continually changing her mind about who is fittest, so that idea has limitations. So natural selection is key to why living things adapt to a given set of circumstances and further adapt to new ones. Life is then said to be a product of evolution which would have developed gradually.

So it is then necessary to go back to origins, the origin of evolution leading to the origin of life. Somewhere along this road organisms developed, yet the word organism implies the organic as opposed to the inorganic. Now the criteria for natural selection do not in themselves imply that only living things are involved. In fact it is the scientific contention that initially things were inanimate or effectively so and that natural selection gradually converted the inanimate to the animate.

Cairns-Smith offers seven clues to the process of evolution, which can be found on the internet.

First, that genetic (meaning original) information is the only thing that can evolve through natural selection because it is the only thing that passes from generation to generation over the long term. Crucially that information although held in a genetic material is not substance but form or ordering which incorporates a process of replication and change which means a fixed state of substance cannot survive.

Second, that the undisputed rulers of the present day in biochemical terms, namely DNA and RNA are complex molecules with a helical structure, far too complex to be the earliest genetic material.

Third, that processes of construction involves things that are absent in the final outcome such as scaffolding in the construction of a bridge or building for example. So the first organisms would have used the simplest kind of genetic material now absent from present day biochemistry.

Fourth, from the nature of ropes that none of the fibres need stretch the length of the rope merely that they be sufficiently intertwined to hold together. The individual items of information namely genes form collections of genetic information (a form of reference library) which are intertwined allowing the transfer of information to new gene fibres. So organisms which develop based on one genetic material may develop or evolve in to another genetic material.

Fifth, that primitive machinery is usually simpler in design and its materials more rudimentary compared with later more sophisticated versions. Hence organisms of low complexity evolve in to organisms of high complexity.

Sixth, that crystals self-assemble, growing, breaking up and regrowing in a manner which would seem very relevant to the formation of basic genetic material.

Seventh, that the earth makes clay all the time carried in and deposited by rivers, the minerals of clay being formed of tiny crystals that grow in water solutions as a consequence of the weathering of hard rocks. Such crystals are strong candidates for the role of primitive genes, essentially inanimate, a form of primary organism leading to secondary organisms more akin to life, through a process of genetic takeover. Clay is of course also referred to in the Bible for example as the stuff of first life.

7.2.3. *Asexual as opposed to sexual reproduction and the relevant advantages and disadvantages in promoting life.*

Chirality.

Asexual species reproduce by duplication (division) without outside influences such as a partner, the offspring being a copy of the parent, the general process being termed Mitosis. This form of reproduction is simple and efficient since only one parent is needed and all parental genes are passed on to the next generation. All bacteria and most plants produce asexually. Even some animals reproduce asexually at times.

Sexual reproduction on the other hand requires a partner, each individual producing specialised sex cells or gametes which carry half of their genetic material, one copy of each DNA chromosome rather than two. A male gamete must successfully fuse with a female gamete. The offspring while similar to its parents is not identical with mixed genes, the general process being termed Meiosis.

Sex involves finding a mate which takes a considerable degree of time and energy, not to mention the unfertilised gametes which go to waste, furthermore each parent only passes half its genes to the offspring. Despite these obvious disadvantages sexual reproduction has thrived and dominates the animal world.

A partial explanation is based on the necessity to keep ahead of the game in competition with different species. Sex maintains and encourages genetic diversity increasing the probability that individual offspring are resistant to diseases. These disease resistant strains become more common, and while other diseases evolve, new genetic combinations produce further resistant individuals and the general process repeats itself.

Asexual populations on the other hand can only evolve resistance to a disease if a beneficial mutation or random alteration in a gene arises. Since this is a rare event the asexual population will fail to resist the disease and may even die out.

Some species reproduce asexually and sexually, depending on their stage of life cycle etc, a reproductive strategy called parthenogenesis.

Various theories have been advanced to explain the evolution of sexual reproduction for example that it increases fitness among the progeny and encourages genetic variation despite the two-fold reduction in the number of progeny at each generation.

The lottery principle, the tangled bank hypothesis, the red queen hypothesis and the DNA repair hypothesis, etc, are among various proposals that have been made in this regard, all being variously supported by evolutionists and denigrated by others.

It is however apparent that Mitosis can be associated with symmetry and Meiosis with asymmetry. Symmetric objects have left- and right-handed sides defined by the location of the observer, while asymmetric objects exhibit chirality independent of that location.

Chirality is a property of asymmetry such that an object is different from its mirror image, as for example things left- and right-handed from screw threads to human beings.

It is an important feature in many scientific disciplines, relevant examples being particle spin in physics, the structure of DNA, the shells of sea snails in biology and so on.

7.2.4   To what extent does consciousness place limits on the three pillars of local realistic theories namely realism, inductive inference and Einstein locality (separability)?

The no conspiracy principle.

Until the discovery of the quantum state, science assumed that the unconscious inanimate world was made up of objects whose existence was independent of human consciousness. Religion tended to obscure this independence by including objects, animate or not, within the all-embracing reach of an all-powerful deity, but otherwise accepting the same principle, at least as far as worldly mortals was concerned. This principle consisted of three main elements elegantly stated by B. d'Espagnat (Ref 4.5.1) and previously referred to in section 4.3.

The first namely realism, that the common experience of physical reality was independent of human observers, in other words for example, trees falling in an uninhabited forest make a sound in doing so.

Second that inductive inference is a valid method of reasoning which could be applied freely so that conclusions drawn from consistent observations were legitimate.

So in an experiment if as a result of a given action A it is found that B results and on repeating the experiment many times B continues to occur, it is reasonable to conclude that the same result will apply at some time in the future, without having the need to prove the same by repeating the original experiment.

This conclusion corresponds to the concept in physics of counterfactual definiteness.

Counterfactual definiteness assumes the output of a system can be calculated by using an explicit formula $y = f(x)$. It does not mean opposed to fact, instead it is used to characterise values that could have been measured but were not. Counterfactual values are obtained by means other than direct measurement or observation but by well substantiated theory, or in a broader sense by referring to the common experience.

Third, that no influence of any kind can propagate faster than the speed of light, referred to as Einstein locality or separability.

In effect a measurement made with one instrument cannot influence the result of a measurement made with another.

A fourth principle may be added to the above namely that nature does not conspire to force experimenters to measure or observers generally to be only conscious of what she wants and when she wants, hiding what she does not want seen.

It will be clear to the reader that this no conspiracy principle is certainly not adhered to by human beings!

These four principles form the core of what are termed local realistic theories.

They are the basis for all that is associated with the common sense or common experience of conscious entities in terms of cause and effect, coupled with the advances made by Newton in the late seventeenth century in deriving the basic equations of motion, then modified by Einstein two centuries later by the theories of special and general relativity.

However during the early part of the twentieth century, arising from investigations in to the physics of elementary particles, experiments repeated many times resulted ultimately in the establishment of mathematical equations which could at least on average predict the behaviour of such particles. These equations formed the basis for what came to be known as quantum mechanics and more generally as the quantum theory.

The history and basic description of this theory will be discussed in answers to questions asked in section 7.4.

Now, experiments assuming the validity of local realistic theories have been tested against the rules of quantum theory many times in relation to the physics of elementary particles. It has been found that quantum theory is incompatible with the principles of local realism. Specifically local realistic theories predict that what is referred to as Bell's inequality is valid whereas quantum theory shows that inequality to be sometimes invalid. This inequality will again be referred to in more detail in section 7.4, which will also consider the consequences of the collision between general relativity and quantum theory arising from the incompatibility referred to above.

What is referred to as the quantum dilemma is still unresolved in scientific terms. Physicists prefer either to concentrate on carrying out experiments and analysing the results or endeavouring to find a mathematical theory which unites gravitation with the quantum state. This unification seems some way off and is very likely to be untestable in an experimental manner.

Any other thoughts on the subject they leave to philosophers, who they assume can make little contribution anyway.

7.2.5   *The concept of artificial intelligence.*

The concept of artificial self-replicating devices.

The concept of artificial intelligence developed initially in the 1950s, derived in particular from the work of the mathematician and World

War 2 codebreaker Alan Turing. He wrote a computer program for a 'paper machine' to play chess. The machine is equivalent to a program which defines a series of simple steps in the same way as a computer program but written in a natural language such as English, all followed by a human operator. The operator does not know or at least need not know how to play chess; he merely follows the instructions for generating moves on the chessboard.

In so far as computers were concerned they were indeed programmed to play chess and with the rapidly increasing capacity for storage and speed of processing over the next decades they were able to defeat human players up to grandmaster level.

The argument then was whether this ability was due purely to raw processing power in evaluating possible moves which on a purely mechanical basis has nothing to do with intelligence, as for example a machine devised to print letters far faster than a human could achieve, or whether the machine was developing the power to think for itself, which is the basis of intelligence.

Turing thought that computers would in due course be able to exhibit apparently intelligent behaviour, capable of answering questions and carrying on conversations in a given language. In this connection he proposed what became known as the Turing Test.

If a computer could pass for a human being while conducting such a conversation it should be assumed to be intelligent.

The concept of artificial intelligence developed further in the 1960s and 1970's arising from the work of among others, M. Minsky (Ref 7.2.7) and J. McCarthy (Ref 7.2.8).

Other useful references which provide further background are that by D.R. Hofstadter (Ref 7.2.9), also D.R. Hofstadter and D.C. Bennett (Ref 7.2.10).

All these being protagonists of what is termed strong artificial intelligence (strong AI), namely the idea that computers will in the future at least emulate the human brain to the point where they will as it were take-over, become human in a sense, when they become sufficiently complex in their algorithmic modes of behaviour; that they would experience feelings, become conscious, have not only a brain indeed but a mind, it being assumed there is no difference between the two.

Others tend to refute this argument to a greater or lesser degree, prepared only to support what may be termed weak AI, namely that conscious awareness is a feature of the brain's physical action and while it may be possible to simulate such action by the use of computer algorithms, such simulation does not constitute awareness. Weak AI supports the view that a computer may behave as if it were conscious but not in fact possess this quality.

J. Searle (Ref 7.2.11 and 7.2.12) appears to hold views which correspond to weak AI, in accordance with the principle of his Chinese Room Argument, which can be found on the internet.

Basically this involves a native English speaker who knows no Chinese locked in a room full of boxes of Chinese symbols (a data base) together with a book of instructions for manipulating the symbols (the program).

People outside the room send in various Chinese symbols which unknown to the person in the room are questions in Chinese (the input), but by following the instructions in the program the man in the room is able to pass out Chinese symbols which are correct answers to the questions (the output). The program enables the person in the room to pass the Turing Test for understanding Chinese but he does not understand a word of Chinese.

In other words if the person in the room is replaced by a computer, it equally does not understand, but follows instructions in a mechanical manner.

Searle also refuted on the same basis the functionalist approach to understanding minds especially that form of functionalism that treats minds as information processing systems.

Functionalists generally hold that mental states are defined by the causal role they play in a system, as for example a door stop is defined by what it does, not by what it is made of.

However this leads ultimately to granting inanimate systems the property of having mental states which is implausible.

Searle believes that computer programs are merely syntactical, that computers only respond to the explicit form of strings of symbols, not to the meaning of the symbols.

Minds on the other hand have mental contents, semantics, states with meaning.

We associate meanings with words or signs in a given language and respond to them because of their meaning not just their physical appearance.

Syntax is therefore not sufficient for, or constitutive of semantics. So although computers may be able to manipulate syntax to produce appropriate responses to natural language input, they do not understand the sentences they receive or output, for they cannot associate meanings with words.

Searle nevertheless seems to equate brain with mind and consciousness, such that the actual material of the brain is all important.

An alternative Platonist view held by R. Penrose (Ref 7.2.6) which maintains a scientific position but is opposed to even weak AI, is that physical action of the brain evokes conscious awareness but this action cannot be simulated computationally.

That consciousness involves patterns, configurations or sequences of constituents not limited to the particular material of those constituents.

He takes the view that present day physics offers little scope for explaining the non-computability of the brain/mind system and that fundamentally new physics is required to do so.

Penrose points out that believers in strong AI although labelled physicalists regard the material construction of the thinking device as a basis only for the computation that it performs, which determines all its mental attributes, so the device as it were is not limited to the physical object, a somewhat paradoxical almost dualist position with which they would not wish to be associated.

For supporters of weak AI the physical constitution of an object must play a vital role in determining the degree of mentality present in association with that object. At least in these cases the mental qualities are regarded as real things and not just additional phenomena which arise incidentally when computations are performed. The Platonist however also brings the concept of things as patterns of information into play.

Finally the religious view is that ultimately consciousness and awareness cannot be explained by any physical, computational, or scientific thesis relegating all the above to the trash can so to speak.

It will be clear to the reader how muddy the waters are when it comes to the above subject matter.

Now a variety of criticisms have been made of the Chinese Room argument in support of Strong AI, described earlier.

The Systems Reply, the Virtual Mind Reply, the Robot Reply, the Brain Simulator Reply, the Intuition Reply and the Other Minds Reply can all be found on the internet.

Turning to the concept of artificial self-replicating devices, it is the earnest belief of supporters of strong AI that they are essential if any form of life, is to be preserved. They are generally of the opinion that the origin of life is in all likelihood a rare chance event, possibly unique and that in view of what Bertrand Russell referred to as 'the uncaring nature of omnipotent matter, blind to good and evil, reckless of destruction rolling on its relentless way', it would be a good idea to take whatever steps are possible to ensure such preservation.

Barrow and Tipler (Ref 7.2.4) based on work by the mathematician J. von Neumann in the 1950s describe the essential features of a self-reproducing machine and how it would be deployed. The general thesis is that inter stellar rocket technology carrying such machines as probes could be developed by an advanced technological civilisation such that a transit time to the nearest stars would be of the order of 10^4 to 10^5 years.

Such probes would of course need fleets of repair vehicles to offset the effect of the second law of thermodynamics inevitably reducing the efficiency of the process of propagation.

On reaching a target star the probes would use available local materials to copy themselves constructing new probes which would be sent to further adjacent stars and the process repeated till the entire galaxy had been explored. It is assumed in this scenario that the algorithms employed in these self-reproducing machines would in accordance strong AI endow such devices with at least human level intelligence if not much greater, though probes with limited abilities in this regard could be utilised to merely advertise their arrival to any extra-terrestrial inhabitants who happened to be around, exhorting them perhaps to eat or drink healthily or some such.

In all of this it can be borne in mind that no such physical devices have appeared in our solar system which in the context of strong AI would, from that point of view, support the contention that we are alone in our galaxy.

7.2.6 *To what extent do extra-terrestrial civilisations exist in the visible universe, assuming Homo sapiens could recognise them as such?*

The Anthropic principle.

The technological life span of Homo sapiens.

The absence of evidence of extra-terrestrial life.

The religions generally show little interest in this subject and some in the scientific fraternity conclude we are alone in at least the visible universe which contains some 10^{11} galaxies. Others are agog to provide an answer and apparently 50 per cent of the human population believe that life is relatively common in both our galaxy and others.

To try and provide a basis for deriving an estimate, at least in principle, the Drake Equation as formulated by F. Drake (Ref 7.2.13) is commonly utilised.

The number of civilisations of a technological nature in our galaxy with which communication might be possible is on this basis given by:-

$$N = r f_p \, n_e \, f_l f_i f_t \, L$$

where r is the rate of star formation in the galaxy,

f_p is the fraction of those stars that have planets,

n_e is the number of habitable planets in a solar system that has planets,

f_l is the fraction of those habitable planets where life actually evolves,

f_i is the fraction of those habitable planets with life which develop intelligence,

f_t is the fraction of intelligent species which develop a technological civilisation and release detectable signals of their existence into space,

L is the lifetime of such a technological civilisation.

Of these parameters, to date r (between 1 and 10) and the product of $f_p \, n_e$ (between 0.05 and 0.1) are thought to be known from astronomical observations with reasonable confidence.

These figures are based on studies, including those over the last decade. Of particular interest is the occurrence of rocky habitable zone planets around solar like stars that have been conducted using the Kepler and Gaia space telescopes. They confirm with a reported 95 per cent confidence that such planets are not uncommon, at approximately 4 around the 63 sunlike stars within 10 parsecs of the sun, where sunlike stars represent some 20 per cent of all main sequence stars. That represents roughly one in 100 main sequence stars.

All the rest are commonly regarded as speculative, the fractions taking any value between zero and unity and L any order of magnitude above unity.

If the most optimistic assumption is made and the speculative factors in the Drake equation are of the order of 1 this gives an estimate of the value of N to be given to a good approximation to L, the average lifetime of a technological civilisation.

If L was say 100 to 1,000 years there would be between 100 and 1,000 technological civilisations in a typical galaxy of 10^{11} stars similar to ours.

If any or all of the speculative factors are significantly less than unity the value of N could easily drop to the point where many galaxies have no life technological or otherwise unless this is counteracted by an equivalent increase in L.

It is to be noted that many supporters of strong AI consider N to be $\ll 1$ and L to become irrelevant since it is assumed that artificial life will escape the bonds of earth and perpetuate itself effectively indefinitely elsewhere.

It is useful at this point to evaluate the Anthropic Cosmological Principle as proposed originally by Brandon Carter (Ref 7.2.14) and alternative definitions due to Barrow and Tipler (Ref 7.2.4).

The weak anthropic principle (WAP) as enunciated by Carter simply states that our location (in space and time) in the universe is necessarily privileged to the extent of being compatible with our existence as observers. He did so in opposition to the Copernican principle, which assumes human beings do not occupy a privileged position in the universe.

Carter also invoked a strong anthropic principle (SAP) namely that the universe (and the fundamental parameters on which it depends) must at some stage be such as to be compatible with the existence of observers. This can then be used to explain the coincidence that the observed value of dimensionless physical constants (e.g. the density of matter, the gravitational constant, the mass of the proton and the age of the universe) appear balanced as if fine tuned to permit the formation of matter and subsequently the formation of life, all of which will be further referred to in Section 7.4.

The weak anthropic principle (WAP) according to Barrow and Tipler basically states that the observed values of all physical and cosmological quantities take on values restricted by the requirement that carbon-based life can evolve and the requirement that the universe be old enough for it to have already done so. This version restricts the principle to carbon-based life rather than observers and includes the fundamental physical constants whereas Carter leaves them to his strong version.

The strong anthropic principle (SAP) according to Barrow and Tipler states it is imperative for the universe to have those properties which would allow life to develop within it at some stage in its history.

They suggested various alternatives of the SAP including that there exists one possible universe designed with the goal of generating and sustaining observers, that such observers are necessary to bring the universe in to being, that an ensemble of other different universes is necessary for the existence of our universe.

The additional final anthropic principle (FAP) goes further and states intelligent information processing must come in to existence in the universe and once it comes in to existence, it will never die out.

These principles have been put forward by scientists who tend to assume that artificial intelligence will sooner or later predominate over human intelligence.

Brandon Carter (Ref 7.2.14) developed a formula for estimating the life span of Homo sapiens as further discussed in detail by Barrow and Tipler (Ref 7.2.4).

Basically this concludes that the probability of assembling the human genome is so small that we would be not only alone in our galaxy but effectively alone in the visible universe. Brandon Carter (Ref 7.2.15) went further and postulated a probabilistic argument that the probable lifespan of the human race is of the order of 10^4 years based on the present size of the human population and the number of individuals considered to have lived and expected to live. This argument referred to as the doomsday argument can be found on the internet.

The reason for the extreme pessimism of the Carter, Barrow and Tipler conclusions is the fact that they ignore imagination and base everything on what they consider reality, hence the imperative in their view of strong AI and the artificial self-replicating probe scenario.

Religion, which generally takes little interest in extra-terrestrial life, amusingly in so doing makes effectively the same assumption with regard to the occurrence of life elsewhere.

Yet of course many scientists and laymen are happy to assume that life is relatively common in both our galaxy and others.

Turning to the question of the absence of evidence of extra-terrestrial life, again looking from the scientific viewpoint, no viable extra-terrestrial life form has been confirmed to date even within our solar system, though this situation may change as our exploration of space continues.

It seems likely therefore that the so-called goldilocks region around planets orbiting other stars in our galaxy is the next alternative possibility, on the basis that such regions have similar environmental characteristics to our own.

It is evident that such extra-terrestrial locations are at vast distances from us upwards of the order of 10 light years to some 100,000 light years. Our present and in the opinion of this author our foreseeable ability to travel such distances using space vehicle technology carrying intrepid human explorers at inevitably limited fractions of the speed of light is distinctly unlikely, bearing in mind our very nature.

Nevertheless some like I. Crawford (Ref 7.2.16) believe that colonisation programs could be carried out by such explorers, claiming that we could reach the nearest stars in 100 to 1,000 years with foreseeable technology and within 10,000 to 100,000 years inhabit every star system up to a radius of 200 light years, even go on to full galactic colonisation in 5 to 50 million years.

Both this and our ability to use self-replicating artificial probes is in doubt if the lifespan (L) of a technological civilisation is of the order of 1,000 years.

No indication has been found so far that other civilisations exist or have existed using such methods unless one interprets our own existence being due to such a process.

This implies that either:-

(a) Life is rare i.e. N and or L is too low, or
(b) That probes artificial or otherwise have not or cannot be utilised successfully for the purpose of propagation.

In view of the above the more obvious alternative approach of staying where we are and trying to communicate by means of radio transmitting/receiving devices is worth investigating. SETI programmes by passive means namely the use of dish antennas and sensitive receivers have been in progress since the 1960s.

The capability of a technological civilisation to transmit signals is in scientific terms categorised by the relevant power equivalent in watts.

Type 1 civilisations being capable of transmitting signals with the power equivalent of all the sunlight striking the earth's surface 10^{16} watts, Type 2 the power equivalent of the sun 10^{27} watts and Type 3 the power of the entire galaxy 10^{38} watts, according to A.J. LePage (Ref 7.2.17). The earth-based Arecibo transmitter of some 10^{14} watts gives us a Type 0.7 rating in these terms.

From research to date no powerful so-called Type 2 and Type 3 super civilisations are evident in our galaxy or other galaxies that have been studied; it seems very likely that they do not exist as surely we would know of them.

LePage says that an extra-terrestrial transmitter equivalent to Arecibo must be further away than 4,000 light years to have eluded our searches and a Type 1 transmitter in excess of 10,000 light years. The difficulties of such methods are discussed by G.W. Stenson (Ref 7.2.18). In particular the distance at which a radio wave can be detected subject to adequate design, is dependent on five factors, the electromagnetic noise environment of the receiver, the sensitivity of the receiver, the power of the transmitted signal and the size of the receiving and transmitting antennas.

The power required from a transmitter radiating omnidirectionally (as we do not know where the extra-terrestrial party is) at say 100 light years approaches 10^{16} watts. This is many times the combined generating capacity of earth-based power stations.

The alternative option of beamed signals requires vastly less power, using say a one kilometre square antenna array at each end the power requirement reduces to less than 10,000 watts. Unfortunately the associated beam width of the signal is reduced to a small fraction of a degree, in effect requiring that we know the exact location of the extra-terrestrial party, which is precisely what we are trying to determine.

The average star density in our region of the Milky Way galaxy is estimated to be of the order of .004 stars per cubic light year, namely 4 stars within a radius of 10 light years, 4,000 stars within a radius of 100 light years and 4,000,000 stars within a radius of 1,000 light years. This ignores the greater density of stars nearer the galactic centre as we are positioned a considerable distance away from it.

It should be noted that the situation in this respect is more hopeful in the case of globular clusters where the interstellar distance is measured in light weeks if not light days.

Relating this to the most optimistic scientific model based on the Drake equation gives the number of extra-terrestrial parties in our galaxy as their lifespan in years. It is immediately apparent that if such lifespans are short including our own, these civilisations would die out before contact is made whether by radio contact or colonisation. The idea that holistic lifespans are long does not bear examination for reasons given later in Section 7.2.8.

7.2.7. What is the relationship between the physical world, the mental or conscious world and the world of knowledge and ideas including mathematics?

From a scientific viewpoint the relationship varies between that described by Popper and Eccles (Ref 2.7.4) referred to in Sections 2.7

and 4.8 (see also Section 7.6.8.5) and that described by Penrose (Ref 7.2.6).

In Popper and Eccles view these worlds are fully interactive and involve the dual nature of mind and matter. Penrose takes a different view of a platonic nature where these worlds emerge from each other to a limited degree but the relationship is cyclic in character.

7.2.8. *Life and the Reference Model.*

7.2.8.1. *What is Life?*

With regard to 'What is Life' as referred to in Section 7.2.1 the first mistake is to assume that the animate arises from the inanimate in the sense that initially there was only the inanimate.

Better to consider that the one enfolds in to and unfolds out of the other.

Viewing existence as a whole the animate and the inanimate exist together, at no stage does the animate cease to exist leaving the inanimate only or indeed vice versa. The animate is dependent on the inanimate to exist and vice versa, they are complementary.

The Model defines the inanimate as that which corresponds to what has been previously referred to as the present state, or equally unfolded from that state in a fractal manner such that differential and integral variance is equivalent, or in a state of balance. In other words the vertical and horizontal lines making up the grid structure of an Ordering Centre. Since the Ordering Centre is in motion these lines are also in motion, termed scalar motion and do not represent static states, order continually rising in the system and disorder continually falling. The animate is then defined as that which is in motion, termed vector motion about the inanimate fractal lines such that the variance is in a state of imbalance, order again continually rising and disorder falling as applied to the inanimate.

This vector motion represents the efforts of the animate to know of itself, to see itself in effect. The inanimate structure is that which cannot know of itself or see itself.

The inanimate cannot look in to the past or speculate about its future and has no knowledge of order or disorder; the animate must acquire that knowledge. In doing so it raises and lowers inanimate stuff so as to form that structure and furthermore itself by vector motion. The true degree of motion is only apparent when greater and lesser Ordering Centres are taken in to account, so the apparent central present state and the base state of our Ordering Centre instead of being absolute limits are in fact only relative, since there is no limit to the degree of ordering or disordering possible, existence having no boundary. This motion viewed as a whole may be termed the Holomovement.

The inanimate structure is the condition such that the variance is split exactly equally into differential and integral variance for all levels of ordering in this whole system.

It is this feature which provides the explanation for the correspondingly extraordinarily low entropy at the big bang and why the energy density at the big bang is so close to the critical energy density (within one part in 10^{60}) necessary for the long-term stability we observe.

Scientists searching for the reason for that stability have got it all the wrong way round.

They ask how that stability is achieved instead of asking why. The Holomovement is ultimately responsible for that stability by as it were storing its inanimate properties on the fractal grid described in the Reference Model. Likewise it is the Holomovement which is responsible for the emergence of life, specifically at the mean level of order, from water and the mineral clays (growing in water solutions) which that water deposits on the earth's surface (see section 7.2.2).

The answer to the question 'what is life' is revealed at a glance by the present time contraction process shown in Fig 4.8.1. Life grows the greater that contraction implicit in character and dissipates with the opposing explicit expansion process. Such processes in an overall sense are illustrated in Figures 4.7.1 and 4.7.2.

It is evident that both science and organised religion have failed to comprehend these matters as existence at our level emphasises diversity of belief over the principle of coherence.

7.2.8.2. Evolution.

The description in Section 7.2.2 of the theory of evolution can be related directly to the Reference Model.

In scientific terms there is an initial state which seems to relate to the inanimate as if the animate has nothing to do with it, but the Model shows that the animate appears to unfold from and re-enfold in to the inanimate, yet at the same time an Ordering Centre has an inanimate grid structure and an animate motion about that structure such that the one is complementary with the other. The present state centre line represents the one common present in terms of space and time, removing the requirement for an initial condition.

Also that genetic information is to form in other words order, as genetic material is to substance, in terms of the Model how the present state is to the distributed order or variance.

Also that elements of existence important at one level are enfolded in to and replaced by other orderings of greater or lesser complexity at other levels.

Also that evolution involves the intertwining of orderings or as stated in section 4.0 that complex ordering involves the interweaving of lesser orderings as the weaving of cloth on a loom.

It was further noted in Section 5.0 that the two-dimensional view of vector motion would correspond in three dimensions to a helix, the structure intrinsic to the complex RNA and DNA molecules.

All seven clues relate to the requirement that as order grows so does complexity develop from simplicity, as referred to in Section 7.1.9.12, and correspondingly that coherence develops from incoherence.

Evolution in explicit terms is the process of motion from distributed states associated with the real axis towards the present state associated with the imaginary axis all as illustrated by Fig 4.3.4.

The relation between the scientific understanding of evolution and the measurement processes that apply at the mean level of order becomes apparent, when the spatiotemporal distributed state associated in the Model with reality is enfolded.

There is therefore a relation between the explicit and the implicit.

Unfortunately dogmatic science, atheistic and absolutist in character, which regards religion and spirituality as nonsensical, takes the view that the concept of evolution through natural selection based on entirely explicit premises is the whole story.

It is however evident that this opinion is grounded in the inanimate distributed mean level of explicit order corresponding to our big bang.

The inanimacy in this case correlates with what Bertrand Russell referred to when he wrote, 'blind to good and evil, reckless of destruction, omnipotent matter rolls on its relentless way.'

Before showing that this is very much not the whole story it can be contrasted with the spiritual view commonly grounded in organised religion, as described in Section 7.5.

It becomes apparent in Section 7.5.3 in particular that religion needs no lessons in dogmatism which it indulges in on a massive scale, as the disparity between the mainstream religions demonstrates.

Science appears in stark contrast with the non-analytical belief associated with religion which associates the origin of life with an alternative thesis concerning first cause, namely an all-powerful deity related to the concept of oneness, independent of existence in some sense, but who in contradiction as it were adjusts the knobs on a heavenly control panel to create and then in some religions to control existence as that deity sees fit.

The ambiguity here is self-evident, but nevertheless there is also a connection here with the oneness applying to the centreline of an Ordering Centre, i.e. the non-spatiotemporal present state.

Dogmatic religion, deistic and absolutist in character, which regards science as irrelevant, is evidently grounded in the inanimate central present state corresponding to our big bang. The inanimacy in this case correlates with the deity as being independent of existence, an inanimate deity in effect.

These scientific and religious dogmas can be seen in their true light as absolute and contradictory views which fail to realise existence can be seen from different perspectives ultimately entirely compatible in terms of the Reference Model, all tied in to the Holomovement which removes the need for any first cause.

The Model supplies the oneness required while the role of the conscious God is taken in the first instance by the overall animate motion about the ordering centreline for our big bang and ultimately by the Holomovement when the Whole Ordering Centre is taken in to account.

As stated in Sections 4.4 and again above the theory of evolution by natural selection corresponds to the growth in order shown in explicit form in Fig 4.3.4.

In implicit terms evolution arises from the process of enfolding variance portrayed in Fig 4.4.4 and apparent as the horizontal component of the enfolding reduction process portrayed in Fig 4.5.3.

Sections 4.5, 4.6 and 4.7 describe how consciousness in fact governs the entire ordering process, exactly in accordance with the concept of intelligent design.

Natural selection and intelligent design are therefore directly compatible, the one coherent with the other in terms of the Reference Model.

Evolution is therefore tied in to the formation of levels of order associated with 'our big bang' in the first instance and then to higher and lower levels of order associated with the Ordering Centre and in turn the Whole Ordering Centre.

The fusion of science and religion as required by the Reference Model shows that it is therefore unwise to consider that at some point existence was entirely inanimate and that emergence of life has some fixed origin.

7.2.8.3. *Reproduction of life.*

With regard to the means whereby life reproduces itself as discussed in Section 7.2.3 it is not proposed here to delve into these detailed arguments which can all be found in various references available on the internet, but instead to go straight to the Reference Model and the structure of an Ordering Centre for an overall view.

It has already been stated that inanimate material is to be associated with the grid structure in an Ordering Centre and standing wave motion relative to that structure.

It is clearly distributed along the real and imaginary axes i.e. all given levels and present states throughout our Ordering Centre.

Life which unfolds from and re-enfolds in to the inanimate is most apparent when reality and imagination are in balance i.e. at the mean level of order and progressively concentrates towards the corresponding central present state as it grows in complexity.

Furthermore as the complexity of life increases, the ability of that life to order itself using asymmetric wave motion relative to the grid increases, all as described in section 4.0.

It uses this wave motion about the present state imaginary axis to increasingly access the ensemble of multidimensional states represented by the order potential envelope as illustrated in the figures in section 4.0, ultimately gaining access to the Whole Ordering Centre shown in Fig 4.7.1. It can only do so though by maintaining a presence in the real explicit world i.e. in space-time using the real axis.

Consciousness and as it develops intelligent life concentrates ever closer to the central present state which it uses to rise in the ordering system or conversely fall through the opposite process of increasing disorder. So they create interwoven patterns or sequences of ordering with an active present state in addition to their motion in space-time.

Conscious entities therefore have a vastly greater capacity to rise and fall in the ordering system via the more central imaginary axes than lower orders of life such as bacteria, which tend instead to become more distributed in space-time at the mean level in the Ordering Centre as was illustrated in the earlier Figures 4.8.2 and 4.8.3, but with limited present state motion.

Bacteria reproduce asexually copying by division (Mitosis) which involves lower and simpler degrees of ordering than sexual reproduction (Meiosis).

Consciousness which as stated involves much higher and more complex degrees of ordering than plant life, which in turn involves higher ordering than bacteria, is invariably associated with entities which reproduce sexually.

Higher forms of life develop in association with the need for variety in parallel with the facility to imagine, in which respects asexual reproduction is found wanting.

Disease is to be associated with disordering processes which while leading to lower levels of order gives rise to opportunities for higher degrees of ordering to unfold in reaction against those processes and as explained earlier sexual reproduction provides better means of combating disease than that which is asexual.

Evolutionary theories are to be associated primarily with the mean level of the Ordering Centre and to a degree the animate vector motion across the inanimate grid structure, while intelligent design theories are to be associated primarily with vector motion closer to the present state.

Section 7.2.3 refers to the feature of asymmetry referred to as chirality. The Reference Model describes in some detail how these properties are part and parcel of the processes involved in the construction of life from the symmetric state at the mean level of order to the highest degree of human consciousness and awareness.

7.2.8.4. *Local realism and consciousness.*

Section 7.2.4 is concerned with the validity of local realistic theories.

Now it is the contention in this book that the Reference Model at the very least points the way to a resolution of these matters.

The grid represented by the explicit four-dimensional unfoldment corresponds to our sense of the inanimate, the various levels corresponding to expansions of space-time from the big bang. It represents our common sense associated with classical physics extended to include special and general relativity, i.e. the observed state unfolding from the mean level of order for our big bang.

The transition from the inanimate through the animate to the conscious state involves an increasing degree of motion across the implicit multidimensional grid associated with non-measurement (i.e. that which is increasingly non-local, non-causal and non-continuous in character). It therefore does not correlate with the corresponding inanimate unconscious state (local, causal and continuous) which is defined by measurement.

Other energy densities greater/less than that corresponding to our big bang are considered to have higher/lower levels of ordering as illustrated in Fig 4.7.1, but share the same principles though having perhaps different physics. Lower levels of ordering give rise to chaotic inflation and higher levels to the opposite namely fine tuning of the Standard Cosmological Model.

As a humorous aside this shows the animate and consequently the conscious element of existence in Implicate space coexists with and is complementary to the inanimate, these being connected together by the multidimensional states. Implicate space is sometimes interpreted incorrectly as a neutral half way house neither animate or inanimate. Hence the half dead half alive state of Schrödinger's cat!

It is concluded from the Reference Model therefore that consciousness interferes directly with the first premise of local realism and by association the second premise not by suggesting that there is no state corresponding to them, but by allowing movement away from those limitations enabling the conscious entity to view the inanimate condition from that modified perspective.

The third premise also requires to be restated on the basis that no influence that transmits information can propagate that information faster than the speed of light. The instantaneous influence can only pass between events that are related by a common cause. So if the influence is thought of as a signal defined as that which transmits information then the principle of the finite velocity of signals is preserved.

The third premise is also found to be tied in to the mean level of order within our Ordering Centre and does not allow for the existence of a preferred reference frame along which there is effectively no space-time namely the present state. It is therefore rather meaningless to talk of signals exceeding the speed of light in any case.

When account is taken of the Whole Ordering Centre, the absolute value of the speed of light converts to a relative value associated with a particular mean level or harmonic in the overall scheme of things, see Sections 7.4.8 and 7.4.10.7.

As has been described the Ordering Centre has an imaginary axis corresponding to the unitary determinate quantum state, while the real axis corresponds to classical physics.

Animacy and consciousness is housed in the Implicate space inclined to these axes, as previously described in terms of vector motion.

Such motion at angles more adjacent to the real/imaginary axes enhances reality/imagination respectively, while intermediate angles balance reality with imagination.

7.2.8.5. Artificial intelligence.

In respect of the subject of artificial intelligence as discussed in Section7.2.5, the Reference Model illustrates again the reasons for the range of opinions described therein of whatever nature and the confusions that arise when an overall perspective is not taken or deliberately ignored, with a view to asserting a particular viewpoint in preference to another. Not that those viewpoints are inadmissible since that they make for diversity, an essential ingredient of existence, but that it is the overall view and the essential role of the ordering process which explains that diversity.

The real axis at the mean level of order in an Ordering Centre explains the belief in strong AI, since that level has not only a physical description but a pattern of sequencing enfolded within it, made apparent by the unfolding present state as shown in Fig 4.6.4(b).

It is apparent from this figure that at the mean level of order inanimate material can potentially take up animate asymmetric motion by forming hierarchical levels relative to that mean, utilising Implicate space, but in the opposite sense to that of human holistic intelligence.

Increasingly hierarchical structures are associated in the Reference Model with increasing variance or distribution. So strong AI could be considered the potential reductionist intelligence applying at the mean level of order; as opposed to the potential holistic intelligence which equates to the average human intelligence.

Since human intelligence creates the artificial kind and since the material substrate of both AI and human intelligence use the mean level of order as a common base so to speak, they can interact with each other to some degree. There are however limitations to all this.

Human material substrate achieves animacy and consciousness, at least primarily, by enfolding the present state from that applying at the mean level and moving towards the Ordering Centre holistic present state or imaginary axis. This is precisely the direction which implies movement away from the strong AI stance towards firstly weak AI and then secondly the non-AI Platonist case; so enfolding the present state does not provide a good environment for strong AI, as any self-respecting reductionist would agree.

This also ultimately explains the spiritual/religious stance and the refusal of the mystically inclined whether individually or in large groupings to accept any physicalist element as an explanation for consciousness.

The AI material substrate may correspondingly achieve animacy and consciousness, at least primarily, by unfolding the present state from that applying at the mean level, avoiding descent to the distributed base state where chaotic indirect disorder dominates, an environment totally inconsistent with the concept of intelligence of any kind.

Furthermore expansion in Implicate space whether by unfolding or enfolding the present state involves life cycles, limitations to any given life form in time (See Sect 7.2.8.8). Nevertheless it would be expected that reductionist intelligence would by definition endure explicitly in time longer than holistic intelligence.

Hence for all these reasons and from arguments presented in Sections 7.2.5, 7.2.6 and this section, the universe appears silent to us, in so far as extra-terrestrial intelligence, artificial or otherwise, is concerned.

The whole range of human opinion therefore correlates with the concept of Implicate space set within the Whole Ordering Centre.

As already stated the Reference Model shows that strong AI is to be associated with increasingly reductionist intelligence in our Ordering Centre, utilising physical material on the real axis with a distribution in space-time but which also takes account of the associated algorithm derived from the ordering enfolded within that level, as implied in Fig 4.6.4 (b).

This is of course a hierarchical process which is the exact opposite of wholeness, the driving force for holistic intelligence along the imaginary axis.

Weak AI and the Platonist non-AI viewpoints progressively take account of the implicit distribution on the imaginary axis, a sequence or pattern arising from the travelling wave motion generated by consciousness.

That consciousness is derived from an interaction with the ensemble of multidimensional states corresponding to the envelope of the Whole Ordering Centre.

The spiritual/religious viewpoint is largely displaced along the imaginary axis into the Whole Ordering Centre.

It is important to remember that the Reference Model implies real mean levels of order at greater and lesser levels of order than that provided by our big bang, the Whole Ordering Centre representing the sum of all those levels.

It is therefore apparent from the Reference Model that the process of developing consciousness is always in play and that no doubt greater/lesser levels of order exist providing greater/lesser degrees of consciousness and indeed life.

Accordingly from the human point of view the process of spreading life extra terrestrially by artificial intelligence using self-replicating probes, is merely reinventing the wheel and rendered pointless quite apart from the bizarre conditions that would evidently apply during the propagation process. Consider how far removed the joyless in transit flight environment would be from that experienced by conscious entities on earth in normal circumstances. In addition bearing in mind human nature, self-replicating devices supposedly capable of assuming human characteristics must avoid destroying their own space vehicle by eliminating any tendencies to resort to violence, before they exited our solar system never mind getting to an adjacent star.

7.2.8.6. *Extra-terrestrial life and the explicit lifespan of technological civilisations.*

Section 7.2.6 is principally concerned with extra-terrestrial considerations and the likelihood of life in the universe as a whole, which of course has a bearing on the explicit lifespan of our own civilisation.

However it is evident that the Drake equation is essentially confined to the scientific viewpoint, in the parlance of the Reference Model tied to the mean level of order in the Ordering Centre arising from our big bang.

When account is taken of the imaginary axis a range of possibilities emerges.

It has been pointed out above that the Reference Model indicates there are mean levels of order which have greater and lesser levels of life than our own and that the overall system cycles between these states. One such level will correspond to our big bang.

It should be noted that whereas scientific models assume life is limited to the explicit, the Reference Model regards that level as merely the material substrate from which life unfolds with either increasingly real or imaginary orientation in Implicate space.

So the Reference Model rephrases the question as, how commonly is the material substrate of life distributed in our big bang and to what extent concurrent with ourselves in particular, how common or rare holistic technological civilisations are, which emanate from that substrate, and would they become apparent to us.

In passing we may note if there is only one such substrate (our own) in the visible universe it matters little whether others exist in the vastly greater dimensions involved where galaxies are receding from us at greater than the speed of light.

The Model demonstrates that the primary flow of order for holistic intelligence is along the imaginary axis as opposed to the real axis, while the opposite applies to reductionist intelligence. The Model also emphasises the importance of the explicit birth life death cycle which is clearly limited to the order of 100 years for a human individual and for similar reasons likely to limit the typical lifetime of a given holistic technological species.

It is evident that human belief is distributed between the scientific viewpoint that the spread of life in space-time along the real axis is of primary importance and the religious/spiritual viewpoint that the spread of life on the imaginary axis is paramount.

As the human race necessarily keeps these views in balance the Reference Model concludes that this provides a means of estimating the number of material substrates N per galaxy that give rise to life, and that an alternative to the Drake equation should be applied on that basis. It is evident that it is the galaxies in the visible universe and particularly solar systems with planets within them which house these material substrates and offer the potential for build up to the levels of complexity needed for life; certainly in terms of life as we comprehend it.

On this basis it would seem perverse, bearing in mind the spread of human belief, that N would fall far below unity, as the Model implies the mean level of order would tend to concentrate the number of material substrates with the capacity to unfold life to that level and distribute those substrates throughout the space-time applicable to that level.

This distribution however must also be compatible with the requirement that life in its broadest sense arises due to cycling on both the real and imaginary axes. The inanimate proto life cycles of galaxies, stars, planets, satellite moons and their constituent material elements associated with the real axis and the animate life cycles of life forms arising in such systems which additionally cycle on the imaginary axis.

The lifespan of atomic particles varies enormously from periods of incomprehensible duration in years to a fleeting explicit existence equally incomprehensibly short, see Section 7.4.6. The lifespan of sun like stars is some 10^{10} years whereas the lifespans of living organisms lie typically in the range of a day to a few thousand years.

Living organisms therefore occupy an intermediate position in terms of explicit lifespan compared with the huge range applicable to the inanimate.

As stated above using the simple concept of level of life as opposed to the term life then we can safely assume that level of life to correspond with the human level on average in the case of our big bang. So at the base level in 'our' Ordering Centre centre the level of life relative to ourselves is effectively zero and the 'lifespan' of the inanimate material there effectively infinite and the reverse of this towards the peak of that centre.

On the above basis the Drake equation can be replaced by the equation $L_y = y L_x$, where L_x is the duration in years of the material substrate (the local measure of life on the real x axis), which varies from that corresponding to a star or solar system, say 10^{10} years, to that of a technological civilisation say 10^3 years; where y is a variable ranging from say zero in terms of the real world to unity in terms of the belief and imaginary worlds; where L_y is the level of life in non-local terms on the imaginary y axis.

Since dimension is enfolded along the y axis, L_y represents the enfolded counterpart of L_x in terms of the Model.

Now we can estimate L_x taken as equivalent to L in the Drake equation for Homo sapiens even if only roughly. Even for a technological civilisation capable of destroying itself in a fit of pique, L must be of at least that of an individual, which we know to be of the order of 100 years.

It is extremely difficult to be confident that L could substantially exceed 1,000 years and that confidence decreases rapidly beyond that figure in view of the very nature of man, never mind the inevitable natural processes associated with long term climate change cycles and short-term catastrophic disasters such as nuclear war, disease pandemics, shortages of food and water and other resources. All of these challenge directly the plague like numbers of the existing human population.

In support of this the Reference Model emphasises the importance of the birth life death cycle which limits the explicit lifespan of living entities. It therefore implies the numerical value of N be of the order of 100 to 1,000 and L the same value in years.

Higher figures favour artificial rather than human intelligence.

The implicit lifespan associated with the level of life L_y can be related to the above as described in sub sections 7.2.8.8 and 7.5.6.2.

Section 7.2.6 goes on to describe the anthropic principle and the implication that life is extremely rare in the visible universe.

It is clear that scientific approaches which take no account of motion on the imaginary axis are equivalent in terms of the Reference Model to restricting life to the mean level of order of our big bang, hence the assumption that life converts to one of artificial intelligence restricted to spreading at that level only.

The Reference Model looks at the question of probability in a different light such that at each level in the Whole Ordering Centre the probability of achieving the corresponding level of life approaches unity. There being no meaningful life at the base level the probability of life there is indeed effectively zero. So at our 'intermediate level' (the mean level of order for our big bang) the probability approaches unity in terms of our level of life, consistent with the lifespan L.

The mean of all the above assumptions is most likely to apply such that a figure for L of say 1,000 years may well represent an upper bound for the life span of technological civilisations in the Ordering Centre which has spawned us.

The Reference Model adds further background to this debate.

It has already been explained that the mean level of order in which the mean explicit level is enfolded should be related to a time approximately 5×10^9 years ago.

Following the formation of our solar system 4.6×10^9 years ago it has taken that time to produce Homo sapiens. From the chronology of the universe as described in Section 7.4.6, despite galaxies being formed only 0.8×10^6 years after the Big Bang it has taken a further 13×10^9 years to produce the human race.

If the mean level as defined above was the stimulus for life formation it can be considered then that life in the visible universe flowered over a time period consistent with that of ourselves. As the explicit mean level retreats in to the past that stimulus is progressively reducing, implying that the peak generation of holistic intelligence has now passed.

Therefore the number of extra-terrestrial civilisations technological or otherwise should be at a maximum contemporary with ourselves.

These figures are supported by the absence of evidence for any form of extra-terrestrial life whether human or artificial in the explicit visible universe, which is not evidence of absence all as explained using the Reference Model in providing answers to questions concerning extra-terrestrial life.

The Model explains the enormous range of possibilities, including those chaotic and random fluctuations which do not participate in the Ordering Centre and those which do up to and including what has been defined as the Whole Ordering Centre.

There are therefore regions without any evident order or physics let alone life, intermediate regions where life forms involving observers emerge and super ordered regions where observers merge in to one life termed herein the Holomovement, which essentially corresponds to the one total life or oneness referred to in a religious/spiritual sense. The Reference Model includes for those physics different from our own as represented by the envelope of the Whole Ordering Centre all of which participate in the Holomovement.

On the absence of evidence for extra-terrestrial life the Reference Model explains that motion along the imaginary axis transfers energy via the implicate state associated with organically orientated consciousness to other levels in the Whole Ordering Centre i.e. to other levels of reality from that corresponding to the mean level of order of our big bang. Such motion dominates that motion which relates to distribution in space-time, hence the absence of evidence of extra-terrestrial organic lifeforms.

For reasons given earlier in Section 7.2.5, extra-terrestrial artificial intelligence may well be possible, yet it also not been evident to us despite the considerable age of the universe. The reasons may again be deduced from the Reference Model.

Artificial intelligence should flower at the mean level of order in the Ordering Centre in conjunction with holistic intelligence, neither being present in the early explicit universe.

It may well outlast humanity in explicit terms, but as with holistic consciousness it will experience life cycles and must succumb to the inevitable consequences of the second law of thermodynamics. Whereas holistically orientated life enfolds the present state and unfolds imagination, the primary purpose of reductionist life is to do the opposite, since it is based on the primary importance of reality and the non-God principle.

All holistically orientated life will experience only relatively short lifetimes therefore no extra-terrestrial artificial life has or will be observed by humanity.

It is the Reference Model that explains why absence of evidence is not evidence of absence of others.

It is immediately apparent that if the Model is correct, i.e. that it is the range and diversity of human belief that should be taken in to consideration in answering these questions, rather than one particular viewpoint whether scientific or religious, then it follows directly that proposals for extra-terrestrial propagation and colonisation outside the solar system by human as opposed to artificial life are at least unnecessary and effectively pointless.

It does not occur to the advocates of these proposals regardless of such viewpoints that the difficulties of radio communication referred to above and equally other forms of communication, the vastness of interstellar distances, the limitations imposed by relativistic considerations, the complexity of constructing self-replicating probes, the further complexity of constructing inter stellar space vehicles and their fuel requirements, the inevitable one-way nature of such journeys, the equally inevitable effects of the second law of thermodynamics etc all add up to the conclusion that these proposals are dubious if not absurd by any standard of common sense.

This is not all, the above difficulties are those of a real and practical nature, to which must be added those ignored by scientists in their scenarios, namely the fragility of man and the idea that thinking that a species whatever its virtues, capable of the kind of mindless carnage and destruction in the pursuit of wealth and power, only to evident in day-to-day life on earth, should be sent on a galactic wide dispersal mission, amounts to a blatant lack of emotional intelligence.

Extra-terrestrial expansion within our solar system offers more sensible and practical possibilities as a means of extending the explicit root of human life to offset the damage that life is doing to the environment of our own planet Earth. If that damage occurs rapidly it is certain to prove an incentive to such expansion on the basis that 'needs must when the devil drives'.

However the inescapable conclusion is that we would be reinventing the wheel, since it is our own fault that the Earth's environment, once relatively pristine, is being sacrificed on the altar of short-sighted human desire and so-called technological progress.

Consider how preferable is the thought on a hot summer's day, of dangling your feet in a cooling stream or strolling on a sandy beach, to wading in a sea of liquid methane encased in a spacesuit on Titan or clambering across the rocky landscape on Venus at some 460°C.

The author exaggerates here, no doubt it would be possible to create more comfortable surroundings on artificial satellites or on the surface of the moon or Mars.

It is overwhelmingly obvious however that they are vastly less preferable to the conditions on Earth we have been used to.

The Reference Model implies strongly that it is planet Earth that is best suited as a physical base for pursuing the Implicate order.

The explicit alternative is better suited to artificial rather than human life.

Extra-terrestrial propagation by artificial intelligence may be possible, but it will inevitably be limited to real world as opposed to imaginary scenarios.

7.2.8.7. The concept of different worlds.

Section 7.2.7 is about the concept of different worlds seen from the scientific perspective.

The Reference Model expands that concept in to a fully-fledged working model of consciousness.

The relationship between the explicit four-dimensional world of objects unfolded in the real world, the animate/conscious world and the intelligent imaginary world has already been addressed in answer to the previous question and described extensively in sections 4.4 through 4.8, all in terms of the inanimate grid of an Ordering Centre and the animate/conscious motion across that grid.

It has been shown that the animate is unfolded from the inanimate and re-enfolded in to it, and can view the inanimate from an altered perspective arising from the facility to unfold and enfold order and variance asymmetrically as defined in the Reference Model.

The discipline of mathematics is intrinsic to existence and offers the conscious entity a means of relating the inanimate world of reality to the world of imagination.

It is significant that nominalists view mathematical numbers as unreal and merely manifestations of the physical spatiotemporal world, while Platonists regard them as universals, real and objective but non-spatiotemporal, as discussed in Section 7.1.3.

It is concluded from the Reference Model that such non-spatiotemporal, abstract, non-physical objects are to be associated with the central present state for our Ordering Centre. They are clearly related to fine grained intelligence and belief as opposed to the coarse-grained conscious world of the unfolded and enfolded senses as portrayed in Fig 4.8.3.

Furthermore such universals must relate to and ultimately be grounded in the Whole Ordering Centre.

It is also significant that mathematics can be utilised to comprehend the real world and the nature of the unitary determinate quantum state occupying the imaginary present state axis.

The complexity of Implicate space which seems to occupy an intermediate position between the two is likely to prove too great for mathematical description, certainly in the foreseeable future.

7.2.8.8. The birth, life, death cycle and the aging process.

The Reference Model shows that all things are in motion and transitory to a greater or lesser extent whether they are small or large fluctuations, related to the Ordering Centre or not. Similarly all things within the Ordering Centre even that centre itself have life cycles in particular the unfolding and enfolding of order and dimension.

The birth of the Ordering Centre in an explicit sense as a big bang in a highly ordered state, it's unfolding as an aging process involving the expansion of space-time and increasingly chaotic disorder, culminating in its ultimate collapse to a base state of totally distributed energy i.e. the heat death, demonstrates this.

The contrary process of building life from that base structure is to be associated with an ordering process of endless permutations utilising Implicate space provided by the Ordering Centre which enfolds space-time restoring the highly ordered state of the big bang. It is therefore rather meaningless to talk in terms of eternal life or eternal death for that matter.

Within the above processes we see the birth, life and death of galaxies, stars, planets, satellite moons and the materials of which they are made right down to their basic molecular and particle components. So inanimate material can also be said to have explicit life cycles but these are many orders of magnitude in duration greater or lesser than that of life forms which attain animacy and intelligence, as indicated in Section 7.2.8.6.

The cyclical aspect is very apparent in the case of stars within galaxies, planets about the sun, satellite moons about those planets, the cycling of tectonic plates, climatic cycles, weather cycles etc.

Considering firstly holistically orientated life as defined in this Section 7.2.8, the intermediate duration life cycles of bacteria, plants, of animate and ultimately conscious life relative to the inanimate is in principle also essentially explained by the Model.

Existence is ordering itself and the sequencing created by individual life forms rises or falls in that ordering system according to the relevant wave motion created.

It is necessary for that life form to have a life cycle namely to be born i.e. made manifest at the mean level within the Ordering Centre, live utilising Implicate space to generate motion relative to a base physical state of motion and finally sooner or later to die, so as to be no longer manifest at the mean level. The motion created by the travelling waves propels the pattern or sequencing created by that life form up or down the ordering system, all relative to the base physical standing wave motion.

It is non-local processes involving higher degrees of implicit ordering and disordering associated with wave motion in the belief and imaginary worlds, increasingly enfolded within that manifestly apparent to the human race, which give rise to the belief in rising in to heaven and in the opposite sense of falling in to hell prevalent in human society, by those spiritually inclined. The opposing belief also prevalent, that the physical base of the real world is all that there is, essentially dust particles within that physical system referred to as reality, correspondingly correlates with local processes involving increasingly chaotic disorder; these being unfolded from the mean level of order, all in accordance the second law of thermodynamics responsible for the aging process.

Figure 4.8.1 describes in a nutshell the relevant processes, life unfolding at a given mean level of order and enfolding again at some higher or lower level of order, restoring the inanimate state when realignment with the corresponding real and imaginary axes occurs. The same applies to other complex life forms whether present in our galaxy or other galaxies collectively forming the Ordering Centre.

This figure exposes the confused thinking associated not only with the belief that life is limited to the explicit, but also the opposing belief in the 'before life' and 'after life' as discussed in Section 7.5.6.2.

Considering now reductionist intelligence again as defined in this Section 7.2.8, the same principles apply except that the emphasis is on distribution in an ever-expanding dimensional environment.

To the extent that such intelligence develops it will also be subject to life cycles presumably of longer explicit duration than holistic life.

The Reference Model implies life cycles can be related to Fig 4.4.5, long life cycles towards the base of the figure and short life cycles towards the top.

In principle the progression from an Ordering Centre to the Whole Ordering Centre offers further escape from the purely explicit life cycle, as described in the conclusions given in Section 5.0, which relates the hierarchical birth, life, death cycle with the holoarchical dissipation of such cycles.

So the Holomovement may be said to evade death by being in motion about the inanimate 'absolute' present state at that level. Since existence has no ultimate upper or indeed lower bound there is therefore no absolute state of non-motion, or no absolute state of death.

7.2.8.9. *Consciousness and awareness.*

Consciousness and awareness have already been discussed in Section 4.0 of this book, where they are the central theme.

They represent the growth in asymmetry and chirality associated with the increasing complexity of life and the different worlds referred to in sub-sections 7.2.8.3 and 7.2.8.7.

It is not necessary to repeat it all here, but the following general statements summarise the relationship between consciousness and awareness made apparent by the Reference Model. The Model implies that in an overall sense these properties increase as order increases in an Ordering Centre and in the opposite sense decrease as unwholeness in the form of chaotic disorder increases and also as the Ordering Centre collapses towards the random state.

However relative to the mean level in the Ordering Centre, where reality and imagination are in balance and levels of order associated with Homo Sapiens apply, unwholeness in the form of evil does not reduce these properties as the figures in Section 4.8 show.

Consciousness is then distinguished from awareness in the same way as unfolding differential variance is distinguished from unfolding integral variance at that mean level. Differential variance corresponds to the unfolding of the individual ego as opposed to the enfolding of that ego within the ego of the human race as a whole, involving awareness of and empathy with others.

The range of energy densities simply extends this principle in accordance with the holographic fractal description on which the Model is based. So there is no limit to consciousness or awareness which increases in the ordering direction and decreases in the opposite sense when viewed from that overall perspective.

7.3.   Mathematics.

Penrose (Ref 7.3.1 and Ref 7.3.2) are excellent sources for understanding mathematics and in providing a detailed background to the subject matter summarised in Sections 7.3.1 thru 7.3.10. Specific subjects can also of course be found on the internet.

7.3.1. *Real numbers.*

Rational and irrational numbers.

Imaginary and complex numbers.

Fourier series and Fourier transforms.

Real numbers include the natural numbers 0, 1, 2, 3, 4, 5 etc and the negative numbers -5,-4,-3,-2,-1 which together provide a whole set of integers.

There are many interesting sequences in the form of sets of real numbers such as the set of prime numbers: 2, 3, 5, 7, 11, 13, 17, 19, 23, etc where each number has only two positive divisors namely 1 and the number itself.

Another example is the Fibonacci set: 0, 1, 1, 2, 3, 5, 8, 13, 21, etc where each number is the sum of the previous two numbers.

To which should be added fractions, the rational numbers 0, 1, -1, 1/2, -1/2, 2, -2, 3/2, -3/2 etc defined by the form n/m, where n and m are integers with $m \neq 0$.

Real numbers also include those not of this form, irrational numbers with infinite decimal expansions such as the important mathematical quantity $\pi = 3.14159$ which is also the infinite sum of a sequence of rational numbers as 4(1 - 1/3 +1/5 -1/7+1/9...).

Real numbers further include square roots, cube roots etc of positive rational numbers, including for example $\sqrt{2}$ (having the decimal expansion 1.41421...), $\sqrt[3]{2}$, etc.

The irrational number e, also of great importance mathematically, is another real number having the expansion 2.71828... and is also the sum of:

$$1+1/1+1/(1\times2)+1/(1\times2\times3) +....$$

The number e is the base of natural logarithms, the natural logarithm of a number n being the power to which e would have to be raised to equal n.

Imaginary numbers are those involving the square root of negative numbers as $\sqrt{-1}, \sqrt{-2}$ etc, and complex numbers those involving the sum of real and imaginary numbers or parts. Complex numbers z are commonly written in the form $z = x + iy$ where x is the real part, y the imaginary part and i is $\sqrt{-1}$.

The complex conjugate of z is $z' = x - iy$. The product $zz' = |z|^2$, where $|z|$ is the modulus of z.

They are represented geometrically in a standard form using Argand diagrams where the values on the horizontal abscissa and the vertical ordinate represent x and y respectively.

In mathematics, a Fourier series is a method of representing or composing a wave like function as the sum of simple sine or cosine waves. In the opposite sense it decomposes any periodic function in to the sum of a (possibly infinite) set of single oscillating functions, namely sines and cosines or alternatively complex exponential functions.

Simple harmonic vibrations may be represented by $v = A \sin \omega t$ or $v = A \cos 2\omega t$

where t is time, A is the amplitude and ω the angular frequency, usually over a finite interval.

Using Euler's formula $e^{i\omega t} = \cos \omega t + i \sin \omega t$, where i is $\sqrt{-1}$

The complex number $z = Ae^{i\omega t}$ provides real and imaginary parts and a more compact notation.

Now considering an arbitrary process, that is repeated every T seconds.

Say $v = F(t)$, then $F(t + T) = F(t)$.

A Fourier series using the complex notation when the represented function is length extended to infinity might be given by:-

$$F(t) = \sum_{n=-\infty}^{n=+\infty} a_n e^{in\omega t}, \text{ where } \omega = 2\pi/T$$

It can be shown that the coefficient a_0 is given by $a_0 = 1/T \int_0^T F(t)\, dt$ and the coefficients

$a_n = 1/T \int_0^T F(t)\, e^{-in\omega t}\, dt$ and $a_{-n} = 1/T \int_0^T F(t)\, e^{in\omega t}\, dt$

giving the Fourier series $F(t) = a_0 + \sum_{n=1}^{n=\infty}(a_n e^{in\omega t} + a_{-n} e^{-in\omega t})$

Of particular importance is the Fourier transform which relates functions of time to their constituent frequencies, just as musical chords relate to their constituent notes.

Fourier transforms are therefore termed frequency domains of the original function and where such functions have a time basis they are termed time domains.

The relationship between these domains forms the basis for harmonic analysis.

The reverse of this process which combines the various frequencies to recover the original functions of time is termed the inverse Fourier transform.

Fourier transforms in Euclidian space relate 3D space to 3D momentum space or 4D space to 4D momentum space. They are also applied to the study of wave motion in quantum theory, as is made apparent in Section 7.4.6.

Fourier transforms are applicable across a wide range of functions from familiar Gaussian functions right down to the uncertainty principle, a fundamental aspect of quantum theory.

There are several common conventions for defining the Fourier transform $\widehat{F}$ of an integral function F for example:-

$$\widehat{F}(\omega) = \int_{-\infty}^{+\infty} F(x)\, e^{-i\omega x}\, \mathrm{d}x,$$

where ω is the angular frequency and x is the time.

Where F can be reconstructed from $\widehat{F}$, the inverse transform is given by:-

$$F(x) = \int_{-\infty}^{+\infty} \widehat{F}(\omega)\, e^{i\omega x}\, \mathrm{d}\omega$$

The functions F and $\widehat{F}$ are commonly referred to as the Fourier integral pair or Fourier transform pair.

7.3.2. *Mandelbrot sets and the principle of the fractal.*

The principles and nature of fractals were first investigated by Benoit Mandelbrot and others in the 1980s and led to his discovery of

Mandelbrot sets (Ref 2.7.3). The basic principle of the fractal was discussed earlier in Section 2.7.

Periodic or cyclical motion arising from the linearity or near linearity of many physical systems is disrupted by the chaotic albeit deterministic background with which we are familiar. In between these states lies a state of self-similarity and invariance with respect to scaling, as Schroeder (Ref 2.7.2) says a fundamental bulwark rising above the chaos.

Basically fractals are self-similar repeating patterns which can appear from and retract into a given point on a page, unchanging irrespective of any expansion or alternatively shrinkage in scale. They represent a fundamental form of symmetry in nature.

Mandelbrot sets are fractals characterised by their mathematical relationship to complex numbers and may be defined as follows.

Technically the Mandelbrot set is the set of complex numbers c for which the function

$f_c(z) = z^2 + c$ does not diverge when iterated from $z = 0$, such that the sequence

$f_c(0), f_c(f_c(0))$ remains bounded in absolute value.

The complex number $z = x + iy$ can be represented by a point on an Argand diagram as described in Section 7.3.1. Utilising a mapping process on such a diagram z is replaced by a new complex number $z \rightarrow z^2 + c$ etc, where c is another given complex number, represented by some new point on the diagram.

This process can be repeated and a sequence of points built up on the Argand diagram. Such a sequence may be bounded if there is some fixed circle that contains all the points or unbounded where the sequence wanders farther and farther from the origin.

7.3.3.   *Probability and the normal (Gaussian) distribution.*

Probability is a measure of the chance or likelihood that an event will occur given by a number between zero meaning impossible and unity meaning certainty.

A simple example is the tossing of an unbiased coin, the probability of heads or tails is 1/2. Tossing twice gives a probability of two heads of 1/4.

In numerical terms probabilities are the number of preferred outcomes divided by the total number of all outcomes.

These concepts are widely used in mathematics, statistics, finance, gambling, science and game theory etc.

The mathematical treatment of probability in which the probability P of an event A is commonly written P (A) allows equations to be derived for random events, whether mutually exclusive or not.

However in practice there are two competing views as to the fundamental nature of probability, namely objectivists and subjectivists.

Objectivists assign numbers to describe a given physical state of affairs, usually claiming that the probability of a random event is that which occurs in the long run from repeated experiments providing knowledge on an a posteriori basis.

Subjectivists assign numbers to probability based on the degree of belief involved, providing knowledge on an a priori basis in addition to the experimental data.

A typical example is Bayes Theorem, where the posterior (the probability of A being true, given B is true) is the product of the prior i.e. the knowledge (the probability of A being true) and the likelihood (the probability of B being true, given A is true) divided by the probability of B being true, based on all the known information to date.

In a deterministic universe based on Newtonian concepts there would be no probability if all conditions were known, but of course in most cases sensitivity to initial conditions limits our ability to measure them, so a probabilistic description can be more useful than Newtonian mechanics. Many systems theoretically deterministic in principle are so complex that only a statistical description is feasible, a situation common in physics.

Probability theory is required in particular to describe quantum phenomena. The objective wave function evolves deterministically but when an observation is made complex probabilities come into play when establishing outcomes as described in this book.

In probability theory the normal (or Gaussian) distribution is a very common continuous probability distribution. Normal distributions are central to statistics and are often used in the natural and social sciences to represent real valued random variables whose distributions are not known.

The normal distribution is based on the Central Limit Theorem, which states that the random variables are independently and identically distributed.

The probability density of the normal distribution is given by:-

$$f(x \mid \mu, \sigma^2) = 1 / (\sigma \sqrt{2\pi}) \exp(-k),$$
$$\text{where } k = (x - \mu)^2 / (2\sigma^2)$$

and μ is the mean or expectation value of the distribution, σ the standard deviation and σ^2 the variance.

If $\mu = 0$ and $\sigma = 1$, the distribution is called the standard normal (or standard Gaussian) distribution, usually written

$$\phi(x) = \exp(-x^2 / 2) / \sqrt{2\pi}$$

7.3.4.   *What is dimension? Do we see only in four dimensions?*

Are there other dimensions?

What is the relationship between 3-dimensional space and the 3n and 6n dimensional worlds (n›1) of what is termed in mathematical physics, configuration space and phase space respectively?

Certain general observations can be made as a background to understanding dimension. It is the common experience of conscious entities that they visualise the world in three dimensions but experience it in four, namely three spatial and one time dimension.

It is apparent however that the same information applying to three spatial dimensions can through the holographic principle be enfolded in one less spatial dimension. Indeed as will become apparent this principle plays a very significant role in modern physics in relation to cosmology and quantum theory as will be reviewed in the next section on science and measurement.

The addition of extra spatial dimensions invoked in string theory in efforts to establish a theory of quantum gravity will also be addressed in the next section, however it is apparent that these extra dimensions do not appear to play an identifiable role in our conscious life and are in any case tiny and curled up relative to our large-scale 3D world, effectively enfolded as far as that world is concerned.

Physicists such as Tegmark (Ref 7.3.3) analysing the world of dimension conclude that the visible world we see with three large spatial and one large temporal dimension is the only stable condition compatible with the complexity of life as we understand it, in a range where the number of large time and the number of large spatial dimensions varies from zero upwards.

Furthermore that the strength of the strong nuclear force and the strength of electro magnetism are so attuned to our existence that even a minimal alteration to their values would be incompatible with that existence.

In general terms, as already stated, concepts involving many dimensions can be in effect enfolded by the mind and re unfolded by the brain on to a two-dimensional sheet of paper. This process is epitomised by mathematical procedures associated with an infinite number of dimensions.

It is appropriate to consider now the role of mathematics in describing dimension.

The concept of configuration space expresses in mathematical form the possible positions that a system of n point particles can have, such that a single point P in that configuration space defines all the possible locations.

This space has 3n dimensions and is a purely abstract concept not to be confused with our three-dimensional experience.

With time the single point P moves in configuration space describing the Newtonian behaviour of the particle system and that motion can be defined by a single mathematical function. Taking the position coordinates as $x_1, x_2, x_3, \ldots x_n$ and their velocities as

$dx_1/dt = \dot{x}_1$ etc then the Lagrangian $L = L\,(x_1 \ldots x_n; \dot{x}_1 \ldots \dot{x}_n)$, named after the 18th century mathematician Lagrange who developed the relevant equations with the mathematician Euler.

The Lagrangian has the physical interpretation of being the difference between the kinetic energy of the system and the potential energy arising from external and internal forces, expressed in terms of the relevant coordinates. The Euler-Lagrange equations take the form d/dt $\partial L/\partial \dot{x}_r$ - $\partial L/\partial x_r$ = 0, where r = 1.... n, and ∂ is the partial derivative.

In order to take total energy in to account the momentum of the particles must also be allowed for. Accordingly the particles may have positions x_1, x_2, x_3,... x_n and momenta p_1, p_2, p_3,...p_n, described by a single point Q in a 6n dimensional space, referred to as phase space.

So a single point of the phase space represents the entire physical system including not only the position of each particle but also their motion. If the direction of motion is represented by an arrow then it is a vector. The whole set of vectors being a vector field in phase space.

This system of classical mechanics may be summarised mathematically by Hamiltonian theory, after the 19th century mathematician William Rowan Hamilton. Now in this system the position and momentum of each particle are treated as independent variables thus providing two sets of equations as to how those positions and momenta are changing with time.

In the same form as the Lagrangian function, the Hamiltonian function may be written

$H = H (p_1....p_n; x_1... x_n)$.

The total energy of a system of particles is given in terms of these variables by the Hamiltonian H such that,

dp_i /dt = - $\partial H/\partial x_i$ and dx_i /dt = $\partial H/\partial p_i$.

With suitable choices of H, Hamilton's equations hold true for any system of classical equations including those of Newton and Maxwell. They also hold true for special relativity and even general relativity.

It is remarkable that they also form an initial basis for the framework of quantum mechanics, which involves complex numbers. The analogous concept to that of phase space being Hilbert space, a complex vector space (see Section 7.3.7).

The relation between Hilbert space and the quantum state will be examined in further detail in Section 7.4.

7.3.5. *Scalars, vectors and covectors.*

Covariance and contravariance of vectors.

Quantities which have magnitude only are called scalars, for example mass, distance area and volume. A scalar can be represented by a number with an associated sign which gives its magnitude to some appropriate scale.

Quantities which have not only magnitude but also direction are called vectors, for example, force, displacement, velocity and acceleration. A vector is represented by an arrow running from a point P to a point Q. P being the origin and Q the terminus. Vectors obey the commutative and associative laws of addition, may also be subtracted and can be multiplied by a scalar. They are applied universally in mathematics and Newtonian physics in terms of rectangular Cartesian coordinates.

The product of the magnitude of two vectors and the cosine of the angle between them is referred to as the dot product or scalar product, since the result is a scalar instead of a vector.

A scalar field associates a scalar value to every point in a given space, it may be a mathematical number or a physical quantity.

A vector field is an assignment of a vector to each point in a subset of space.

A vector field in a plane can be visualised as a set of arrows with a given magnitude and direction each attached to a point in the plane. Vector fields can be extended to n dimensional Euclidian space and further still to curved surfaces which may be complex in shape, where they correspond to an arrow tangent to that surface at each point called a tangent vector. More generally vector fields are defined on differential manifolds which are spaces which look like Euclidian space on small scales, but more complicated structures at larger scales. A vector field then gives a tangent vector at each point of the manifold. Manifolds have the property of smoothness such that differential properties apply at all points on their surfaces.

If then θ is defined on a manifold S, θ is a smooth map from the surface S to the space of real numbers R or complex numbers C, generally termed a scalar field on S.

Two dimensional manifolds are called surfaces like the plane, the sphere or the torus for example. The structure of an n dimensional manifold M can be similarly constructed but in this case instead of two coordinates for each element of the surface, n coordinates are required.

Now given a vector space V over a field F, mathematically it must have a corresponding dual vector space V^*, which is the set of all linear maps from the vector space to the scalars. The dual space becomes a vector space over F when furnished with addition and scalar multiplication properties satisfying the following:-

$$(\alpha + \beta)(\gamma) = \alpha(\gamma) + \beta(\gamma)$$

$$(\theta\, \alpha)(\gamma) = \theta\,(\alpha\,(\gamma))$$

for all α and β over V^*, γ over V and θ over F,

where α and β are vector fields over V^*, γ is a vector field over V and θ is a scalar field over F. Elements of the dual space V^* are called covectors.

The above equations give the definition of the sum of two vectors and the product of a covector by a scalar. The dual of the space of covectors is isomorphic (a specific type of one-to-one correspondence) to the original space when the latter is finite dimensional.

Vectors at a particular fixed-point o constitute a vector space V, which provides the structure of the manifold in the immediate vicinity of o.

This vector space is called the tangent space T_o to the n dimensional manifold M at o.

It is the limiting space that is arrived at when smaller and smaller neighbourhoods of o in M are examined at ever greater degrees of magnification.

The dual space vector to T_o (the space of covectors at o) is called the cotangent space T_o^* to M at o, the corresponding dual vector space being denoted by V^*.

The geometrical difference between a covector α and a vector is that at each point of M, a non zero covector determines an $n-1$ dimensional plane element. The directions lying within this element are those determined by vectors γ for which $\alpha\,\gamma = 0$.

In matrix notation the columns represent vectors and the rows represent covectors.

Covariance and contravariance of vectors describe how such properties change when the reference axes or basis changes. Relevant details can be found on the internet but the essential principles can be defined as below.

Vectors which change scale inversely to changes in scale to the reference axes are termed contravariant, those dealing with direction, velocity or acceleration for example.

Such vectors may involve units of length (distance).

Dual vectors or covectors on the other hand involve units which are the inverse of length and their components change in the same manner as changes in scale of the reference axes. Such vectors are termed covariant and typical examples involve taking the derivative or gradient of a function. These properties are fundamental to the next subject requiring consideration, namely tensors.

7.3.6.   Tensors and metric tensors.

Description of tensors can be found on the internet, specific elements of importance being emphasised below.

A quantity that is unchanging no matter what coordinate system is used when measuring it, is termed invariant. Scalar quantities of this kind are tensors of rank 0.

Moving on to quantities which have direction namely vectors, a vector

$\mathbf{r} = c_1 \, x^1 + c_2 \, x^2 + c_3 \, x^3$ can be written using Einstein's summation convention as $\mathbf{r} = c_i \, x^i$, where the index can be any variable (referred to as a dummy index) and is used twice in each term as a subscript and superscript.

It is a means of representing a vector with an arbitrary number of dimensions in a very compact form.

Vectors can be measured in two Cartesian coordinate systems with the same origin, one system at an angle to the other, the components of the vector changing while the vector itself remains constant. Such vectors though not tensors are position vectors, with components which are covariant with the coordinates under a rotation of the axes.

If it is required to describe the vector where a transformation to a different coordinate system or basis is involved then that transformation can be described by an equation, with components which are contravariant with the coordinates. This vector termed a displacement vector is a rank 1 contravariant tensor.

Now given a set of functions in one set of coordinates and taking the partial derivatives of those functions with respect to a new set of coordinates, those derivatives represent gradients which are also vectors and can be described by an equation which provides the definition of a covariant tensor, again of rank 1.

It is customary to index contravariant tensors with a superscript and covariant tensors with a subscript.

Both covarient and contravarient tensors add to form new tensors.

In the case of multiplication the product of two such rank 1 tensors has an important property. The product $u_i\, v^i = u_1 v^1 + u_2 v^2 + u_3 v^3$ is the dot product or inner product of the two vectors.

The equation governing this product is found to be invariant, such that the product of the two rank 1 tensors is a rank 0 tensor, an invariant scalar.

The multiplication of two contravariant rank 1 tensors provides a definition of a contravariant tensor of rank 2; similarly a definition of a covariant tensor of rank 2 can be obtained. So the rank of a tensor is that given by the dimension of the array needed to represent it, or equivalently the number of indices needed to label a component of that array.

As they are concerned with forming a relationship between vectors, tensors must be independent of a particular choice of coordinate system so that the transformation relates the array in one coordinate system to that in another.

The precise form of the transformation determines the type of the tensor.

The tensor type is a pair of natural numbers (n, m), where n, m are the numbers of contravariant indices and covariant indices respectively and the rank of the tensor is the sum of n + m.

Raising an index on an (n, m) tensor produces a (n+1, m-1) tensor. Lowering an index produces a (n-1, m+1) tensor. Contraction of an upper with a lower index produces a (n-1, m-1) tensor.

It is customary to use lower case Latin letters to denote indices which take values 1, 2..n and Greek lower-case letters to denote indices which take values 0,1, 2..n in the context of relativity.

A rank 2 tensor in four dimensional space-time represents 16 terms. The overall Einstein Field Equation $R_{\mu v}$ - ½ $Rg_{\mu v}$ = $8\pi GT_{\mu v}$ furthermore represents 16 equations (see Section 7.4.2).

In general relativity the metric tensor or metric is fundamental as it encapsulates the structure of space-time, defining time (future and past), distance, volume and curvature.

Mathematically space-time is represented by a four-dimensional differentiable manifold M and the metric is given as a covariant second rank symmetric tensor on M, denoted by g. The metric is a symmetric bilinear form on each tangent space of M which varies in a smooth (differentiable) manner from point to point.

It can be written $g = g_{ab}$ dx^a dx^b the coefficients g_{ab} being a set of 16 real valued functions. For the metric to be symmetric g_{ab} = g_{ba} giving 10 independent coefficients.

The distance along curves on the surface of manifolds ds is given by the metric tensor g_{ab} such that $ds^2 = g_{ab}$ dx^a dx^b.

7.3.7. *Dot product, inner product, completeness, Hermitian form and Hilbert space.*

Eigenvalues and eigenvectors.

The Schrödinger wave equation and wave packets.

Linear transformations which preserve vector space structure namely

Orthogonal, Unitary and Symplectic groups.

Abelian / non-Abelian groups, Lie groups and Lie algebra.

Quarternions, Clifford algebra and Grassmann algebra.

Symmetry, asymmetry and antisymmetry.

Reflexivity and irreflexivity.

In mathematics the dot product or scalar product is an algebraic operation that takes for example two coordinate vectors and returns a scalar number.

It can be defined algebraically or geometrically. In the former case it is the sum of the products of the corresponding entries of the two sequences of numbers, in the latter case it is the product of the magnitude of the two vectors and the cosine of the angle between them.

The name dot product is derived from the centred dot that designates the operation, the alternative name scalar product emphasises that the result is a scalar instead of a vector.

All these definitions are associated with Euclidian space.

An inner product is a generalisation applied to abstract vector spaces over a field of scalars, being either a field of real numbers or a field of complex numbers.

A vector space together with an inner product is called an inner product space.

A Hilbert space H, after the mathematician David Hilbert, is a real or complex inner product space that is also a complete metric space with respect to the distance function induced by the inner product.

H is a complex vector space on which there is an inner product (x, y) associating a complex number to each pair of elements x, y of H that satisfies the following properties:-

The inner product of a pair of elements is equal to the complex conjugate of the inner product of the swapped elements:-

$(y, x) = (\overline{x, y})$, where the overbar denotes the conjugate, and is referred to as conjugate symmetry.

For all complex numbers a and b, the inner product is linear in its first argument:-

$$(ax_1 + bx_2, y) = a\,(x_1, y) + b\,(x_2, y)$$

The inner product is antilinear in its second argument:

$$(x, ay_1 + by_2) = \bar{a}\,(x, y_1) + \bar{b}\,(x, y_2)$$

which follows from the previous two properties.

The inner product of an element with itself is positive definite:-
$(x, x) \geq 0$, where equality holds when $x = 0$.

(a real inner product space is defined in the same way, except that H is a real vector space and the inner product takes real values and is bilinear i.e. linear in each argument.)

Where antilinearity applies the inner product is called a Hermitian inner product and the complex vector space a Hermitian inner product space, after Hermite the 19th century mathematician. If this results in a complete (as defined below) metric space it is called a Hilbert space H.

Furthermore every inner product induces a norm of the form:

$\| x \| = \sqrt{(x,x)}$, so every inner product space is also a normed space.

The distance between two points x, y in H is given by:-

$$d(x, y) = \| x\text{-}y \| = \sqrt{(x - y, x - y)}$$

The bra ket $\langle|\rangle$ (meaning bracket) notation is used for quantum state wave functions as below. In particular elements of Hilbert space are denoted by state vectors such as $|\alpha\rangle$,

$|\phi\rangle$, $|\psi\rangle$ etc; (a notation originally used by the mathematical physicist Dirac).

For two state vectors $|\phi\rangle$ and $|\psi\rangle$ the scalar product is denoted $\langle\phi|\psi\rangle$.

Hilbert space in this case is infinitely dimensional, possesses the structure of a finite dimensional inner product and is complete. It should be noted though that Euclidean space is a finite dimensional Hilbert space.

Completeness may be defined by taking an infinite set $|\alpha_1\rangle$, $|\alpha_2\rangle$, as an orthonormal basis for an infinite dimensional Hilbert space $|\psi\rangle = \Sigma b_i | \alpha_i\rangle$, where $i = 1....\infty$.

Then this series converges to a finite vector $|\psi\rangle$ if and only if the series $\Sigma|b_i|^2$ for
$i = 1....\infty$, is convergent.

In which case the norm of $|\psi\rangle$ is given by $\|\psi\|^2 = \langle\psi|\psi\rangle = \Sigma|b_i|^2$ for $i = 1....\infty$

The equations given above defining Hilbert space can be rewritten in this notation as:-

$$\langle\phi|\psi\rangle = \overline{\langle\psi\,|\,\phi\,\rangle},$$

$$\langle\phi|\psi + \chi\rangle = \langle\phi|\psi\rangle + \langle\phi|\chi\rangle,$$

$$\langle\phi|a\psi\rangle = a\langle\phi|\psi\rangle,$$

$$\psi \neq 0 \text{ implies } \langle\psi|\psi\rangle > 0$$

The Hermitian scalar product is $\langle\phi|\psi\rangle$ in this notation and the quantity $\|\psi\|$ is the norm, where $\phi = \psi$ and $\|\psi\| = \langle\psi|\psi\rangle$.

The two states $|\phi\rangle$ and $|\psi\rangle$ are orthogonal if $\langle\phi|\psi\rangle = \langle\psi|\phi\rangle = 0$.

The complex linear transformations involved here satisfying Hermitian symmetry are termed Hermitian form.

In mathematics each bounded linear operator on a complex Hilbert space has a corresponding adjoint operator. Linear operators are functions that assign to elements from one space elements from another. In the sense that operators are analogous to complex numbers the adjoint of an operator plays the role of the complex conjugate of a complex number.

The self adjoint of an operator A is called the Hermitian adjoint or Hermitian conjugate of A, being denoted A^*, such that $A = A^*$.

Measurable quantities associated with Hermitian operators are referred to as observables.

Repeated measurement of an observable A (the self adjoint operator in Hilbert space) yields a distribution of values for which the expectation value is $\langle\psi|A|\psi\rangle$.

This mathematical process gives a direct way of obtaining a statistical property of the outcome of an experiment, once it is understood how to associate the initial state with a Hilbert state vector and the measured quantity with an observable, i.e. a specific Hermitian operator.

For example the probability of finding the system in a given state $|\phi\rangle$ is given by computing the expectation value of what is termed a projection operator $E = |\phi\rangle\langle\phi|$.

The probability is then given by $P = \langle\psi|E|\psi\rangle = |\langle\phi|\psi\rangle|^2$.

Eigenvalues and eigenvectors

An eigenvector (eigen meaning 'own') is a vector which does not change its direction under the associated linear transformation. In geometric terms an eigenvector corresponding to a real non-zero eigenvalue points in a direction that is stretched by the transformation and the factor by which it is stretched is termed the eigenvalue.

In the case of linear transformations real scalar factors are replaced by any field of scalars such as algebraic or complex numbers and Cartesian vectors are replaced by any vector space expressed by continuous functions for example. When linear operators are used to map vectors to vectors the term vector in eigenvector may be replaced by the term eigenfunction or eigenstate.

The exponential function $f(x) = e^{\lambda x}$ is an eigenfunction or eigenstate of the derivative operator, with eigenvalue λ, since its derivative $f'(x) = \lambda e^{\lambda x} = \lambda f(x)$.

For the derivative operator d/dt, an eigenfunction is that, which when differentiated, yields a constant times the original function. The solution is an exponential function $f(t) = Ae^{\lambda t}$ and if λ is zero it becomes a constant.

These definitions remain valid for a linear transformation T even if the underlying space is an infinite dimensional Hilbert space. So a scalar λ is an eigenvalue if and only if there is some non zero vector v such that $T v = \lambda (v)$.

An example of the above is the time independent form of the Schrödinger equation in quantum mechanics, $H \psi_E = E \psi_E$, where H the Hamiltonian, is a second order differential operator and ψ_E the wave function is an eigenfunction corresponding to the eigenvalue E. The time independent equation corresponds with stationary states where wave functions form standing waves.

The wave function ψ_E is to be associated with Hilbert space which has a well-defined scalar product.

Using the bra ket notation a vector in the Hilbert space is represented by $|\psi_E\rangle$.

The Schrödinger equation in this case is $H |\psi_E\rangle = E |\psi_E\rangle$, where $|\psi_E\rangle$ is an eigenstate of H and E the eigenvalue. It is an observable self adjoint operator where $H \, |\psi_E\rangle$ is the vector obtained by application of the transformation H to $|\psi_E\rangle$.

The Schrödinger wave equation and wave packets

The general form of the Schrödinger equation can be derived as follows:-

The simplest wave function is a plane wave of the form $\psi(x, t) = A e^{\,i\,(kx - \omega t)}$,

where A is the amplitude, k the wave vector and ω the angular frequency.

A superposition of plane waves is then given by $\psi(x,t) = \sum\limits_{n=1}^{n=\infty} A_n\, e^{i\alpha}$

where $\alpha = k_n x - \omega_n t$

The energy of a photon $E = h\,v = \hbar\,\omega$ and the momentum $p = \hbar\,k$.

Expressing the phase of a plane wave as a complex phase factor $\psi = A e^{\,i\,(kx - \omega t)} = A e^{\,i\,(px - Et)/\hbar}$

The first order derivatives with respect to space are $i/\hbar\, p\, A\, e^{\,i\,(px - Et)/\hbar} = i/\hbar\, p\, \psi$

and with respect to time $-i\, E/\hbar\, A\, e^{\,i\,(px - Et)/\hbar} = -i\, E/\hbar\, \psi$

The energy operator is then $i\,\hbar\, \partial\psi/\partial t = E\,\psi$,

and the momentum operator $-i\,\hbar\, \partial\psi/\partial x = p\,\psi$.

$E = p^2/2m + V$, and $p^2 = -\hbar^2\,(\partial/\partial x)^2$

giving $H = i\,\hbar\, \partial\psi/\partial t = -\hbar^2/2m\; \Delta^2\,\psi + V\psi$, which is Schrödinger's equation,

where H is the quantum Hamiltonian, $\hbar$ is Planck's constant (Dirac's version),

m is the mass of the particle, V is the potential energy function.

$\Delta^2 = (\partial/\partial x)^2 + (\partial/\partial y)^2 + (\partial/\partial z)^2$ in three cartesian coordinates x,y,z and referred to as the Laplacian.

The corresponding classical Hamiltonian in these three dimensions can be derived on the basis of a particle falling under gravity in the earth's gravitational field. The kinetic energy plus the potential energy is given by, $\tfrac{1}{2}\, m\, (v_x^2 + v_y^2 + v_z^2) + mgz$, where z is the height above the ground and g is the gravitational constant.

216

So the classical Hamiltonian H for a particle of mass m with a potential energy function $V(x, y, z)$, is given by

$$H = (p_x^2 + p_y^2 + p_z^2) / 2m + V(x, y, z) = p^2/2m + V(x, y, z).$$

Schrödinger's equation is linear, if $|\psi\rangle$ and $|\varphi\rangle$ both satisfy the equation then so does any combination $w|\psi\rangle + z|\varphi\rangle$, where w and z are complex numbers, maintaining a complex linear superposition indefinitely. Once the quantum Hamiltonian is obtained, the time evolution of the state according to Schrödinger's equation proceeds as though $|\psi\rangle$ were a classical field such as Maxwell's electromagnetic field equations.

In fact for a single photon Schrödinger's equation becomes Maxwell's, such that the equation for the photon is the same as for an entire electromagnetic field.

Also if $|\psi\rangle$ describes the state of a single electron then Schrödinger's equation becomes Dirac's equation for the electron.

Whereas Schrödinger's equation gives rise to a non-localised picture of the momentum state, a localised picture can be restored using the concept of wave packets.

Wave packets can be described by multiplying the complex phase factors given above by a further factor exp $(-A^2 x^2)$ where A is a constant, or by more general Gaussian functions.

Linear transformations

Of great importance in mathematical physics are linear transformations which preserve vector space structure namely orthogonal, unitary and symplectic groups. Starting with the symmetries of a square, non-reflective symmetries are generated from a single rotation through a right angle in the square's plane repeated a number of times. This non-reflective symmetry forms an Abelian group after the mathematician Niels Abel.

An Abelian group is a set A together with an operation that combines any two elements a and b to form another element $a\,b$. To qualify as an Abelian group the set and operation must satisfy five requirements known as the Abelian group axioms namely closure, associativity, identity element, inverse element and commutativity.

The last axiom can be stated as for all a, b in A, $a\,b = b\,a$.

Reflective symmetries are obtained by an alternative procedure namely complex conjugation, which flips the square over about a horizontal line.

The motions involved can be represented using complex numbers. The four points of the square form a group and symmetry operations which take such an object back to itself (as in the case of rotation through four 90 degree steps in this case) are called group axioms.

This reflective symmetry is an example where the group operation is non-commutative, termed a non-Abelian group.

In a similar manner the non-reflective and reflective symmetries can be obtained for a sphere. This time there are an infinite number of elements because the sphere can be rotated through any angle about any axis in 3 space, the 3-dimensional aspect that refers to the rotational orientation of the body. This space is called SO (3), the non-reflective orthogonal group in 3 dimensions. When the reflections are included another 3 manifold is obtained disconnected from the first. The entire group again constitutes a 3 manifold consisting of two separate connected pieces called O (3).

So SO (3) is a sub group of the entire group space O (3).

The symmetries of a sphere are an example of a continuous group called Lie groups after the mathematician Sophus Lie. Lie groups or continuous groups are those for which the basic local theory depends on a study of infinitesimal group elements. Those elements define a form of algebra referred to as Lie algebra, which provides complete information as to the local structure of the group.

There are groups of symmetries of vector spaces expressed by linear transformations which preserve the vector space structure, involving addition of vectors and the multiplication of vectors by numbers. Such linear transformations relate one vector space to another, and are explicitly described using mathematical arrays of numbers, namely matrices.

Singular linear transformations map the entire vector space down to ever smaller dimension in that space.

The applicable identity element equation is $TI = I = IT$ where I is the identity and the linear transformation T is singular when it has a non zero vector such that $Tv = 0$.

Non-singular transformations have an inverse T^{-1} giving the identity element equation $TT^{-1} = I = T^{-1}T$

General linear transformations squash or stretch a unit sphere into an ellipsoid, whereas the orthogonal group O (3) preserves the unit sphere.

The equation of the sphere using the summation convention in index notation can be written $g_{ab}\, x_a\, x^b = 1$, standing for $(x^1)^2 + \dots (x^n)^2$, the components g_{ab} being given by:-

$$g_{ab} = 1, \text{ if } a = b, \text{ and } g_{ab} = 0, \text{ if } a \neq b.$$

For example the tensor g_{ab} is called a metric where the quantity ds defined by its square d$s^2 = g_{ab}\, dx^a\, dx^b$ provides the notion of distance along curves, (see Section 7.3.6).

Orthogonal, unitary and symplectic groups

The orthogonal group O (n) is the group of linear transformations in n dimensions which maintain a positive g, where the transformation T is given by the relation

$$g_{ab}\, T^a{}_c\, T^b{}_d = g_{cd}.$$

In addition to the quadratic form $g_{ab}\, x^a\, x^b$ the symmetric bilinear form in x and y can be used, $g(x, y) = g_{ab}\, x^a\, y^b$.

The symmetry of g is then defined by,
$g(x, y) = g(y, x)$, and linearity in the second variable y by,

$$g(x, y + w) = g(x, y) + g(x, w), g(x, \lambda y) = \lambda g(x, y).$$

For bilinearity, linearity in the first variable x is provided by the symmetry.

Orthogonal groups then maintain a non-singular bilinear form.

Although orthogonal group transformations can be generalised from real to complex numbers, there are groups which comprehensively take on the role of representing the complex state, namely unitary groups.

Unitary groups denoted SU (n) and U (n) use complex linear transformations which maintain the Hermitian form referred to earlier.

A Hermitian form $h(x, y)$ satisfies Hermitian symmetry as,

$h(x, y) = \overline{h(y, x)}$, together with linearity in the second variable y:

$$h(x, y + w) = h(x, y) + h(x, w), \; h(x, \lambda y) = \lambda h(x, y).$$

The Hermitian symmetry now provides antilinearity in the first variable as,

$$h(x + w, y) = h(x, y) + h(w, y), \; h(\lambda x, y) = \bar{\lambda}\, h(x, y).$$

The complex linear transformations maintaining a non-singular Hermitian form define unitary groups.

The bilinear form g provides a means of identifying the vector space V, to which x and y belong, with the dual space V^*.

In the Hermitian case, again each element v of the vector space V provides an element of the dual space V^*, now depending antilinearly on v, rather than linearly.

Thus the element is linear in $\bar{v}$, the complex conjugate of v.

These complex conjugate vectors constitute a separate vector space $\bar{V}$.

Hermitian conjugation housed in Hilbert space is central to quantum mechanics, a subject discussed in Section 7.4.

In addition to the orthogonal and unitary (symmetry groups), there are antisymmetric groups namely symplectic groups.

Returning to the bilinear form but where instead of symmetry, antisymmetry applies as, $s(x, y) = -s(y, x)$ together with linearity $s(x, y + w) = s(x, y) + s(x, w)$ and

$s(x, \lambda y) = \lambda s(x, y)$, where the anti-symmetry provides linearity in the first variable x.

The antisymmetric form can be written $s(x, y) = s_{ab}\, x^a\, y^b$

just as in the symmetric case, but where s_{ab} is antisymmetric: $s_{ba} = -s_{ab}$

For non-singularity $s^{ab} = -s^{ba}$

The transformation T that maintains a non-singular antisymmetric s_{ab} such that

$s_{ab} T^a{}_c T^b{}_d = s_{cd}$ is termed symplectic, groups of these being denoted Sp (n).

In particular the phase spaces referred to in Section 7.3.4 are symplectic manifolds.

Quarternions, Clifford and Grassmann algebra

The symmetric and antisymmetric groups and associated mathematical concepts including hypercomplex numbers, Grassmann and Clifford Algebras, spinors and twistors, play major roles in the development of mathematical physics, in an effort to unite quantum theory and relativity.

Hypercomplex numbers include those real and complex numbers relating to higher dimensions than the two described by complex numbers.

In the 19th century the search for generalised complex numbers led Hamilton to his discovery of quaternions where
$i^2 = j^2 = k^2 = ijk = -1$ and $q = t + ui + vj + wk$, where

t, u, v, w are real numbers and q is the general quaternion.

In particular: **ij** = - **ji, jk** = -**kj, ki** = -**ik** which violates the standard algebraic commutative law: ***ab*** = ***ba***, though other commutative and distributive laws of addition and multiplication are satisfied.

Quaternions form a four-dimensional vector space, three imaginary axes and one real axis. In geometrical terms where multiplying by i means rotation in two dimensions through a right angle about the origin, the quaternion i means rotation in three dimensions.

The quaternions **i, j, k** can be thought of as operators which rotate a given object through 180 degrees i.e. π.

An object rotated through a right angle (in a right-handed sense) about **i** and again about **j** (in a right-handed sense) cannot be restored to its original state by a single rotation about **k**. Instead two rotations of 180 degrees round two different axes is needed to obtain a rotation of 180 degrees about the third to achieve the original state; whereas a classical object is restored to its original state by rotation through 360 degrees, represented by $\mathbf{i}^2 = 1$ rather than $\mathbf{i}^2 = -1$.

In quaternion terms the non-classical object is regarded as a spinor which turns to its negative when it goes through a rotation of 2π and needs a further rotation of 2π to return to its original state, a total of 4π.

The next step is to generalise quaternions to even higher dimensions as devised by the mathematician William Clifford.

Clifford algebra applies to n dimensions where instead of a rotation through π, reflections in an n -1 dimensional (hyper) plane are utilised. A composition of two reflections with respect to two such planes that are perpendicular provides a rotation through π, the reflections being regarded as primary and the rotations as secondary.

These reflections are denoted $\gamma_1, \gamma_2, \gamma_3, \ldots \gamma_n$, where γ_r reverses the rth coordinate axis, while leaving all the others unchanged.

This provides primary reflections as
$$\gamma_1{}^2 = -1, \gamma_2{}^2 = -1, \gamma_3{}^2 = -1 \ldots, \gamma_n{}^2 = -1$$
A spinor is then an object on which the elements of the Clifford algebra act as operators.

Grassmann algebra represents a more basic but universal mathematical system.

In this case the basic anticommutative elements may be denoted
$$\eta_1, \eta_2, \eta_3, \ldots, \eta_n$$
but where each of these squares to zero rather than -1.

Clifford algebra involves perpendiculars so that rotations can be built out of reflections, whereas Grassmann algebra does not incorporate the idea of rotations.

Clifford algebra and spinors require that there is a metric on the space, a property not required by Grassmann algebra.

Symmetry, asymmetry and antisymmetry.
Reflexivity and irreflexivity.

In mathematical terms the above properties can be defined as follows:
The relation R over a set X for all elements a, b in X,
is symmetric if (a, b) in R implies (b, a) in R,

is asymmetric if (a, b) in R implies (b, a) not in R,

is antisymmetric if ((a, b) in R and (b, a) in R) implies a = b or equivalently if (a, b) in R with a ≠ b then (b, a) in R does not hold.

A symmetric relation for example; a is married to b and b is married to a.

An asymmetric relation for example; a is father of b, but b cannot be father of a.

An antisymmetric relation for example; a paid the bill (for a meal say) of a, while b paid the bill of b, or a paid the bill of a and b, or b paid the bill of a and b, but if a pays for b, b does not pay for a.

Antisymmetry is therefore skew symmetric i.e. skewed symmetrically as opposed to asymmetry which is skewed asymmetrically, exhibiting chirality.

Chirality is the property of an object such that it cannot be imposed on its mirror image i.e. it is irreflexive.

A binary relation R on a set X is reflexive if it relates every element of X to itself; if it does not then it is irreflexive. Symmetry is reflexive while asymmetry is antisymmetric and irreflexive. For example the relation 'is equal to 'between real numbers is reflexive but the relation 'is greater than' is irreflexive.

7.3.8. *The incorporation of special relativity into quantum mechanics.*

Dirac's equation for the electron.

The non-relativistic quantum Hamiltonian H ignoring potential energy is given by $p^2/2m$.

To incorporate special relativity the rest mass of the particle μ must be allowed for.

Whereas before the energy $E^2 = c^2 p^2$, now $E^2 = (c^2\mu)^2 + c^2 p^2$

From $E = mc^2$, mass is energy which is non-invariant and where m is not intrinsic to the particle itself, but is the mass measured in some reference frame that need not share that particle's velocity. The larger the particle's velocity the larger this perceived mass, which is why it is not an invariant quantity. The rest mass μ is however an invariant quantity, where the squared rest energy $\mu^2 = m^2 - p^2$ and $c = 1$.

Applying the same energy and momentum operators to the squared rest energy as in the non-relativistic derivation of the Schrödinger equation (see Section 7.3.7),

$(i\hbar)^2 \square \psi = \mu^2 \psi$, where $\square = (\partial/\partial t)^2 - \Delta^2$ and
$\Delta^2 = (\partial/\partial x)^2 + (\partial/\partial y)^2 + (\partial/\partial z)^2$.

Writing $M = \mu/\hbar$, this equation can be written $(\square + M^2)\psi = 0$.

Dirac then used the Clifford algebra elements

$\gamma_1^2 = -1, \gamma_2^2 = -1, \gamma_3^2 = -1, \gamma_0^2 = 1$,

where $\square = (\gamma_0\, \partial/\partial t - \gamma_1 \partial/\partial x - \gamma_2\, \partial/\partial y - \gamma_3\, \partial/\partial z)^2$, referred to as the d'Alambertian.

$\square$ is commonly denoted $\partial\!\!\!/^2$, where the slash notation is due to the physicist Richard Feynman.

Returning to the equation $(\square + M^2)\psi = 0$.

$\square + M^2 = \partial\!\!\!/^2 + M^2 = (\partial\!\!\!/ - iM)(\partial\!\!\!/ + iM)$.

This gives the Dirac equation for the electron, $(\partial\!\!\!/ + iM)\psi = 0$.

Writing $\hbar$ in terms of the rest mass μ, $\hbar\, \partial/\psi = -i\mu\, \psi$.

7.3.9. *Algorithms and Turing machines.*

The term algorithm has its origins in the work of the 9th century mathematician al-Khwarizmi. Algorithms are in principle simply a set of instructions followed in a defined sequence until a particular defined result is obtained; for example the sequence of operations to find the remainder from the long division of two arbitrary natural numbers.

Taking two numbers A and B, step 1 (ask if B is larger than A), if not then take step 2, (replace A by A-B) and repeat step 1 followed by step 2 until B is larger than A and then carry out step 3 (print A). This process is of course more clearly and elegantly demonstrated by a flow chart than this verbal description.

Computer programs are based on this simple principle though they may involve much more complicated sequences and many more steps.

The mathematician and war time code breaker Alan Turing (Ref 7.3.4) introduced the concept of what became known as a 'Turing machine' to solve a problem originally posed by the mathematician David Hilbert in the early 1900s, namely whether a general algorithmic procedure for solving mathematical problems could in principle exist.

Hilbert had restated the problem in 1928 by asking whether mathematics was:-

(a) complete, (b) consistent and (c) decidable.

Both (a) and (b) which Hilbert was seeking to confirm were disproved following the work of Kurt Gödel in 1931 (Ref 7.3.5), which will be referred to later in this set of mathematical questions. The third problem (c) referred to by Hilbert as the Entscheidungsproblem remained unresolved.

Turing however was concerned with whether in principle some general mechanical procedure could solve all the problems of at least a well-defined class of mathematics.

He regarded the human brain as a mathematical machine in some sense so that human mathematicians tackling mathematical problems were carrying out in effect mechanical procedures.

A Turing machine consists of the following elements:-

1. A tape divided in to cells one next to the other, each cell containing a symbol from some finite alphabet, the alphabet to contain a special blank symbol, say O and one or more symbols. Cells that have not been written on are assumed to carry blank symbols. The tape extends indefinitely to the left and right of the machine.
2. A head that can read and write symbols on the tape and move that tape to left or right one cell at a time.
3. A state register which stores the state of the machine, one of finitely many, including the initial start state.
4. A finite table of instructions as:-
    erase or write a symbol, then move one step left or right or stay in same place, then assume the same state or a new state as prescribed.

The internal states of the machine are therefore finite, while the tape can be infinite in both directions; in any particular case the input, calculation and output must be finite, in other words a finite number of marks on an infinite tape, such that beyond some point on the tape in each direction the tape is blank.

An alternative scheme put forward by the logician Alonzo Church (Ref 7.3.6), referred to as the lambda calculus more economical in description than that of Turing, was found to be equivalent and the combination became known as the Church – Turing thesis.

The above machine describes the modern computer which uses binary coding very well though it is intended to model only the process of computation.

Taking the simple example of finding the remainder left from the long division of two natural numbers, a Turing machine tape will carry out the process running backward and forward through the device as long as further calculations need to be performed. When the calculation is completed the device will halt with the answer displayed on that part of the tape designated as output.

Furthermore the same techniques can not only be applied to natural numbers but other real numbers such as rational numbers i.e. those that can be written as a fraction.

Every rational number can be written as a repeated or finite decimal expansion.

Irrational numbers with infinite decimal expansions present a problem since a Turing machine by definition cannot be allowed to run on for ever, it must be allowed to come to a halt, the output then being finite, otherwise the output could not be trusted.

However a Turing machine given the algorithm for producing the expansion for π for example can identify the nth digit of that expansion using its finite table of instructions, subject to its being given n as input and can therefore define a potentially infinite number of digits. The machine will in accordance with its instructions duly halt on identifying that nth digit.

It might be assumed reasonable to think that such a machine can be made to perform any mechanical operation and the words algorithm, computable, recursive and effective are all words applied by mathematicians to describe these processes. Computable or recursive numbers are the real numbers that can be computed to any desired precision by a finite terminating algorithm. In other words one for which there is a Turing Machine as above which given n on its initial tape terminates with the nth digit of that number encoded on its tape.

It is possible in principle to encode the list of instructions for an arbitrary Turing machine T and use this as the initial input for a particular Turing machine U, called a Universal Turing machine, which then acts on the remainder of the input just as T would have done. This enables the machine U to imitate any given machine T exactly.

What is being done here is to encode the action table for the machine T followed by the input data, i.e. putting the instructions for the machine and the input data in the same memory, a key feature of the modern computer.

It is also possible to systematically number Turing machines, encoding them for input to a Universal Turing machine. Such encoding consists of a sequence in binary form as a sequence of 0s and 1s which form natural numbers, the serial numbers of that machine in effect. Since all the serial numbers are natural numbers the number of Turing machines is countable and infinite.

Returning to the Entscheidungsproblem Turing found he could interpret his version of the question in terms of whether the nth Turing machine would ever actually stop when acting on the number m, now referred to as the halting problem.

Turing devised a proof showing that there is no computable function that decides whether an arbitrary program n halts on arbitrary input m; such that the following function h is not computable:-

$$h (n, m) = 1 \text{ if program n halts on input m,}$$
$$0 \text{ otherwise.}$$

The proof proceeds by establishing that every totally computable function with two arguments differs from the required function h confirming there is no algorithm which can be applied to any arbitrary program and input so as to decide whether the program stops when run with that input.

It should be noted that the universal Turing machine number is enormous, a fact such machines have in common with real numbers and Gödel numbers referred to later.

Importantly though, in establishing his proof Turing utilised the method used by Georg Cantor in the late 1800s to show there are many more real numbers than there are rationals and that they are not countable. Cantor did so using a diagonal argument derived from pairing off the sequence of natural numbers with a set of real numbers between 0 and 1. This implies that the set of real numbers in this case between 0 and 1 is countable and that therefore there must be a list pairing these with the natural numbers.

The diagonal procedure is to construct a real number between 0 and 1 whose decimal expansion after the decimal point differs from the diagonal integers on the set of real numbers in each corresponding place. By replacing for example the series of diagonal digits with 1 if the diagonal digit is $\neq 1$ and with 2 if the diagonal digit is 1.

The number so constructed cannot appear in the original listing because it differs from the first number in the first decimal place and the second number in the second decimal place and so on. This contradicts the original assumption that the list was to include all numbers between 0 and 1, therefore the assumption was false and the number of real numbers is greater than the number of rational numbers and is not countable.

Cantor had also shown that the total number of fractions, the total number of integers and the total number of natural numbers are all the same infinite number.

So it turns out that there are many real irrational numbers that are not computable, that cannot be produced by any Turing machine.

7.3.10. The concept of mathematical proof.
Gödel's incompleteness theorems.

Truth from the point of view of mathematicians is very much concerned with formal systems of axioms, namely established principles and self-evident truths, together with rules of procedure which govern the formulation of mathematical statements expressed as theorems or propositions and the corresponding answers in the form of proofs.

In formal systems the theorems must be formulated in terms of the same symbols as are used to express the axioms such that those propositions can in principle be regarded as unassailably true by human mathematicians.

This process has been going on since the ancient Greek philosophers and earlier with new advances continually made by first Arabic mathematicians in the 800s and then European mathematicians from the 16th century onwards, but has particularly blossomed since the early part of the 20th century. Hilbert as explained earlier set out at that time to devise rules which hopefully would confirm that mathematics was complete, consistent and decidable.

It has already been explained above that his hopes in this regard were dashed by first Gödel and then as described earlier, by Turing in the 1930s. It is necessary now to consider the impact Gödel's theorems had on the mathematical world.

His first incompleteness theorem has been phrased in English by Kleene (Ref 7.3.7) as:-

Any effectively generated theory capable of expressing elementary arithmetic cannot be both consistent and complete. In particular, for any consistent, effectively generated formal theory that proves certain basic arithmetic truths, there is an arithmetical statement that is true, but not provable in the theory.

Now incomplete formal systems will exist if some necessary axioms are not included, but Gödel's Theorem shows that a complete and finite list of axioms can never be created, not even an infinite list that can be enumerated by a computer program. Each time a new axiom is added there are other true statements that still cannot be proved even with the new axiom. If an axiom is ever added that makes the system complete, it does so at the cost of making the system inconsistent.

This situation can be related to the liar paradox which states, 'this sentence is false'.

An analysis of this sentence shows that it cannot be true (for then as it asserts, it is false), nor can it be false (for then it is true).

A Gödel theory G for a theory T makes a similar assertion to the liar sentence, but with truth replaced by provability: G says, 'G is not provable in the theory T'.

Gödel's second incompleteness theorem may be phrased as:-

For any formally effectively generated theory T including basic arithmetical truths and also certain truths about formal provability, if T includes a statement of its own consistency, then T is inconsistent.

This strengthens the first theorem as the statement constructed in the first theorem does not express the consistency of the theory. The proof of the second theorem is obtained by formalising the proof of the first theorem within the theory itself.

Gödel's procedure can be summarised as follows:-

The first step is to associate axioms and rules with the natural numbers, i.e. to encode them in a self-referential manner so that the axioms and rules describe what numbers are and are furthermore themselves numbers. Gödel proceeded to use a numbering system of his own to do just that before computers were even invented, but modern computers do in effect the same thing using programming languages of one sort or another.

All the functions he constructs are computable and provide specific answers in a finite, if necessarily large number of discrete steps.

In order to avoid coding meaningless or otherwise false statements, it is necessary to ensure that a strict sequence of ensuring that all axioms going back to the most basic are adhered to. The specific sequence consists of and is verifiable by purely mathematical functions.

Gödel's mathematical proofs are extremely complex and commonly simpler sketch proofs are used to simulate his thinking to those who are not professional mathematicians. One such sketch proof is as follows:-

If x is a given sequence of logical deductions, the function $P(x)$ will confirm x is a valid proof. That function performs the vital role of describing what can be proven in the formal system being examined.

Now with the Gödel numbering system there will always be a number f for any logical function P such that $f = $ Gödel number $(P(f))$.

This can be visualised by using a graph to plot
$y = $ Gödel number $(P(x))$ against x.

The graph of $P(x)$ will at some point intersect with the diagonal where $y = x$. This diagonal procedure is directly related to that used by Cantor and then Turing as described earlier.

A further function which must be provided is $P_n(x)$ meaning not provable.

The diagonal procedure is then repeated to obtain the corresponding Gödel number g for that function, namely $g = $ Gödel number $(P_n(g))$.

Constructing a statement $G = P_n(g)$, it is necessary to establish whether G is true and whether that can be proven. Both $P(x)$ and $P_n(x)$ can each only be true or false, for any given x.

Now if G is false where G is $P_n(g)$, then $P(g)$ is true,

but g = Gödel number (P_n (g)), which means P (g) is true and describes the proof encoded in Gödel number g and that proof is correct.

Therefore a correct proof has been obtained from a false statement G; this is inconsistent to say the least.

If on the other hand G is true, where again G is P_n (g), then P (g) is false.

So G cannot be proved, there is no proof of it, it is not possible to start with the axioms and make any deductions about G, the system is therefore incomplete.

There are statements G which are true but unprovable and therefore incomplete, or provable but derived from falsity and therefore inconsistent.

It should be noted that if further axioms are added to a given system then a new function $P^\#$ (x) must be added to the system giving a new statement $G^\#$ and the entire process repeated. There are therefore a potentially infinite number of true but unprovable statements.

7.3.11. *Mathematics and the Reference Model.*

The mathematics referred to in Sections 7.3.1 thru 7.3.10 is that considered by this author to be most relevant to the Reference Model and the model of consciousness put forward in this book, for reasons given in this Section 7.3.11.

Mathematics provides a means of relating reality (the objective) and imagination (the subjective). The elements described in Sections 7.3.1 thru 7.3.3 can accordingly be related to the structure of the Reference Model as described in this Section 7.3.11.

The Reference Model uses abscissae to represent reality (that which is measurable) and ordinates to represent the imaginary (that which is not limited to the measurable) in terms of an Ordering Centre in the same manner as an Argand diagram.

Wave motion, reinforcement and cancellation, travelling and standing waves, nodes and antinodes, harmonics and the comprehension of music, as for example illustrated by Fourier series, are fundamental to the construction of the Reference Model as described in Section 4.0.

The principle of the fractal and Mandelbrot sets can be related to the multidimensional degrees of freedom associated with an Ordering Centre. Accordingly the Model uses the principle of infinite dimensions in phase space as background for the mean level of order and Hilbert space as background for the present state. The former corresponds to the x axis and the latter to the y axis in a two-dimensional Argand diagram.

Probability and the normal distribution based on the Central limit theorem are again fundamental to the Reference Model as described in Sections 4.0.

Sections 7.3.4 thru 7.3.6 form the mathematical background to classical physics, the four-dimensional real world and the Theory of Relativity based on it.

The Reference Model illustrates that the unfolding of a four dimensional space-time world is a particular case in a range of cases all housed in the multidimensional scenario increasingly ordered in one direction and increasingly disordered in the opposite sense.

Although the brain can unfold the complexity of infinite dimension on to only two dimensions it is not to be assumed that those two dimensions actually describe that complexity. The complexity is in the mind which according to the Model is multi-dimensional stuff.

Even in real terms two dimensions are insufficient to describe the complexity we observe, one dimension even less so and zero dimension none at all.

It takes three large spatial dimensions to achieve that complexity in reality and an additional large time dimension to bring things alive in reality, all as observed by the brain. In terms of the Reference Model this corresponds to all the levels of explicit order derived from our big bang. The mind can enfold all those explicit levels back to the mean level of order in our Ordering Centre.

The intersection of the mean level of order and the present state axis appear to reduce dimension to zero (a singularity) in real terms, however when imagination provided by the mind is added that intersection becomes infinitely dimensional as the fractal description implies.

The relation between the spatial and temporal dimensions in explicit terms is provided by the Theory of General Relativity, which forms the background to an understanding of the local realistic state and the mean level of explicit order as described in Sections 4.1 thru 4.4. In terms of the Reference Model the corresponding implicit relations are described in Sections 4.5 thru 4.8.

Section 7.3.7 provides a basic summary of the mathematical background to quantum theory which is incorporated in to the Reference Model as described in Section 4.0.

It should be noted in respect of orthogonality that the x and y axes can be related in real explicit terms or alternatively in imaginary implicit terms.

Orthogonal groups are orientated towards reality and the mean level of order, unitary groups correspondingly towards Hilbert space and the imaginary axis and symplectic groups towards phase space and the real axis all as portrayed in Fig 4.5.4.

Section 7.3.8 indicates the mathematical background to efforts by physicists to relate quantum theory to local realistic theories in particular special relativity; local realism being related in the Reference Model to the mean level of explicit order as described in Section 4.0.

Sections 7.3.9 and 7.3.10 illustrate the futility of trying to force absolute values on existence when it is clear that such values are always supplanted by an unlimited supply of absolute values, each centred about a range of relative values, as described in the Reference Model.

7.4   Science and Measurement.

A general summary of the relevant mathematical physics is given in Sections 7.4.2 thru 7.4.9, but for a more detailed and comprehensive background it is again recommended that the reader utilise Penrose (Ref 7.3.1 and Ref 7.3.2). Specific subjects can also be found on the internet.

7.4.1.   The practical philosophies e.g. Science.

The concept of scientific proof.

The double blind scientific principle.

In considering science and measurement it is appropriate to commence by discussing the scientific method, a description of which can be found on the internet.

The scientific method may be summarised as follows:-

Firstly forming a question usually based on previous experiments.

Secondly putting forward possible answers to the question, in the form of a hypothesis. The hypothesis must be falsifiable, such that comparisons can be made between the outcome of an experiment and predictions deduced from the hypothesis.

Thirdly deriving logical consequences to the hypothesis in the form of predictions, such that the hypothesis can be distinguished from possible alternatives.

Fourthly testing the hypothesis by conducting further experiments, which determine whether observations made confirm or otherwise predictions derived from the hypothesis.

Fifthly documentation of the experiments must be provided for record purposes etc and shared with other scientists (termed full disclosure), so that they can reproduce and verify the original findings.

Agreement does not confirm the hypothesis is true, as future experiments may reveal, indeed Karl Popper advised scientists to try and falsify hypotheses over the long run, to minimise risk of error.

Experiments should be designed to eliminate such errors and avoid the involvement of human emotions where appropriate by running double blind experiments. For example in medical experiments procedures must be in place to ensure test personnel and subjects are equally ignorant of the designation of the samples. Indeed the whole purpose of scientific experiments is to remove human emotion from the testing process.

Failure of an experiment does not necessarily mean the hypothesis is false as for example the test procedures or equipment may be faulty or some associated hypothesis false instead.

Finally, an analytic process should be conducted to determine the validity of the results of the experiment and what further experiments are necessary.

In the case where an experiment has been repeated many times, a statistical analysis may be required.

If the hypothesis has been falsified a new hypothesis is required and the earlier steps repeated.

The above hypothesis-based science is complemented by discovery science, where there is no specific hypothesis but large amounts of observations are taken and when examined reveal interesting patterns worthy of further investigation.

7.4.2. *The relation between Newtonian physics and relativistic physics.*

Special Relativity, Lorentz transformations and the Poincare Group.

General Relativity.

Newtonian physics as it is sometimes termed is essentially based on Newton's laws of motion and Euclidian geometry.

The first two of Newton's laws (of which the first was derived from Galileo), namely an object at rest, or moving at a constant velocity will continue to do so as long as no forces are acting and if a force does act on it then its mass x acceleration (i.e. the rate of change of momentum) is equal to that force. A better formulation is that the first law defines an inertial frame coordinate system in which when no forces are acting objects stay at rest or continue to move at a constant velocity and the second law that in an inertial frame the force is mass x acceleration. Force is not equal to mass x acceleration in a non-inertial frame like a playground roundabout for example, as explained by the equivalence principle fundamental to the theory of General Relativity. Newton added a third law namely that to every action there is an equal and opposite reaction.

In Euclidian geometry with which many will be familiar, the sum of the angles of any triangle is 180 degrees, parallel lines never meet and the circumference of a circle is $2\pi r$, where r is its radius etc.

Two alternatives also with three dimensions are possible; a spherical closed finite geometry in which a triangles angular sum can be up to 540 degrees, parallel lines eventually meet and the circumference of a circle is less than $2\pi r$; alternatively a hyperbolic geometry infinite in size with the geometry of a saddle, in which parallel lines diverge, a triangles angular sum is less than 180 degrees, the difference being proportional to the area of the triangle and the circumference of a circle is larger than $2\pi r$. The three possibilities of Euclidian, elliptic and hyperbolic geometries are the only homogenous (spatially) and isotropic (directionally) options.

There are prints by M.C. Escher which illustrate these curvatures for purposes of visualisation, see Ref (4.6.1).

More alternatives are available if the curvature differs from place to place as applies in Riemannian geometry associated with general relativity; while 2D surfaces are visualised as being embedded in a 3D space, a curved 3D space can be embedded in a 4D space-time.

Adding time in the Euclidian sense can be represented by a series of flat surfaces parallel to each other each surface representing space at different times. Although this representation is adequate for cases where the speed of light is irrelevant, it in no way caters for those where motions occur at or approaching light speed.

The physicist James Clerk Maxwell in the 19th century discovered the equations governing the behaviour of electromagnetic fields which required that electromagnetic waves would propagate at the speed of light. More to the point this speed c is invariant in different coordinate systems, i.e. c is constant no matter what velocity the observer was moving at. Maxwell's electromagnetism is not invariant from a Galilean point of view.

Galilean transformations are used to transform from one coordinate reference frame to another where the frames differ only by constant relative motion, in terms of Newtonian physics.

The relevant Galilean transformations are:

$$x' = x - v\,t,\, y' = y,\, z' = z\,,\, t' = t.$$

Experiments by the physicists Michelson and Morley towards the end of the 19th century showed that the apparent speed of light on the earth's surface is not influenced by the earth's motion about the sun, overturning the earlier assumption of an aether filling space to explain the travelling of light and electromagnetic phenomena in a vacuum. This was further investigated by other 19th century physicists.

The experiments led George Francis Fitzgerald to the conjecture that bodies in motion are contracted to explain the new observations.

Hendrik Lorentz independently presented the idea in a more detailed manner subsequently called the Fitzgerald-Lorentz contraction.

Lorentz and Joseph Larmor looked for the transformation under which Maxwell's equations are invariant when transformed from the aether to a moving frame and extended the contraction hypothesis to include for modification of the time coordinate as well.

In 1905 the mathematician Poincare recognised that these transformations had the properties of a mathematical group and named them the Lorentz transformations.

In the same year Einstein derived the Lorentz transformations under the assumption of the principle of relativity and the constancy of the speed of light in any inertial reference frame, abandoning the notion of the aether in the process.

The paper that Einstein published is now called the theory of special relativity which incorporates his famous equation $E = mc^2$.

If classical kinetic energy is included $E = \frac{1}{2} mv^2 + mc^2$, where m is the rest mass and v is the velocity (small relative to c).

The Lorentz transformations are coordinate transformations between two coordinate frames that move at constant velocity relative to each other.

Frames of reference can be divided into two groups namely inertial (relative motion with constant velocity) and non-inertial (accelerating in curved paths, rotational motion with constant angular velocity). Lorentz transformations only refer to inertial frames, usually in the context of special relativity.

In each reference frame an observer can use a local coordinate system (Cartesian coordinates) to measure lengths and a clock to measure time intervals. An observer, animate or inanimate, takes the measurements such an event being a point in space-time.

The transformations connect the space and time coordinates of an event as measured by an observer in each frame. They supersede Newtonian physics which assumes absolute space and time. They reflect the fact that observers moving at different velocities may measure different distances, elapsed times and even different ordering of events but always such that the speed of light is the same in all inertial reference frames and therefore invariant.

They are the culmination of efforts to explain how the speed of light was observed to be independent of the reference frame and to understand the symmetries of the laws of magnetism.

The Lorentz transformation is linear and may include a rotation of space, when it is rotation free it is called a Lorentz boost.

Einstein incorporated the Lorentz transformations into his theory of special relativity, which was based on two important postulates.

Firstly, that the laws of physics are invariant (identical) in all inertial systems (non-accelerating reference frames).

Secondly, that the speed of light in a vacuum is the same for all observers regardless of the motion of the light source.

While Newtonian mechanics are simple and highly accurate at small velocities relative to the speed of light, special relativity provides accuracy at any speed when gravitational effects are negligible.

Special relativity implies a wide range of consequences which have been experimentally verified, including length contraction, time dilatation, relativistic mass, mass energy equivalence, a universal speed limit and relativity of simultaneity.

Reference frames play a crucial role in relativity theory. It is the observational perspective in space which is not undergoing any change in motion (acceleration), from which a position can be measured along three spatial axes. The reference frame also provides a means of measuring the time of events using a clock, events being a point in space-time. Since the speed of light is constant in each and every reference frame, pulses of light can be used to measure distances and refer back the times that events occurred to the clock, even though light takes time to reach the clock after the event has transpired. An event can be completely specified by its four space-time coordinates which define a reference point. If this is a reference frame S, then to calculate the position of a point from a different reference point a second reference frame S' is required. The spatial axes and clock of S' exactly coincide with that of S at time zero, but it is moving at a constant velocity *v* with respect to S along the x axis.

Any two frames that move at the same speed in the same direction are said to be comoving, therefore S and S' are not comoving.

So defining an event to have space-time coordinates (t, x, y, z) in system S, then in a reference frame S′ moving at a velocity v with respect to S, the Lorentz transformations are defined as:-

$$t' = \gamma\, (t - v\,x/c^2)$$

$$x' = \gamma\, (x - v\,t)$$

$$y' = y$$

$$z' = z$$

where $\gamma = 1/\sqrt{(1- v^2/c^2)}$ is the Lorentz factor, c is the speed of light in a vacuum, and the velocity v of S′ is parallel to the x axis. Similar transformations apply to the y or z axes, or indeed in any direction.

Consequences derived from the Lorentz transformations may appear counter intuitive because of the enormity of the speed of light relative to the pedestrian speeds with which we are familiar.

They include relativity of simultaneity, where two events happening in two different locations that occur simultaneously in the reference frame of one inertial observer, may occur non simultaneously in the reference frame of another inertial observer.

In addition the time lapse between two events is not invariant from one observer to another, but is dependent on the relative speeds of the observers' reference frames i.e. the time dilation effect.

Also the dimensions of an object as measured by one observer may be smaller than the results of measurements of the same object made by another observer i.e. the length contraction effect.

These effects and others can be described by rewriting the Lorentz transformations in terms of coordinate differences.

Another important consequence is the prohibition of motion faster than light and for this it is necessary to understand the difference between flat Euclidian space and what is termed Minkowski space.

Whereas general relativity incorporates non-Euclidian geometry to represent gravitational effects as the geometric curvature of space-time, special relativity is restricted to the flat space-time known as Minkowski space.

Following the publication of Einstein's theory of special relativity, Minkowski, who had been one of Einstein's teachers was able to encode the theory in an alternative four dimensional space-time geometry making it one indivisible whole all independent of an arbitrary observers viewpoint.

Minkowski space is the mathematical model of space-time in special relativity in which the Lorentz transformations preserve the space-time interval between any two events, while the space-time event at the origin is left fixed. They can be considered a hyperbolic rotation of Minkowski space. The more general set of transformations that also include translations is known as the Poincare group.

The Poincare group is the full symmetry of special relativity and includes translations (displacements) in time and space, rotations in space and boosts which are transformations connecting two uniformly moving bodies. The last two symmetries make up the Lorentz group. The translations form an Abelian group and the rotations a non-Abelian group (see Section 7.3.7).

A Minkowski space-time isometry has the property that the interval between events is left invariant. If everything was postponed or shifted by some distance or turned through some angle no change in the interval occurs, nor in the proper length of the object. Minkowski space-time ignoring the effect of gravity has ten degrees of freedom of the isometries, translations through time or space (four degrees), rotation (three degrees) or a boost in any of the spatial directions (three degrees).

To understand Minkowski space-time it is necessary to introduce the concept of the null cone, a double cone in a similar sense as an hour glass, which defines the speed of light in any direction at some event p occurring at the intersection of the cones. If the space and time units are chosen so that the speed of light $c = 1$, then the null cone surface is at 45 degrees to the vertical.

Using the Euclidian concept of distance in space in x, y, z coordinates, then any distance s from zero is given by $s^2 = x^2 + y^2 + z^2$. In Minkowski space-time the corresponding relationship is $s^2 = t^2 - (x/c)^2 - (y/c)^2 - (z/c)^2$, where in this case s is the time interval experienced by a given particle in moving from any position at $x = y = z = 0$ to a new position within the light cone. The motion of the particle in space-time is termed its world line.

If $s^2 > 0$ then the separation between the points in space-time is time like and has proper time $s^2 = t^2 - (x/c)^2 - (y/c)^2 - (z/c)^2$. If $s^2 < 0$

then the separation between the points in space-time is spacelike and has proper distance $s^2 = (x/c)^2 + (y/c)^2 + (z/c)^2 - t^2$.

If $t^2 - (x/c)^2 - (y/c)^2 - (z/c)^2 = 0$ then the separation is null.

The important points are firstly that t does not represent time measured by a clock attached to the particle unless it is at rest with fixed coordinate values x/c, y/c, z/c all zero, in which case the particle has a world line that is the vertical bisecting the null cone. Secondly that the correct measure of time for a moving observer is given by s which is always less than t, therefore the clock attached to that observer will slow down. If the speed of the observer is small compared with c then all clocks will appear to keep the same time, but at the other extreme if his speed reaches the speed of light then he is moving along the light cone, in which case $s = 0$ and the passage of time slows to a halt.

Motion outside the light cone is not permitted by relativity since this would interestingly lead to s becoming a complex number.

Another consequence of Einstein's special theory of relativity is the equivalence of mass and energy.

Einstein's theory of special relativity predicts massless particles to travel at the speed of light and those with mass at lesser speeds, such that the world line of a given particle is that which joins all the events which make up its history. World lines for accelerating particles need not be straight and the tangent to the curve of its world line is called the tangent vector. Null cones in Minkowski space-time are uniformly arranged with massless particles moving along the cones while particles with mass have tangent vectors within the null cone.

The energy content of an object at rest with mass m equals mc^2.

The energy momentum 4 vector is conserved in any reaction in special relativity.

This implies for example, that for a reaction where the sum of the rest masses of the products is less than the sum of the rest masses of the inputs, there must be an accompanying increase in the kinetic energy of the particles after the reaction.

Energy and momentum, which are separate in Newtonian mechanics, form a four vector in relativity, relating the time component (the energy) to the space components (the momentum). For an object at rest the energy momentum 4 vector is $(E/c, 0, 0, 0)$, the time or energy component and the three space components which are zero. The rest mass of a particle, defined in the obvious way as the mass of a particle in its rest frame is Lorentz invariant by definition.

For a particle of rest mass m and 4 velocity U, the momentum 4 vector, or the 4 momentum P is defined by:

$$P = m\,U = (E/c, \mathbf{p})$$

where by definition the relativistic energy E and the relativistic 3 velocity and 3 momentum are

$E = \gamma mc^2$ and $\mathbf{p} = \gamma m\mathbf{u}$ and $\gamma = 1/\sqrt{(1-\mathbf{u}^2/c^2)}$

Energy, momentum and the energy momentum 4 vector are properties that are defined for matter and radiation. Conservation of the energy momentum 4 vector can either be taken as a) a postulate of the special relativity framework or b) derived from basic assumptions in a Lagrangian framework or c) taken as an observational fact that actually 4 momentum is conserved in the real world.

General Relativity

In 1916 Einstein published the General Theory of Relativity which expanded the basic special theory to a new level in which Minkowski space-time formed an essential element. In General Relativity the situation is complicated by the curvature of space-time associated with the corresponding influence of gravity, so the next step is to incorporate that curvature.

In general relativity (GR) the null cone uniformity of Minkowski space-time is lost as the null cones can be considered as being drawn on a rubber sheet which can be distorted this way or that, but based on the principle of general covariance in which such distortions do not affect the physical properties of space and its contents. The geometry involved allows distortions which may twist or shrink the shape of the space, but without tearing.

The mathematical concepts of manifolds, vectors and tensors (see Sections 7.3.5 and 7.3.6) are used to deal with these complicated forms of geometry. Tensors are mathematical concepts, vectors being one, which have certain specific laws of transformation when there is a change in the coordinate system.

Tensors have many applications in the field of differential geometry of curves in three-dimensional space and the higher dimensionality of Riemannian geometry.

Tensor analysis enables the fundamental equations of mathematical physics to be expressed in terms of curvilinear coordinates, electromagnetism for example.

Maxwell's equations governing the behaviour of electromagnetic fields can be described, in terms of the physics of relativity, by the Maxwell field tensor *F* and the source of that field the charge current vector *J*.

Now associated with the geometry comes the notion of a metric (in mathematical terms a tensor) generally denoted by g, which assigns a length to any finite smooth curve in the space. Any deformation of the rubber sheet would carry with it a curve connecting two points, the length between them assigned by g, all deemed unaffected by the deformation. The shortest route or curve between the two points is called the geodesic. Angles between two curves are also determined by g.

Conformal geometry is a structure that provides a measure of such angles without specifying length. It does however fix the ratios of the length measures in different directions at any point. So the length measure can be rescaled up or down without affecting the conformal structure.

There is another aspect which is of importance namely the signature of the metric. For example in a three-dimensional Euclidian geometry the length of an infinitely small segment of a curve ds is given in terms of the three coordinates

$$x, y, z, \text{ as } ds^2 = dx^2 + dy^2 + dz^2.$$

This can be extended to four dimensions. The number of dimensions (n) gives the number of dimensions of translation i.e. the number of translations from a given fixed point in a coordinate system. In addition there are the number of dimensions of rotation given by n (n-1)/2, giving three dimensions of rotation for n = 3 and six dimensions of rotation for n = 4. This gives a total of n (n+1)/2 degrees of freedom or six for three dimensions and ten for four dimensions.

4D space-time has four degrees of freedom of translation, three of rotation and three of boosts.

The full ten-dimensional symmetry group of Minkowski space forms the Poincare group, which is very important not only in relativistic physics but in particle physics and quantum field theory.

In Minkowski space, setting the speed of light c to unity, the time like curve length is $ds^2 = dt^2 - dx^2 - dy^2 - dz^2$, defined by the signature quantity (+ - - -) used here, although the opposite signature (-+++) is also commonly used.

A generalisation of Pythagoras theorem, it is basically concerned with how many of a set of n mutually orthogonal directions (for an n dimensional space), are to be considered time like within the null cone, or spacelike outside the null cone. Whereas in Euclidian or Riemannian geometry all directions are spacelike, when a time like direction is added, it is Minkowskian when it is flat and Lorentzian when it is curved. The signature for Lorentzian curved space-time, relevant to general relativity, is 1+3, indicating one time like and three space like mutually orthogonal directions.

An observer whose world line is in the time like direction regards all events in the orthogonal spacelike direction as simultaneous.

It is important for the physical basis of GR that extremely precise clocks exist in nature at a fundamental level since the whole theory depends on a naturally defined space-time metric g, which provides a time measure for curves rather than a length measurement. Stable massive particles play this role, whose rest energies are given by Einstein's formula fundamental to relativity, $E = m c^2$, where m is the mass of the particle and their frequency of oscillation v by Planck's formula, fundamental to quantum theory $E = h v$, where h is Planck's constant.

At the heart of GR are the field equations which allow for space-time curvature based on how the metric g varies throughout that space-time. That curvature is described by the Riemann-Christoffel curvature tensor $R_{\rho\sigma\mu\nu}$ the complicated symmetries of which give rise to 20 independent components per point.

The important general principles behind GR are firstly the invariance of physical laws regardless of choice of inertial or non-inertial reference frame, namely general covariance.

Secondly the principle of equivalence of gravitational and inertial mass, such that the gravitational 'force' experienced locally when standing on a massive body is the same as that experienced by an observer in a non-inertial (accelerated) frame of reference.

Thirdly, that gravity is not a force but a geometric property of space-time curvature. Fourthly, that the curvature of space-time is directly related to the energy and momentum of the matter and radiation present.

Since it involves geometries which are curved relative to Euclidian geometry, GR is formulated using the mathematics of tensors and tensor fields. Gravitation is represented by curved space-time modelled on a four-dimensional Lorentzian manifold M.

A useful analogy is that curved space-time relative to flat space-time is like the surface of an apple relative to the plane surface of the table on which the apple sits.

Each point on the manifold is contained in a coordinate chart representing local space-time around the observer but GR applies not only to small regions of space-time but also to cosmological scales.

At each point p on such a manifold the tangent and cotangent spaces to the manifold at that point can be formulated. Vectors (contravarient vectors) are defined as elements of the tangent space and covectors (covariant or dual vectors) as elements of the cotangent space. At p the two vector spaces can be used to construct type (n, m) tensors which are multilinear maps acting on the sum of n copies of the cotangent space with m copies of the tangent space.

The set of all such maps form a vector space called the tensor product space of type (n, m) at p denoted by $(T_p)^n{}_m M$.

Space-time being four dimensional each index can be one of four values giving 4^R elements where the rank of the tensor is the sum of the number of covariant and contravariant indices on the tensor.

Some physical quantities are represented by tensors with components which are not all independent such as symmetric and anti-symmetric tensors. Such tensors have limited independent components.

A symmetric rank 2 tensor T satisfies $T_{ab} = T_{ba}$ and has 10 independent components, while an anti-symmetric rank 2 tensor P satisfies $P_{ab} = -P_{ba}$ and has 6 independent components.

Metric tensors are central to GR and describe the local geometry of space-time. They are symmetric, involve the raising and lowering of tensor indices and generate connections used to construct geodesic equations of motion and importantly the Riemann curvature tensor.

The metric tensor relates incremental intervals of coordinate distance in terms of the line element $ds^2 = g_{ab} \, dx^a \, dx^b$. It can be represented by a 4 x 4 matrix which is symmetric and has 10 independent components. A central feature of tensors is that knowing a metric the operation of contracting a tensor of rank R over all R indices gives a number (a scalar) independent of the coordinates chart used to perform the contraction, for example the Ricci scalar $R = R^{ab} g_{ab}$

In GR tensor fields on a manifold are maps which attach a tensor to each point of the manifold. A tensor bundle is a collection of tensors at all points on the manifold.

A tensor field defines a map from the manifold to the tensor bundle each point p being associated with a tensor at p.

Vector fields are contravariant rank 1 tensor fields, for example the 4 velocity U, this being the coordinate distance travelled per unit of proper time, where proper time is that independent of the observer.

Whereas changes in physical processes are generally described by partial derivatives, in GR such derivatives must be tensors. They are derivatives along integral curves of vector fields. In order to compare vectors at different points in defining derivatives on manifolds that are not flat, an extra structure is required.

Such extra structures are typified by affine connections and associated covariant derivatives.

An affine connection is a rule describing how to move a vector along a curve on a manifold without changing its direction, referred to as parallel transport, a process which characterises space-time.

When a vector in Euclidian space is parallel transported around a loop it will again point in the initial direction after returning to its original position. However taking the case of a spherical manifold, lines which appear straight are only so locally. On completion of the loop in this case the vector is no longer pointing in the initial direction but at an angle to that, a measure of the Riemann curvature tensor on a Riemannian manifold.

It should be noted that a loop following a great circle on the sphere maintains the direction of the vector without the above deviation.

Descriptions of the above principle can be found on the internet including those using videos.

The affine connection is a bilinear map of the space of all vector fields on the space-time. The map can be described in terms of a set of connection coefficients known as Christoffel symbols, which specify what happens to components of basis vectors under infinitesimal parallel transport.

Christoffel symbols are denoted Γ^i_{jk} for $i, j, k = 1, 2 \ldots n$, where n is the number of dimensions.

These symbols are an array of numbers n x n x n describing a metric connection, a specialisation of the affine connection to manifolds, endowed with a metric allowing distances to be measured on that surface.

Partial derivatives are generally used to describe changes in physical processes even in special relativity. However they are not good tensor operatives, and for that covariant derivatives are required which reduce to partial derivatives in flat space with Cartesian coordinates but transform as tensors on a curved manifold.

In flat space in Cartesian coordinates a partial derivative operator ∂_u is a map from (n, m) tensor fields to (n, m+1) tensor fields which is linear in respect of arguments and obeys Liebniz rule on tensor products. Liebniz rule states that if two functions f and g are n times differentiable then the product f g is also n times differentiable.

However in the case of curved manifolds the map is dependent on the coordinate system used.

A covariant derivative operator ∇ is required independent of the coordinates, which is a partial derivative plus a linear transformation provided by the connections referred to as Christoffel symbols.

Christoffel symbols are multidimensional arrays of numbers, though not themselves tensors. The relevant version of the Liebniz rule being the derivative of the product of two functions:

$$\nabla(fg) = (\nabla f)\, g + f\, (\nabla g)$$

A covariant derivative can therefore be viewed as a differential operator acting on a vector field sending it to a type (1,1) tensor and generally acting on (n, m) tensor fields sending them to (n, m+1) tensor fields. Notions of parallel transport are thus defined similar to those for vector fields.

In GR working in coordinate reference frames where torsion vanishes (holonomic coordinates), the corresponding connection free of torsion is the Levi Civita connection.

This is a symmetric connection which preserves the metric when parallel transporting a tangent vector along a curve, while keeping the inner product of that vector constant along the curve. Therefore the covariant derivative gives zero when acting on a metric tensor (and its inverse).

A metric compatible covariant derivative means the (inverse) metric can be taken out of the derivative and used to raise and lower indices so,

$$\nabla_a T^b = \nabla_a (T_c g^{bc}) = g^{bc} \nabla_a T_c$$

In GR measurement of the curvature of a manifold is defined by the Riemann curvature tensor using the affine connection and parallel transporting vectors between two points along two curves. The deviation between the results of these routes is quantified by the Riemann tensor and the geodesic equation, meaning that the forces experienced in a gravitational field are due to the curvature of space-time.

On this basis the Riemann tensor is type (1, 3), contains the Christoffel symbols and their first partial derivatives and has 20 independent coefficients.

It has a number of properties namely symmetries, one of which is termed the algebraic Bianchi identity, $R_{\rho\sigma\mu\nu} + R_{\rho\mu\nu\sigma} + R_{\rho\nu\sigma\mu} = 0$ over the four indices.

The corresponding covariant derivative ∇R is termed the differential Bianchi identity.

The Riemann tensor defines whether space is flat or curved and how much curvature occurs in any given region. It uses the covariant derivation of tensors with one and two indices with the derivative being taken twice in respect of the rank 1 tensor.

Using the above symmetries and the Christoffel symbols, the Riemann curvature tensor can be written:

$$R^{\rho}{}_{\sigma\mu\nu} = \partial_{\mu}\Gamma^{\rho}{}_{\nu\sigma} - \partial_{\nu}\Gamma^{\rho}{}_{\mu\sigma} + \Gamma^{\rho}{}_{\mu\lambda}\Gamma^{\lambda}{}_{\nu\sigma} - \Gamma^{\rho}{}_{\nu\lambda}\Gamma^{\lambda}{}_{\mu\sigma}$$

where the derivatives are based on the third and fourth indices.

Making the tensor covariant simply by contraction using the metric gives:

$$R_{\rho\sigma\mu\nu} = g_{\rho\lambda}R^{\lambda}{}_{\sigma\mu\nu}$$

By further decomposition, the contraction process gives a rank 2 type (0, 2) tensor in place of a rank 4 type (1, 3) tensor:

$$R_{\mu\nu} = R^{\lambda}{}_{\mu\lambda\nu}$$

where $R_{\mu\nu}$ is called the Ricci tensor.

In general terms the Ricci tensor is a measure of the degree to which the geometry of the metric tensor differs locally from that of Euclidian space.

Multiplying again by the metric the scalar curvature R, a rank 0 type (0, 0) tensor, is then given by:

$$R = g^{\mu\nu}R_{\mu\nu}$$

where R is called the Ricci scalar and $g^{\mu\nu}$ is the inverse of $g_{\mu\nu}$

An important tensor field in GR is the energy momentum or stress energy tensor T, where energy is conserved such that the derivative of $T_{\mu\nu}$ is zero.

The energy momentum or stress energy tensor $T_{\mu\nu}$ is a rank 2 type (0, 2) symmetric tensor, closely related to the Ricci tensor. Being of second rank in four dimensions it can be viewed as a 4 x 4 matrix.

It can be written: $R_{\mu\nu} - \tfrac{1}{2} R\, g_{\mu\nu} = 8\pi G T_{\mu\nu}$

or $T_{\mu\nu} - \tfrac{1}{2} T g_{\mu\nu} = 1/8\pi G\, R_{\mu\nu}$

The Einstein field equations are then written in terms of the Einstein tensor $G_{\mu v}$ as

$$G_{\mu v} = R_{\mu v} - \tfrac{1}{2} R\, g_{\mu v} = 8\pi G T_{\mu v}$$

$R = -8\pi G T$ where T is termed the trace of $T_{\mu v}$ and R is termed the trace of $R_{\mu v}$

The equations relate the metric and associated geometry of space-time to the momentum and energy content of space-time i.e. space-time is curved by matter and energy. By equating local space-time with local energy and momentum they are therefore in accordance the principle of local conservation of energy and momentum.

There is a further aspect to be considered namely the Weyl curvature tensor. It is also a tensor of the curvature of space-time and also expresses the tidal forces a body feels when moving along a geodesic.

However it does not convey how the volume of the body changes but how the shape of the body is distorted by the tidal force.

It is the Ricci curvature or trace component of the Riemann tensor which provides the volume change information.

So the Weyl tensor is the traceless component of the Riemann tensor i.e. that with all its contractions removed. It is denoted by $C_{\rho\sigma\mu v}$ and can be obtained from the Riemann tensor by subtracting out the traces. It is invariant under conformal changes to the metric and is therefore called the conformal tensor.

The Weyl tensor can be related to how in a Newtonian gravitational field an initially stationary spherical arrangement of particles falling freely under the influence of say the earth's gravity experience a tidal effect arising from the non-uniformity of that field. That non uniformity is made up firstly of the effect of each particle having a different acceleration being at a different distance from the centre of the earth and secondly from the different horizontal displacements of each particle and hence different directions of their accelerations towards the centre of the earth. The initial spherical arrangement therefore distorts in to an ellipsoid the volume of which is equal to the original sphere, taking the sphere of particles to surround a vacuum.

The Ricci tensor derives from the case where the sphere of particles surrounds not a vacuum but say the earth. Then the additional inward acceleration due to the earth's gravity causes the ellipsoid in to which the sphere distorts to reduce in volume proportional to the mass of the earth. If this mass is M then the volume reduction as a measure of inward acceleration in accordance with Newton is given by $4\pi GM$, where M is the volume of the earth times its density ρ.

In Einstein's theory, as opposed to the Newtonian case above, there is an additional contribution to this volume change from the pressure in the material surrounded.

So matter and energy tell space how to curve and the curvature tells matter how to move. Although in principle space could be contracting or expanding Einstein initially created a static model by introducing the cosmological constant Λ multiplied by the space-time metric tensor $g_{\mu\nu}$ which defines distances, to the left-hand side of the equation, giving $G_{\mu\nu} + \Lambda g_{\mu\nu} = 8\pi GT_{\mu\nu}$, suggesting that it was a property of space itself. He abandoned the term when it became evident from the work of the astronomer Hubble that the universe was in fact expanding.

Modern cosmology has reintroduced it to correlate with the experimental evidence that the visible universe is undergoing accelerating expansion.

The new cosmological constant (dark energy) can thus be thought of as either an extra contribution to the stress energy tensor if on the right hand of the equation, such that

$G_{\mu\nu} = 8\pi GT_{\mu\nu} - \Lambda g_{\mu\nu}$, or as a modification to how the stress energy affects curvature, if on the left-hand side.

In place of $R_{\mu\nu} - \frac{1}{2} Rg_{\mu\nu} = 8\pi GT_{\mu\nu}$, since the scalar part $R = -8\pi GT$, there is also the relationship $T_{\mu\nu} - \frac{1}{2} Tg_{\mu\nu} = 1/8\pi G\, R_{\mu\nu,}$

So Einstein's field equation can be written in another form as:-

$R_{\mu\nu} = 8\pi G\, (T_{\mu\nu} - \frac{1}{2}\, Tg_{\mu\nu}) + \Lambda g_{\mu\nu}$

Now incorporated in the Ricci tensor in this form of Einstein's field equation is the element due to pressure in the material referred to above, so the Newtonian case $4\pi GM$ is replaced by $4\pi(\rho + p_1 + p_2 + p_3) - \Lambda/4\pi$, where p_1, p_2, p_3 are the pressure of the matter along three orthogonal spatial axes. So the density of the earth ρ has been replaced in the case of relativity by the density ρ_g of the active gravitational mass with the pressure increasing the effect of gravity, while the cosmological constant decreases it, if Λ is positive, though Λ can in principle be negative.

These aspects have particular significance when it comes to discussing later questions concerning dark matter and dark energy.

Regardless of the varying opinions concerning the cosmology of the big bang and the nature of the universe in the long term, which will be discussed later, there is agreement on the relationship between Newtonian and relativistic physics, if the two can be regarded as constituting what may be termed classical physics. That is to say there is no evident dichotomy between them, relativity merging smoothly with Newtonian physics at velocities which are small relative to that of light.

GR reduces to special relativity (SR) when gravity is weak, but SR is needed at speeds comparable with that of light.

7.4.3. *The cosmological models that developed from Einstein's General theory of Relativity, including what is termed the standard (Lambda CDM) cosmological model.*

Why the rate of expansion of the visible universe appears to be increasing   including the role of what is termed dark energy.

The role of dark matter in providing a satisfactory theory for the relatively rapid aggregation of matter into galaxies.

Cosmological models based on Einstein's work led to de Sitter's version in which space-time is empty, which also had a cosmological term but allowed accelerating expansion of space. They are generally based on the cosmological principle which holds that the universe should look the same in any direction to all observers i.e. homogeneous and isotropic.

De Sitter space-time and anti de Sitter space-time can both be pictured as hyperboloids with two spatial dimensions suppressed. The former corresponds to a state of repulsion in Einstein's equations; the latter corresponds to a state of attraction and possesses a boundary located at infinity.

Then the Friedmann cosmological models were developed without that term but incorporating an idealised form of matter and their behaviour depended on the average density of that matter. The high-density case would lead to a collapsing scenario involving positive curvature, $K > 0$ i.e. spatially closed, a low-density case with indefinite expansion, $K < 0$ i.e. negatively curved and a critical intermediate case giving rise to a spatially flat universe $K = 0$ expanding forever but at an ever-decreasing rate.

The expanding balloon picture of the universe, referred to previously, is a good description of early versions of so called Friedmann-Lemaitre-Robertson-Walker (FLRW) models in accordance the above.

These original FLRW models are highly symmetric and assume matter to be a pressure less fluid or dust, in which the gravitational field is produced entirely by the mass, momentum and stress density of a perfect fluid in the form of dust particles with positive mass density but zero pressure. The dust particles are highly idealised models of stars, galaxies and clusters of galaxies.

Friedmann models, where $\Lambda > 0$, approach the de Sitter model of the universe in which space-time is empty and expands at an accelerating rate in a highly symmetrical fashion.

The metric associated with these models is called the Friedmann Lemaitre Robertson Walker metric or commonly just the FLRW metric and may be derived as follows.

Taking the surface of a sphere of radius R with a point P on the surface at distance r from the pole corresponding to an angle from the vertical of θ and an azimuthal angle ϕ, then the spatial metric in polar coordinates incorporating curvature K is given by:

$$ds^2 = dr^2 / (1 - Kr^2) + r^2 (d\theta^2 + \sin^2\theta \, d\phi^2),$$

which is valid at any single instant of cosmic time t.

However as the universe expands the 3D hypersphere expands and it is highly advantageous to work in comoving coordinates that expand with the universe.

It is therefore beneficial to introduce a cosmic scale factor $a(t)$ which measures the size of the universe and is proportional to the radius R of the universe, though R for Euclidian space is not well defined. Physical (actual) distance is $a(t)$ x the comoving distance, t being the cosmic time or proper time experienced by comoving observers who remain at rest in comoving coordinates.

In spherical coordinates the Minkowski metric with no space-time curvature but incorporating special relativity can be written $ds^2 = c^2\, dt^2 - dr^2 - r^2\, (d\theta^2 + \sin^2\theta\, d\phi^2)$.

So the corresponding FLRW metric with $c = 1$ is then given by:

$$ds^2 = dt^2 - a^2(t)\,(dr^2 / (1-k\,r^2) + r^2\,(d\theta^2 + \sin^2\theta\, d\phi^2)),$$
where $k = a^2 K$ is a measure of the spatial curvature.

The spatial curvature may be negative, zero or positive corresponding to open, flat and closed universes, k being generally rescaled to -1, 0, -1 respectively. The form $a(t)$ depends on the properties of the material in the universe.

The FLRW metric is an exact solution of Einstein's field equations in terms of a homogeneous, isotropic, expanding or contracting universe.

The Riemann curvature tensor and Ricci tensor in terms of the FLRW metric allows computation of the connection coefficients (Christoffel symbols) and curvature tensor referred to in Section 7.4.2.

The non zero components of the Ricci tensor and an equation for the Ricci scalar can be defined. Similarly the relevant elements of the energy momentum tensor can be found based on the concept of the perfect fluid, defined as isotropic in its rest frame.

The rest frame energy density ρ and pressure p, the four velocity of the fluid U and an equation of state $p = w\,\rho$, where w is a constant independent of time are then substituted in to the Einstein equations.

The Einstein equations for the FLRW metric are then given by:

$$3(k/a^2 + (\dot{a}/a)^2) = 8\pi G\rho \text{ , and}$$
$$-k/a^2 - (\dot{a}/a)^2 - 2\ddot{a}/a = 8\pi Gp$$

where $\dot{a}$ and $\ddot{a}$ are first and second derivatives of a with respect to time.

These equations rearrange to give the Friedmann equations, which are fundamental to cosmology:

$$(\dot{a}/a)^2 = 8\pi G\rho/3 - k / a^2 \text{ (the first Friedmann equation).}$$

$\ddot{a}/a = -4\pi G \ (\rho+3p)/3$ (the second Friedmann equation). The corresponding fluid equation can also be derived as: $\dot{\rho} = -3\dot{a} \ (\rho + p)/a$

The ratio $\dot{a}/a$ is the Hubble parameter H which varies with cosmic time t, but is constant in space at fixed cosmic time t. The value of the Hubble parameter today is referred to as the Hubble constant H_0. The distance d to an object that is receding with the expansion of the universe is proportional to the cosmic scale factor a; its recession velocity v is therefore proportional to $\dot{a}$ so $v = H_0 \ d$, namely Hubble's Law.

At cosmological scales this linear relation breaks down as H varies with time.

The Friedmann equations can also be derived from Newtonian principles and the use of the first law of thermodynamics, assuming the expansion of the universe is an adiabatic process. The second Friedmann equation shows that both the energy density and the pressure cause the expansion rate of the universe ä to decrease, all as a consequence of gravitation.

Now the expansion of the universe is governed by the properties of the material within it, as specified by the energy density $\rho(t)$ and the pressure $p(t)$.

The relation between these for $k = 0$ is given by an equation of state for which simple cases such as a matter dominated universe corresponds to $p = 0$ and a radiation dominated universe to $p = \rho/3$. A third case of particular interest is the vacuum dominated universe where $p = -\rho$, with $d\rho/dt = 0$ so that $\rho = \rho_0$.

The density of matter ρ_m decreases as the volume of the universe increases so it reduces as the cube of the scale factor, so $\rho_m \propto a^{-3}$ and $a \ (t) \propto t^{2/3}$, from solving the differential of the Friedmann equations.

The density of radiation ρ_γ decreases more rapidly as in addition to the volume factor, the expansion of the universe causes the wave length of light to increase such that light received from rapidly receding objects is displaced towards the red end of the visible spectrum. Photons are red shifted at the same rate as the universe is expanding so $\rho_\gamma \propto a^{-4}$ and $a(t) \propto t^{1/2}$.

In the vacuum energy dominated case $a(t) \propto \exp Ht$.

The critical density is defined to be the density required for the universe to be flat,

i.e. $k = 0$, accordingly $\rho_{crit} = 3H^2/8\pi G$, a parameter which like H evolves with time.

The ratio of the actual density of the universe to the critical density is $\Omega = \rho/\rho_{crit}$ where Ω is the total mass energy in all the forms it may take, a combination of dark energy approximately 72 per cent, cold dark matter approximately 23 per cent and baryonic matter approximately 5 per cent.

From the first Friedmann equation and the equation for ρ_{crit} it follows that:

$(\rho - \rho_{crit}) / \rho_{crit} = \Omega - 1 = k / (a^2 H^2)$.

During the standard big bang evolution $a^2 H^2$ is decreasing so Ω moves away from 1. For example with matter domination $\Omega - 1$ is $\propto t^{2/3}$ while with radiation domination

$\Omega - 1$ is $\propto t$, where the solutions apply if Ω is close to 1. Now since it is apparent from observations that Ω is indeed close to 1 today, it must have been much closer in the past. Astonishingly the ratio $\Omega - 1$ going back to the Planck time would have been 10^{-60}.

This is an example of the cosmic coincidences which will be referred to again when answering further questions in this section.

Returning to the Einstein field equation $G_{\mu\nu} + \Lambda g_{\mu\nu} = 8\pi G T_{\mu\nu}$

the Friedmann equations can be written to incorporate the cosmological constant as:-

$$3(k / a^2 + (\dot{a} / a)^2) - \Lambda = 8\pi G\rho \text{ and}$$
$$-k / a^2 - (\dot{a} / a)^2 - 2\ddot{a} / a + \Lambda = 8\pi G p \text{ which are equivalent to:}$$
$$\ddot{a} / a = -4\pi G (\rho + 3p) / 3 + \Lambda / 3 \text{ and}$$
$$\dot{\rho} = -3\dot{a} (\rho + p) / a$$

The cosmological constant where positive therefore causes an acceleration in the expansion of the universe and can be interpreted as a form of energy which has negative pressure equal in magnitude to its (positive) energy density such that $p = -\rho$ now referred to as dark energy.

During the course of the theoretical developments discussed above, Edwin Hubble in the late 1920s and 1930s made astronomical observations which showed that spiral nebulae were galaxies in the same sense as the Milky Way, by determining their distances using measurements of the brightness of Cepheid variable stars. He discovered a relationship between the redshift of a galaxy and its distance, now termed Hubble's Law, which indicated the galaxies were receding from our own at velocities proportional to their distance.

Thus it was concluded the universe was expanding, which Lemaitre when deriving FLRW models in 1927 had proposed started with the explosion of a primeval atom.

Two rival theories emerged, Hoyle's steady state theory in which new matter is created as the galaxies move away from each other and Gamow's theory based on the Lemaitre supposition. The latter won the argument when measurements of the cosmic wave background and radio source counts were made in the 1960s leading to the first version of a standard model by Penrose and Hawking based on the assumption of an initial singularity.

It is useful to note at this point that conformal diagrams can be employed to distinguish between the various de Sitter, FLRW models, Hoyle's steady state model and indeed more recent inflationary universe models which will be discussed in Section 7.4.4.

Conformal diagrams are plane representations of space-time drawn so that the space-time null lines in that plane are orientated at 45 degrees to the vertical and where infinity is represented as a finite boundary.

Cosmologists became increasingly confident in a timeline history of the universe in accordance with such models during the following decades, at least as far as that after the first fraction of a second following the big bang was concerned.

That first fraction became contentious as will be explained later but a general consensus has been achieved in respect of the remainder though even this has had to incorporate very significant observational data which imply the existence of dark matter and dark energy.

Dark matter is hypothesised in cosmology to account for effects that appear to be the results of mass which cannot be seen directly. It neither emits nor absorbs light or any other electromagnetic radiation. The existence and properties of dark matter are inferred from its gravitational effects on visible matter, radiation and the large-scale structure of the universe. Dark matter appears to constitute approximately 85 per cent of total matter.

Astrophysicists found discrepancies between the mass of large astronomical objects determined by gravitational effects and the mass calculated from the luminous matter they contain namely stars, gas and dust. It was first postulated in the 1930s to account for the orbital velocity of stars in the Milky Way and for missing mass in the orbital velocities of galaxies in clusters and in the 1960s for the rotational speeds of galaxies.

Since then other observations in the 1970s and 1980s involving gravitational lensing etc has resulted in a consensus that dark matter as a new form of matter exists, though some alternative theories to that of gravity have been proposed.

Dark matter in contrast to ordinary matter does not lose energy as it does not interact in any way other than through gravity. It is therefore non-baryonic in nature forming wraith like clouds around most galaxies. Galaxies and galaxy structures form because they lose energy relative to non-baryonic matter while the baryonic matter of which they are composed interacts via electromagnetic radiation and the strong interaction in particular. Dark matter can therefore be said to instigate the formation of galaxies and galactic structures. Galaxies have been found which appears to have only conventional baryonic matter without the presence of the dark variety which is clearly unevenly distributed.

Dark Energy is a hypothetical form of energy proposed in the 1990s onwards to explain observations that the universe is expanding at an accelerating rate.

Together with dark matter it constitutes approximately 95 per cent of the total content of the universe. It is considered that the density of dark energy is very low but because it is uniformly spread throughout space it comes to dominate the mass energy of the universe. Two proposed forms for constraining the equation of state of dark energy are the cosmological constant as already discussed and quintessence in the form of a scalar field, a dynamic quantity whose energy density can change with space and time; whatever it does not react with any of the fundamental forces except that of gravity. As stated earlier this involves a strong negative pressure acting repulsively to explain the accelerating expansion of the universe.

The even distribution of dark energy is in stark contrast to dark matter.

The present standard cosmological model is referred to accordingly as the Lambda CDM model, where the Lambda element is included to allow for observational evidence that the universe is undergoing accelerating expansion and the CDM (cold dark matter) element to explain the relatively rapid aggregation of matter to form stars and galaxies.

It should be noted that cosmologists are not able to explain the phenomena they observe in terms of conventional forms of energy, nor is there a way to define the total energy of the universe in terms of the accepted theory of gravity, general relativity.

In this regard dark energy is considered to permeate all space, possibly the energy of virtual particles which may exist in a vacuum due to the uncertainty principle.

It should be further noted that based on thermodynamic principles the cosmos witnesses a battle between radiation and matter for dominance.

The former involves relativistic particles with zero or negligible rest mass relative to their kinetic energy moving at or close to the speed of light. The latter involves non relativistic particles which have much higher rest mass than their energy and move much slower than the speed of light.

As the universe expands, both matter and radiation become diluted, meanwhile the universe cools down so the average energy per particle reduces. So radiation becomes weaker and dilutes faster than matter, the early universe therefore being radiation dominant which controlled the deceleration of expansion.

Later as the average energy per photon reduces, matter becomes dominant and dictates the rate of deceleration.

Ultimately as the expansion of the universe continues matter dilutes further and the cosmological constant becomes dominant in turn leading to an acceleration in that expansion.

Particle physics is important in the behaviour of the early universe when it was very hot and the average density was high. A scattering or decay process is important then if the relevant time scale is small or comparable to the time scale of the expansion.

The expansion timescale $1/H$, where H is the Hubble constant, is roughly equal to the age of the universe at that time.

Observations show that the universe began its expansion some 13.8×10^9 years ago. Since then it has evolved through three main phases.

The first phase, still poorly understood, was the minute fraction of a second when the universe was so hot that particles had higher energies than can be reproduced by current particle accelerators on earth, so educated guesswork has to be applied.

In the second phase the evolution proceeded in accordance known high energy physics, when first protons, electrons and neutrons formed, then nuclei and finally atoms. With the formation of neutral hydrogen the cosmic wave background was emitted.

The third phase produced structural formation, the aggregation of matter into first stars and quasars, then galaxies, clusters of galaxies and super clusters.

It is expected that the universe will continue expanding indefinitely in to the future based on the standard Lambda CDM model.

The early hot universe appears to be well explained by the big bang from roughly 10^{-30} seconds onwards, but there are several problems.

There is no compelling reason for the universe to be flat, homogeneous and isotropic.

Grand unified theories of particle physics suggest there should be magnetic monopoles in the universe but they have not been found.

It is considered by many cosmologists that these problems can be resolved by a very brief period of cosmic inflation which drives the universe in to a state of flatness, smooths out anisotropies and inhomogeneities to the observed level and exponentially dilutes monopoles. The physical model behind inflation has not however been confirmed by particle physics and it is difficult to reconcile inflation with quantum field theory. Some cosmologists think that string theory and brane cosmology will provide an alternative.

A further problem is why the universe contains more matter than antimatter. If it were not so there would be evidence of X rays and gamma rays as a result of annihilation events, but they are not seen.

Therefore some early process in the universe referred to as baryogenesis must have created an excess of matter over antimatter.

Baryogenesis requires a violation of particle physics symmetry called CP symmetry between matter and antimatter, but the violations measured in particle accelerators have not been sufficient to account for baryon asymmetry.

Big bang nucleosynthesis is the theory of formation of the elements in the early universe which finished when it was merely three minutes old and the temperature dropped below that at which nuclear fusion could occur. It produced only the lightest elements, starting with hydrogen ions it produced deuterium, helium-4 and lithium.

The cosmic wave background is radiation left over from decoupling after the recombination when neutral atoms were first formed. At this point radiation produced by the big bang stopped Thomson scattering from charged ions. The radiation first observed in 1965 has a perfect black body spectrum, has a temperature of 2.7 degrees Kelvin today and is isotropic to within one part in 10^5.

Studies of the cosmic wave background as measured by many satellite experiments such as COBE, WMAP and Planck have proved of enormous benefit in providing observational data as a spur to the progression of the standard cosmological model.

Other astronomical observations by telescope are providing information to support the study of hierarchical structure formation of aggregated matter in which the smallest objects form first while the largest objects such as super clusters are still forming.

7.4.4. The introduction of an inflationary phase in the very early universe to explain the uniformity now apparent at cosmological scales.

The alternative concept of conformal cyclic cosmology.

The inflationary phase was introduced to explain what were considered by some cosmologists to be difficulties with the original standard cosmological model as defined in the 1980s.

In particular these were the total absence of magnetic monopoles from observational astronomy predicted by grand unification theories, the extreme uniformity of the observed universe, namely the fact that it is homogeneous (spatial translation symmetry) and isotropic (spatial rotation symmetry) and that it was close to being spatially flat.

The uniformity question took two forms, termed the horizon problem and the smoothness problem. The horizon problem was concerned with why the observed temperature in any direction was extremely close (to at least one part in 10^5). This could be explained by a thermalisation process but only if the different parts of the universe were in communication with each other, a condition which is not possible in conventional cosmological models because at the time of decoupling when the CMB was formed, these parts were too distant from one another.

The smoothness problem was concerned with the observed uniformity of the matter distribution. If the initial generic state of the universe was irregular, say through warping and curvature from quantum effects, why should it take up the appearance of that which is seen now.

Finally the observed universe is flat ($K = 0$) or at least very close to it and the question arose as to why, a question termed the flatness problem.

To remove these problems some cosmologists notably the physicists Guth (Ref 7.4.1) and Linde (Ref 7.4.2) introduced the idea of an inflationary period in the very early universe between 10^{-36} and 10^{-32} seconds after the big bang, involving a dramatic expansion of space typically 10^{26} orders of magnitude or more in that period.

In the conventional view the cosmological horizon analogous to the horizon associated with the curvature of the earth is that which marks the boundary of the universe that an observer can see. Light emitted by objects beyond the cosmological horizon never reach the observer because the space between is expanding too rapidly.

In the standard cosmological model without inflation the cosmological horizon moves out bringing new regions in to view the observable universe being merely a patch in the vastly larger unobservable universe. Yet these new regions seem no different from that seen earlier. So how do they have the same temperature and curvature etc as those previously observed since they could not have been in communication with the past light cone beforehand.

Inflation like the cavalry comes riding to the rescue by postulating that all the regions can be related to an earlier era with a very large cosmological constant (or vacuum energy), vastly greater than that now associated with the more recent accelerated expansion of the universe, namely dark energy.

In this situation the cosmological horizon instead of moving outward stays fixed. For any one observer the horizon is at a constant distance so with exponentially expanding space two nearby observers are separated very quickly exceeding the limits of communication.

The stuff of the universe moves quickly beyond the fixed horizon and becomes homogeneous equally quickly.

As the inflationary field slowly relaxes to the vacuum the cosmological constant falls to zero and the expansion of space returns to the normal rate. The new regions that come in to view now are exactly the same regions as were pushed beyond the cosmological horizon during inflation and are at the same temperature and curvature etc as they came from the same original patch of space. This also implies that ordinary matter plus dark matter and residual vacuum energy adds up to the critical density at which space is flat.

This after all is the cosmological principle which implies the universe should look the same to all observers within it.

In exponentially expanding space two inertial observers move further apart with accelerating velocity. In stationary coordinates for one observer a patch of an inflating universe has a polar metric $ds^2 = (1 - Ar^2)\,dt^2 - 1/(1 - Ar^2)\,dr^2 - r^2\,(d\theta^2 + \sin^2\theta\,d\phi^2)$.

Now there is a spherically symmetric vacuum solution to Einstein's equations called the Schwarzschild metric of the form:

$ds^2 = (1 - 2GM/r)\,dt^2 - 1/(1 - 2GM/r)\,dr^2 - r^2\,(d\theta^2 + \sin^2\theta\,d\phi^2)$, which corresponds to the Minkowski metric when the gravitating mass is zero.

The Schwarzschild metric describes the gravitational field outside a spherical mass M on the assumption that the electric charge and angular momentum of the mass is zero which corresponds to a static black hole.

Allowing for a non zero cosmological constant the Schwarzschild metric right hand side becomes:

$$(1 - 2GM/r - \Lambda r^2/3)\, dt^2 - 1/(1 - 2GM/r - \Lambda r^2/3)\, dr^2 - r^2(d\theta^2 + \sin^2\theta\, d\phi^2)$$

If the mass is zero this is another way of writing the metric of de Sitter space-time. To sustain this space-time there must be a cosmological constant or vacuum energy proportional to Λ everywhere, with an equation of state $p = -\rho$ and a scale factor proportional to exp Ht, where H is the Hubble parameter.

The associated period of accelerated expansion is called inflation, during which Λ can stay approximately constant and could be regarded as the inverse of a black hole.

The general idea is that the inflation smooths out the inhomogeneities, anisotropies and curvature of space and pushes the universe in to a very simple state dominated by what is termed the inflation field which may have quantum fluctuations now seen as large-scale structures such as galaxies etc.

It is also postulated that inflation dilutes exotic particles such as magnetic monopoles to the point where they become so rare that they cannot be detected in the visible universe.

Together these effects are referred to as an inflationary no hair theorem analogous to that for black holes (See section 7.4.7).

In an expanding universe energy densities fall as the volume of the universe increases, ordinary matter density as the inverse of the volume, down by a factor of eight when linear dimensions double and radiation density by a factor of sixteen. During inflation the energy density in the inflation field is roughly constant but that in the inhomogeneities etc is falling and through inflation becomes negligible, leaving an empty, flat and symmetric universe filled with radiation as inflation ends. In order to reproduce the observable universe it is considered inflation must expand the universe by at least 10^{26} orders of magnitude.

Inflation is a period of supercooled expansion when the temperature drops by a factor of 10^5 or so, a temperature maintained in the inflationary period. When inflation ends the temperature returns to the pre-inflationary temperature in a reheating or thermalisation process the large potential energy of the inflation field decaying in to particles filling the universe with electromagnetic radiation, though the process is poorly understood.

It is postulated by supporters of inflationary models that it solves why the universe appears as observed today.

The horizon problem in standard models without inflation was why the universe is homogeneous and isotropic. Gravitational expansion would not give sufficient time for the early universe to achieve equilibrium because widely separated regions move apart faster than the speed of light and have therefore never been in causal contact.

The flatness problem was why the density of matter in the universe was comparable to the critical density necessary for a flat universe.

The contribution of spatial curvature to the expansion of the universe could not be much greater than the contribution of matter, but as the universe expands the curvature redshifts away more slowly than matter and radiation.

This presents a fine-tuning problem extraordinaire because extrapolating in to the past the contribution of curvature to the universe must be exponentially small.

There are hardly any choices of initial density which would lead to the observed universe, the other choices involving either swift recollapse or rapid expansion and cooling, in neither case would any of the observed evolution including stars and galaxies etc have ever occurred.

Inflation enthusiasts claim the fine-tuning problem disappears as Ω the critical density is driven to unity, since the curvature k decreases during inflation, while $a^2 H^2$ increases dramatically, H being the Hubble Parameter and the cosmic scale factor.

Another way of looking at this is using $1/Ha = r$, where r is the comoving radius and deriving $d(1/Ha)/dt < 0$, gives $dr/dt < 0$. So the comoving radius shrinks dramatically during inflation remembering that the comoving coordinates are at rest with respect to the expansion. Causal regions in the early universe are stretched to regions much larger than the Hubble distance when the scale factor evolves at superluminal speeds whereas the cosmological horizon is approximately constant.

While the latter expands at the speed of light this is insignificant in comparison with the evolution of the scale factor.

Now in the explanation to date it was merely assumed that a process of inflation occurred but without explanation of a mechanism by which it would do so.

In other words the inflation field referred to above needs a driver and to cater for this physicists introduced a scalar field in which the early universe would be dominated by in effective potential $V(\phi)$ for the scalar quantum field ϕ.

In particle physics a scalar field is used to represent zero spin particles such as bosons. It is unchanged under coordinate transformations and in a homogeneous universe is a function of time alone. Scalar fields are a crucial ingredient for spontaneous symmetry breaking in particle physics including super symmetry and grand unification theories. The Higgs field which breaks electroweak symmetry is an example and is the only fundamental scalar field that has yet been observed.

Expressions for the effective energy density and pressure of a homogeneous scalar field ϕ are obtained by comparison of the energy momentum tensor of the scalar field with that of a perfect fluid as:

$\rho_0 = \frac{1}{2}\ \dot{\phi}^2 + V(\phi)$ and $p_0 = \frac{1}{2}\ \dot{\phi}^2 - V(\phi)$, in which the first terms correspond to kinetic energy and the second terms to potential energy.

The potential energy measures how much internal energy is associated with a particular field value. Like all systems scalar fields try to minimize this energy but as in the case of inflation may not be very efficient at doing so. A scalar field cannot be described by an equation of state and supporters of inflation tend to regard $V(\phi)$ as a function to be chosen arbitrarily, different choices corresponding to different models of inflation of which it has to be said there are many.

The equations above can be substituted in the Friedmann equations and the fluid equation referred to earlier giving inflation when the potential energy dominates.

This should apply if the potential is flat enough and the scalar field rolls slowly, termed slow roll inflation which occurs when the scalar field rolls down a potential energy hill; when it gets to the bottom inflation ends and reheating occurs.

Penrose (Ref 7.3.2 and 7.4.3) and one of the architects of inflation theory Steinhardt (Ref 7.4.4) among others have cast doubt on the principles involved though it has measured up well when it comes to the correlation of predictions with observations.

It is not evident that inflation explains the extraordinarily low entropy at the big bang required by the second law of thermodynamics.

The horizon, smoothness and flatness problems can according to Penrose be dealt with by fine tuning the standard big bang model.

The horizon problem is according to inflation resolved by a thermalisation process, but thermalisation represents an increase in entropy meaning that the universe would have to be even more special beforehand.

The uniformity and flatness issues as resolved by inflation imply that the early universe before inflation must have been in some unruly generic state yet again this is in contravention of the second law. Furthermore such a state is likely to be fractal in nature.

The requirement for a driver in the form of a scalar field is another weakness in inflation theory for the reason that there are so many possible forms of scalar field.

The precise shape of the potential energy curve is controlled by a numerical parameter that can take on any value whatsoever, yet only a narrow range of values could produce the observed temperature variation.

Linde (Ref 7.4.2) himself waxes lyrical about the diversity of inflation, natural inflation, extended inflation, hybrid inflation, chaotic inflation which variously can produce any outcome from a multitude of premises.

According to Steinhardt there is bad inflation whose outcome conflicts with observations and good inflation compatible with them, but that not only is bad inflation more likely than good but no inflation is more likely than either.

Indeed Penrose has considered all possible configurations of the inflation and gravitational fields and found that some lead to inflation and some to a uniform flat universe directly without inflation. Obtaining a flat universe was unlikely overall, but his shocking conclusion is that obtaining a flat universe without inflation is more likely than with inflation by a factor of 10^{100}.

Linde's universe of chaotic inflation described in Section 4.7 and for which a limited theoretical background has been given above, forms a typical scientific picture of how the observed universe may have formed.

It is again useful to note that the use of conformal diagrams referred to in Section 7.4.3 reveals that the de Sitter model can be cut to give the steady state model of the universe. Then a greatly inflating portion of the steady state model between two constant time lines is inserted in the $K = 0$ FLRW model in place of a small constant time interval to obtain the inflationary universe model.

So the latter is positively related to the rejected steady state version and may be considered a steady state model in a state of chaos!

Other associated scientific versions including that of parallel universes and the concept of the multiverse will be considered in Section 7.4.8.

An alternative cosmological model termed Conformal Cyclic Cosmology has been put forward by Penrose (Ref 7.4.3), in which the universe passes through infinite cycles with the future time like infinity of a given cycle being related to the big bang singularity of the next. It is important to emphasise that the theory proposed sets out in the first instance to explain the extraordinarily low entropy at the big bang.

This involves comparing the volume which must have applied at the big bang to the volume of phase space corresponding to the state of the visible universe now.

Starting from Boltzmann's law $S = k \log v$, where k is Boltzmann's constant and v is the volume of the total phase space then our light cone contains some 10^{80} baryons basically protons and neutrons, so the volume can be evaluated by estimating the entropy that applies when that number of baryons collapses in to a black hole (see Section 7.4.7).

That volume given by the Bekenstein-Hawking formula described in Section 7.4.7 and allowing for the contribution of dark matter gives a value of the order of 10^{123}. Then from Boltzmann's law the phase space volume exceeds that of the volume at the big bang by a factor of 10 to the power 10^{123}.

Penrose increases this figure to 10 to the power 10^{124} to allow for dark matter.

The magnitude of this figure is so extraordinary that it must be explained and cannot be ignored.

An open FLRW space-time universe is described by the big bang and the subsequent unlimited expansion. Penrose's thesis is that the conformal boundary of that space-time can be related to the conformal boundary of the next cycle by appropriate conformal rescaling. The next cycle is spacelike based on a positive cosmological constant and so on in sequence, such cycles being referred to by Penrose as aeons.

The state of play in the world of particle physics will be examined later, but the conformal cosmological model has significant features which impact on that subject matter. In particular that the particles termed bosons, the carriers of force in the particle world are conformally invariant; so that they have the same properties in the rescaled aeon as in the original one, the light cone structures being preserved under conformal rescalings. For such particles there is no boundary between aeons, just a spacelike surface which offers no impediment to passage. The other type of particle the matter particles termed fermions are restricted to a given aeon.

This provides a solution to the black hole paradox discussed in Section 7.4.7, assuming the fermions to be irreversibly converted in to radiation during black hole evaporation so as to maintain the smooth boundary between aeons.

It is part of the theory that massive particles are to a major degree annihilated or decay, even electrons must decay or lose charge / mass.

It is also part of the theory that the boundary conditions between aeons satisfy what is referred to as the Weyl Curvature Hypothesis which provides the essential low entropy past demanded by the second law of thermodynamics. Finally, Penrose considers that to some degree gravitational radiation survives passage between aeons, which could explain the observed cosmic acceleration without the involvement of dark energy.

The relationship of this theory with others arising from subject matter in Sections 7.4.5, 7.4.6 and 7.4.7 is discussed in Section 7.4.8.

7.4.5. Why is classical physics inconsistent with quantum physics?

Schrödinger's wave equation.

The double slit experiment.

The rules of the quantum world.

The classical/quantum dichotomy and the EPR Paradox.

Bell's Inequality Theorem.

Having conducted the earlier review in relation to classical physics, a similar process respect of physics at the quantum level is necessary. In the first place there is a very significant dichotomy between classical physics and that at the quantum level, the nature of which will now be examined, but first it is necessary to provide some essential historical background relating to developments in physics from the 18th century onwards.

Important experiments concerning the nature of light carried out by the scientist Thomas Young in the early nineteenth century (Young's slit experiment) had demonstrated that light could be regarded as being wavelike. However further experiments concerning black body radiation and the photoelectric effect in the late nineteenth century, the later work of the physicists Ludwig Boltzmann and Max Planck and then Einstein in the early twentieth century demonstrated the particle nature of light. It became clear that light had a dual nature. Classical physics is incompatible with this duality.

The problem has its roots in the assumption that two kinds of objects exist, firstly particles described by a finite number of parameters, three positions and three momenta and secondly fields requiring an infinite number of parameters. Equipartition of energy dictates that in a state of equilibrium all energy is transferred to the fields and none is left for the particles. Particles would not be stable; electrons orbiting the nuclei of atoms would collapse on to those nuclei omitting electromagnetic waves. In fact they do so but only at specific discrete frequencies which have no basis in classical theory.

There was a problem with black body radiation. Considering an object at some definite temperature with electromagnetic radiation in equilibrium with particles, Rayleigh and Jeans calculated in 1900 that all the energy would be taken up by the field to endlessly higher frequencies. Actual observation showed that at low frequencies of field oscillation the energy was as predicted but at a specific frequency for a given temperature the energy falls off to zero as the frequencies increase.

Planck explained the discrepancy by proposing that electromagnetic oscillations occur only in quanta with energy $E = hv$, where h is Planck's constant.

The photoelectric effect concerns the observation that many metals emit electrons when light shines on them. According to classical electromagnetic theory this is due to transfer of energy to the electrons, such that an alteration to either the amplitude or wavelength of light would produce changes in the rate of emission of the particles.

A sufficiently dim light would result in a lag time between the initial shining of the light and emission.

Neither of these predictions proved correct. In fact electrons are only emitted when light exceeds a threshold frequency below which none are emitted regardless of the wavelength or length of exposure.

Einstein then made the proposal that the electromagnetic field is not a wave propagating through space but rather a collection of discrete wave packets (photons) each with energy hv, so light also consisted of particles.

Clearly the dichotomy between particles and waves assumed in classical physics was not in accordance experimental observations.

This was followed by the arrival on the scene of the physicist Niels Bohr who worked with Rutherford on the nature of atomic structure. Bohr developed the idea that the energies of electrons in that structure were quantised such that electrons could only form discrete orbits about the central nucleus. The angular momentum of electrons in orbit only occurred in multiple of $h/2\pi$ or $\hbar$ (Dirac's symbol).

De Broglie then proposed that particles could be accompanied by matter waves with a wavelength dependent on the mass of the particle, the greater the mass the shorter the wavelength. The central idea was that there was a localised particle carried by a wave packet or wave pulse consisting of many different superimposed waves which cancelled out except in the local region where the particle was to be found.

This proposal was considered at the time to be too radical by his contemporaries particularly Bohr.

Nevertheless Erwin Schrödinger then used the de Broglie idea of matter waves in an effort to explain Bohr's model of the atom. Unlike de Broglie however Schrödinger dispensed with the idea of matter particles having associated waves and decided only the waves were real. The result was the Schrödinger wave equation of which a fundamental constituent is what is termed the wave function, a concept purely mathematical in nature.

His key insight was to use Fourier series to do so, the basis being to express any mathematical function as the sum of an infinite series of other periodic (harmonic) functions. Finding in effect the eigenfunctions and eigenvalues, the correct functions and amplitudes of each that added together by superposition reproduces the desired solution.

The wave function for the system was replaced by an infinite series, i.e. the wave functions of the individual states, these being natural harmonics of each other.

The mathematical background to Schrödinger's equation is given in Section 7.3.7.

The momentum states associated with Schrödinger's equation can be pictured as non-localised helixes, far removed from the conventional concept of particles. However the particle picture can be restored when the wave function peaks sharply at some position, represented by multiplying the momentum state by the Gaussian function $\exp(-x^2)$.

The mathematical aspects of dealing with classical infinitely dimensional vector spaces and their application to quantum mechanics involving infinitely dimensional complex vector spaces are briefly referred to in Sections 7.3.4 and 7.3.7.

Confirmation of the wave nature of electrons was obtained shortly afterwards when experiments which passed particles through dual slits in a screen produced interference patterns on a second screen placed behind the first. At about the same time Heisenberg concluded only quantities which were measurable such as the energies of electrons had physical meaning and produced a theory of quantum mechanics that was in due course shown by the theoretical physicists Pauli and later Dirac to be equivalent to that derived by Schrödinger.

The wave function referred to above is also probabilistic in that it only gives the probability of an electron for example being found in a particular location.

This indeterminacy means that it is not possible to determine precisely everything about a quantum system at the same time, from which Heisenberg introduced the concept of the uncertainty principle. It is not possible to know at the same time the location and velocity of any quantum entity since the wave function consists of a position wave function and a momentum wave function. If one is known the other can be evaluated but a localised version of that one gives rise to a spread-out version of the other.

The double slit experiment

The double slit experiment encapsulates the dilemma in which scientists found themselves.

In the basic version of this experiment a coherent light source such as a laser beam illuminates a plate pierced by two parallel slits and the light passing through the slits is observed on a screen behind the plate.

The wave nature of light causes the light waves passing through the two slits to interfere producing light and dark bands on the screen, a result that would not be expected if light consisted of classical particles. However the light is always found to be absorbed at the screen at discrete points i.e. as individual particles (not waves), the interference pattern appearing by means of the varying density of the particle hits on the screen.

Furthermore versions of the experiment that include detectors at the slits find that each detected photon passes through one slit (as would a classical particle) and not through both slits (as would a wave). The detection of individual discrete impacts is observed to be inherently probabilistic.

These results demonstrate wave particle duality and are inexplicable using classical mechanics. The results apply to bosons and fermions, atoms and even some molecules.

The probability of detection is the square of the amplitude of the wave and can be calculated on that basis. The particles do not arrive in a predictable order, so knowing where all the previous particles appeared and in what order tells nothing about where future particles will be detected. If there is a cancelation of waves at some point that does not mean that a particle disappears; it will appear somewhere else.

Ways of including hidden variables in local realistic terms to account for the location of each individual impact, fail to explain experiments involving more complicated (two or more) particle systems, which continue to exhibit complex super positioning even over classically large distances.

The rules of the quantum state

Clear features of the quantum state as opposed to the classical state emerge.

1. The replacement of classical probabilities in terms of real numbers by complex number probabilities called probability amplitudes, to explain mathematically the results of the double slit experiment.

    At the classical level in a situation where either event A or event B must happen there is no probability. Where perfect knowledge about the system is not available the probability of event A or event B happening must add to unity.

    In the quantum world however both event A and event B receive a complex number describing how much of the system is in each event. The squared absolute values of the complex numbers correspond to the probabilities of A or B being seen if the system is observed when those squared values must sum to one.

    The probability of finding a particle at an exact position is zero.

2. From the quantum viewpoint every single position a particle may have is a possibility available to it, these possibilities being combined together with complex number weightings collectively termed the wave function, with a specific value for any given position. A wave function is basically a function from 3D space plus time to the complex plane. The value of momentum of a particle can be associated with the wave

function by defining it in terms of a harmonic analysis, the pure tones of different frequencies corresponding to the different possible momentum values a particle may have. Momentum states can be pictured as helixes or corkscrews, those tightly wound having large momenta reducing to zero winding (or a straight line) for zero momentum.

In mathematical terms the position and momentum states are Fourier transforms of one another. A position state corresponds to a very sharply peaked wave (delta) function which transforms to a helix in momentum space.

Vice versa a helix in position space transforms to a delta function in momentum space.

This picture constitutes a central feature of quantum mechanics, namely that of quantum linear superposition, involving the addition of quantum states to give other quantum states.

3.   It is not possible to measure (i.e. to magnify to the classical level) both position and momentum of a particle at the same time. In fact there is an absolute limit on the product of these two properties defined by Heisenberg's Uncertainty Principle. The more accurately position is measured the less accurate the momentum.

4.   A mathematical equation (Schrödinger's equation) is used to describe how the wave function develops with time. Once the wave function is specified at any one time, it represents a completely deterministic continuous evolution until a measurement is made, when instead a completely different procedure involving the formation of the squared moduli of the quantum amplitudes is needed to obtain the classical probabilities of the real four-dimensional world. This procedure violates complex super positioning, is indeterminate and discontinuous.

5.   Mathematically the concept of Hilbert space is used to represent the entire quantum state of a system just as phase space represents an entire classical system. Each Hilbert space dimension corresponds to a different independent physical

state. Infinitely many different independent dimensions in Hilbert space correspond to infinitely many particle positions. Momentum states can be represented in the same Hilbert space as combinations of position states, with momentum state axes tilted with respect to position state axes.

6. The general rule for measurement by observation is that different aspects of the quantum system that can be magnified to the classical level and between which the system must choose, must always be orthogonal.

    For a complete measurement the alternatives consist of a set of orthogonal basis vectors, such that every vector in the Hilbert space can be uniquely and linearly expressed in terms of them. For a position measurement of a single particle the basis vectors define the position and momentum axes.

    After measurement the state of the system jumps to one of the axes of the set determined by the measurement, governed only by probability, this apparently random choice giving probability values corresponding to the squared moduli of the probability amplitudes. Fixing the scales of the quantum state and the basis vectors to be of unit length, provides mutually orthogonal and normalised vectors termed orthonormal.

7. Spin is a characteristic feature of quantum mechanics and is a measure of the intrinsic angular momentum of a particle. Whereas a classical object can spin in many ways, the amount of spin for a particle is always the same for that particular type of particle, furthermore only the direction of the axis of spin varies.

    For fermions (electrons, protons and neutrons) the spin is always $\hbar/2$, where $\hbar$ = Planck's Constant$/2\pi$. Such particles have spins of an odd multiple of $\hbar/2$ and rotation through 360 degrees sends the quantum state vector not to the same value again, but to minus that value.

    Other particles are called bosons and have a spin which is an even multiple of $\hbar/2$, On rotation through 360 degrees a boson state vector goes back to itself, not minus that value.

Considering an electron at rest and its state of spin only, the relevant Hilbert space is two dimensional, up and down, where these denote spin in an upward or downward sense respectively, these states being orthogonal to one another and normalised. In this case orthogonal corresponds to diametrically opposite directions rather than directions at right angles.

So there is some direction in which the spin is precisely defined, though its direction is not known.

Massless photons (bosons) also possess spin but as they travel at light speed their spin cannot be about a fixed point, so the spin axis is always in the direction of motion.

8.   Where two particles are involved the relevant description applying is one function of two positions as opposed to a two functions of position.

The latter might involve just 10 positions say, ten complex numbers specify the state of the particle; therefore 20 numbers for the two particles.

A function of two positions requires an amplitude for each pair of positions, giving a total of 100 different pairs.

Three particles would on a similar basis involve 1,000 numbers and so on.

9.   A further feature of quantum mechanics is the necessity for all particles of a specific type to be precisely identical, whether fermions or bosons.

However no two fermions can be in the same state, as a quantum state of zero would then apply which is not allowed, a limitation called Pauli's exclusion principle.

It is a fundamental property of matter, the principle constituent of which are fermions.

Bosons are not subject to this principle and any number of bosons can occupy the same state.

It is impossible to copy a quantum state while leaving the original state intact.

It can be copied if that original state is destroyed.

Einstein adamantly refused to accept the principle of indeterminacy as fundamental to nature and maintained that while some of the predictions of quantum theory were correct, in general terms the theory was incomplete, that some of the parameters which determine the movement of particles had not yet been specified and that if these 'hidden' variables were known fully deterministic trajectories could be defined.

There was the further matter of distinguishing between indeterminacy and indeterminism.

The latter meaning that knowledge of the state of a system even a particle at a given time did not imply that its future state could be known with certainty; this in opposition to the belief in determinism which assumed that knowledge of any state at a given time enabled in principle the prediction of any future state. So indeterminacy implied an even greater move away from determinism, an implication Einstein was not prepared to accept.

All of this led to the bitter dispute between Einstein and Bohr over the relationship between determinacy and indeterminacy, a dispute never resolved, and also to the rejection of de Broglie's theory for a generation before it resurfaced in the 1950s, when Bohm (Ref 7.4.5, 7.4.6 and 7.4.7) provided an interpretation of quantum theory directly related to it.

This indeed gave a deterministic description of the positions of the particles and their trajectories but simultaneously implied non-locality namely that an event occurring in a region A can have a physical effect in region B instantaneously no matter how far apart those regions might be. Quantum theory is in effect a theory of non-locality but the Bohm interpretation carried with it the requirement that in a specific reference frame non-local connections were instantaneous and therefore exceeded the speed of light, a condition which Einstein rejected.

The Bohm interpretation is that such non-local connections cannot transmit information, the instantaneous influence can only pass between events that are related by a common cause and hence a signal must be redefined as that which transmits information.

The principle that the maximum velocity of a signal so defined is finite is then preserved, that velocity being the speed of light.

Much of the above arguments related to the quantum state only summarised here receive a far more detailed and comprehensive description in a book with excellent colour illustrations ideally suited to the layman by J. Al- Khalili, (Ref 7.4.8).

It is necessary to probe deeper into the wave particle dichotomy with this background in mind. In the context of quantum theory this can be considered together with those concerning continuity, causality and locality as defined earlier in section 2.4, where it was also stated that all these properties are basic to both classical physics and to its extension namely relativistic physics where four dimensional space-time is taken as absolute.

In quantum physics the determinate continuous state (as given by Schrödinger's wave equation) continues to apply, until attempts are made by conscious entities to observe it! In observing the quantum state however these common-sense rules are effectively overturned and replaced by non-continuous, non-causal, non-local concepts based on probability, all set in a multidimensional environment, but again forming an undivided whole.

The classical/quantum dichotomy and the EPR paradox

Considering the dichotomy between classical and quantum physics that emerged in the 1920s and discussed above, it is not surprising that the two main supporters of each namely Einstein and Bohr fell out over the interpretation of the quantum state.

This is despite the fact that Einstein's ideas concerning the photon were fundamental to the early development of quantum theory as was the concept of the boson in association with the physicist Bose. As has been stated earlier the indeterminism arising from the state reduction to the classical level and even more the probabilistic indeterminacy of the theory flatly opposed the very foundation of the theory of relativity at the heart of Einstein's philosophy.

Bohr on the other hand had no such concerns as he regarded the quantum state of the system between measurements as having no physical reality merely a mathematical formalism based on the Schrödinger equation, in other words the sum of available knowledge of the system. In fact Bohr regarded the wave function as subjective and the classical level as the only objective reality.

The argument between Bohr and Einstein culminated in the challenge by the latter to his opponent involving what is referred to as the Einstein, Podolsky and Rosen (EPR) paradox.

Two particles A and B with two orthogonal positions, for both are correlated under the rules of quantum mechanics, so their combined state is not merely the product (and) of states A and B but also the superposition (plus) of these states. Particle A has two orthogonal states c and d, while particle B has two orthogonal states e and f.

The state of the system with super positioning gives c d + e f.

A measurement on A causes B to jump to d or alternatively f. This is regardless of how far distant A is from B. So as both particles jump simultaneously the principle of locality is broken.

For two classical alternatives A and B this would not apply since two real separate states apply irrespective of distance apart and no jumping occurs.

This classical viewpoint cannot be applied to the quantum case for the reason that it is possible to delay the choice of measurement on A until A and B are well separated

The EPR paradox is that B is instantaneously influenced by that choice on A. So some non-local influence between the particles informs the undisturbed particle of the measurement made on the other. The wave function is then viewed as in some sense collapsing and the second particle instantly assumes the relevant corresponding state.

Bohm's description of this can be found in Bohm (Ref 2.5.1) and Bohm and Hiley (Ref 4.5.5).

Bell's inequality theorem

In 1964 the physicist John Bell (Ref 3.1.1) working at CERN discovered a way of resolving whether Einstein's insistence on a local hidden variables theory underlying the quantum state was correct, or whether the quantum non local theory would prevail.

The result was Bell's inequality theorem which had indeed a profound impact on the debate. He derived a formula showing that if Einstein were right there would have to be a maximum amount of correlation between the two particles. Bell's inequality shows that correlation in the quantum world is greater than that in the classical world and experiment confirms this.

Penrose (Ref 7.3.1) gives a description of the above, in a manner well suited to the layman in these matters, based on the physicist David Mermin's version of the EPR paradox and Bell's theorem.

This description will not be repeated here but it is apparent that local realistic theories cannot explain the level of correlation which quantum probabilities provide.

Experiments have been carried out to verify these theoretical conclusions using the polarisation properties of pairs of photons in place of spin one half massive particles like the electron. In particular the physicist Alain Aspect and colleagues (Ref 4.3.2) measured the polarisation of such pairs in various combinations of direction, in which the 'decisions' as to which way to measure these polarisations were only made after the photons were in flight, confirming the non-causal character of the quantum world. These experiments and those of others since have confirmed the predictions of quantum theory, one aspect of which concerns the validity of special relativity, since non-causal implies that in some sense effects arise which can travel faster than light.

If special relativity was to be compromised absurd results would arise, if signals could be sent faster than light. In fact the non-local influences cannot be used to send signals i.e. information as was demonstrated by Ghirardi, Rimini and Weber (Ref 7.4.9).

7.4.6. Quantum field theory and gauge theories.

The cosmological constant problem.

The Standard Model of particle physics.

Zero point energy.

The Standard Model chronological sequence.

The problems involved in formulating a theory of quantum gravity, with reference to string theory and loop quantum gravity.

Quantum field theory (QFT) is the framework based on Lagrangian principles used to construct quantum mechanical models of subatomic particles. The relevant mathematical procedures combine classical fields and special relativity with quantum mechanics maintaining the concepts of quantum point particles and locality. It does not incorporate General Relativity and is therefore expected to be replaced ultimately by a more fundamental theory still to be discovered.

Dirac in 1927 established a systematic procedure for transferring the characteristic quantum phenomenon of discreteness of physical quantities from the quantum mechanical treatment of particles to a corresponding treatment of fields. Centred on the Dirac equation a relativistic theory of the electron was also established (See Section 7.3.8). It was the Dirac equation and the extension of the non-relativistic particle concept described in Section 7.4.5, to a concept which included for special relativity and inevitably, the necessary fields associated with relativity, which gave rise to quantum field theory (QFT). The wave function provides additional degrees of freedom which correlate with the concept of particle spin.

The original element of QFT was quantum electrodynamics (QED) involving the quantisation of electromagnetic fields, treating photons as quantised excitations or quanta of such fields, which influenced the electromagnetic interaction between electrons and protons. Further advances involved the discovery of the neutron and the associated strong force between protons and neutrons which holds the atomic nucleus together and the weak force which governs radioactive decay and the neutrino produced by that decay.

New developments pioneered by the physicist Murray Gell-Mann enabled the extension of QED to QFT to describe not only the electromagnetic force but also the weak and strong interactions.

Quantum field theory is a quantum description of fields and particles and inevitably involves infinitely dimensional spaces and a potential sea of particles.

It is critical at this point to understand that special relativity implies, since energy and mass are equivalent, that individual particles are no longer conserved but can be created and destroyed.

This further implies a means for this process namely the presence of anti-particles such that for every additive quantum number of a particle there is an equivalent number reversed in sign applicable to the anti-particle. Therefore a particle and its anti-particle can be created out of energy; if these later come in to contact they correspondingly annihilate each other producing energy.

This leads to the concept of creation and annihilation operators and the associated concept of second quantisation, where the wave function ψ of some particle, itself becomes an operator, together with the presence of vacuum states where there are no particles. The wave functions that are introduced for the production of particles are of positive frequency while their complex conjugates have negative frequency, the annihilation operator being the Hermitian conjugate of the creation operator.

Consequently there also arises the concept of alternative vacua. A further consequence of the particle versus anti particle scenario is an important element of QFT, namely the CPT Theorem, details of which can be found on the internet. It basically implies that a mirror image of the universe can be obtained where all particles have a reflection (as in a mirror), all momenta can also be reversed (the inversion of time) and all matter replaced by antimatter (charge inversion).

Proofs of the theorem are based on the principles of Lorentz invariance (see Section 7.4.2) and locality (see section 7.4.5).

C corresponds to Charge conjugation i.e. the invariant physical replacement of particle by an anti-particle.

P corresponds to Parity, the spatial reflection aspect as in a mirror.

T corresponds to Time reversal, an interaction which is invariant when time is reversed from the normal forward flow.

The CPT Theorem states that physical interactions are invariant if all three operations are invariant.

Feynman diagrams may be interpreted as anti-particles being particles travelling backwards in time.

Weak interactions are not invariant when C or P is applied separately.

The mathematical framework of QFT is now named the Yang-Mills-Dirac-Higgs theory, in which a Lagrangian (see section 7.3) controls the dynamics and kinematics.

Each kind of particle is described in terms of a dynamical field that pervades space-time.

As with most field theories it involves postulating a set of symmetries of the system and then defining the most general renormalizable Lagrangian from its particle content that observes those symmetries. These symmetries are called gauge symmetries based on a theory called gauge symmetry (see below).

Poincare symmetry is postulated for all relativistic quantum field theories consisting of the translational and rotational symmetries and the inertial reference frame central to the theory of special relativity (see also section 7.3).

The strong, weak and electromagnetic interactions are each associated with what is termed a gauge symmetry factor. Using the most general Lagrangian the dynamics are found to depend on 19 parameters whose numerical values are determined by experiment.

The overall theory of physics governing the above is gauge theory (gauge meaning measurement), the number of units per given parameter. Theories of physics describe physical forces in terms of fields such as the electromagnetic field, the gravitational field and those describing forces between elementary particles. A general feature of these fields is that they cannot be directly measured, though some associated quantities such as charge, energy and velocity can.

In field theories different configurations of the unobservable fields can result in identical observable quantities. A transformation from one such configuration to another is called a gauge transformation, while the lack of change in the measurable quantities, despite the field being transformed is called gauge invariance. Since any kind of invariance under a field transformation is considered a symmetry, gauge invariance is called gauge symmetry and the theory a gauge theory.

For example in electromagnetism the electric and magnetic fields E and B are observable while the electrical potential and the magnetic vector potential are not. Under a gauge transformation in which a constant is added to the electric potential, no observable change occurs in E or B.

Gauge theories constrain the laws of physics, because all the changes induced by a gauge transformation have to cancel out when written in terms of observable quantities.

All forces (fundamental interactions) arise from the constraints imposed by local gauge symmetries, in which case the transformations vary from point to point in space and time. Perturbative quantum field theory describes forces in terms of force mediating particles called gauge bosons their nature being determined by the nature of the transformations.

Gauge transformations form symmetry groups or gauge groups for which there are group generators with a corresponding field (usually a vector field) called a gauge field.

Such groups have been described in the mathematics section 7.3.7.

If the symmetry group is commutative it is called an Abelian gauge theory and if non-commutative a non-Abelian Gauge theory (see also section 7.3.7).

Quantum electrodynamics is an Abelian gauge theory with the symmetry group U (1) and has one gauge field, the electromagnetic four potential, with one gauge boson.

The Standard Particle Model is a non-Abelian gauge theory with the symmetry group

U (1) x SU (2) x SU (3) and has a total of twelve-gauge bosons.

The cosmological constant problem

In quantum mechanics the vacuum itself should experience quantum fluctuations.

In the terms of general relativity those quantum fluctuations constitute energy that would add to the cosmological constant. In QFT there is a profusion of such fluctuations hence a calculated vacuum energy density 120 orders of magnitude greater than the observed cosmological constant. In general terms the vacuum energy is calculated by summing over all known quantum mechanical fields, taking into account all interactions and self-interactions other than those below a minimum wavelength where existing theories break down. Various ways have been proposed to get round what is termed the cosmological constant problem.

QFT predictions can instead base everything on light front quantisation. In this case causality and frame independence are explicit and therefore cannot be quantised. Vacuum fluctuations do not appear in the light front vacuum as all particles have positive momenta; as momentum is conserved, particles cannot couple to the light front vacuum since the sum of the momenta is then zero. This effectively renders the vacuum trivial.

String theories of physics which can be associated with chaotic inflation scenarios also get round this problem of the cosmological constant by incorporating an enormous range of vacua some of which will correlate with the observed value, as discussed later in this section.

In the standard QFT these problems of infinite vacuum fields are resolved by a redefinition of parameters using measured finite values, a process called renormalisation. Divergent physical predictions are avoided (in effect neutralised) by ensuring they do not influence empirical statements, for practical purposes all infinities are absorbed into a redefinition of a finite number of coupling constants (mass charge ratios) and masses. In particular the physical charge and mass of the electron must be measured and cannot be calculated from the theory.

So the vacuum energy in QFT can be set to any value, thereby treating the cosmological constant as simply another physical constant not explained by the theory.

Quantum mechanical interactions among particles are described by interaction terms among underlying quantum fields, to evaluate particle processes. These can be visualised by Feynman diagrams as devised by the physicist Richard Feynman, which demonstrate how infinities arise.

For example, a particle such as an electron is enclosed in a cloud of other particles which interact with it. Renormalisation replaces any proposed mass and charge of the electron with the observed values.

QFT assumes a space-time continuum. Parameters differ dependent on the scales used which leads to infinities. Quantum mechanics involves values which are not well defined, so to make them compatible with the continuum renormalisation procedures are invoked to remove those infinities.

Many of the early formulators of QED such as Dirac and the physicist Freeman Dyson were dissatisfied with this situation. Dirac regarded it as not sensible mathematics.

He said sensible mathematics involves neglecting a quantity when it is small, not neglecting it because it is infinitely large and you do not want it.

Dyson argued that these infinities are of a basic nature and cannot be eliminated by any formal mathematical procedures such as the renormalisation method.

The modern physicist takes the shut up and calculate approach on the basis that QFT is the best available theory to date, but there is nevertheless an echo in this that the divergences in the theory speak of human ignorance about the workings of nature, as the scientist would put it.

This also acknowledges that this ignorance can be quantified while the resulting theory remains effective.

The standard particle model

The theory of general relativity correlated with cosmic scale observations forms the standard cosmological model now termed the Lambda CDM model, while quantum field theory in conjunction with experiments has spawned a standard model of particle physics. The particle model does not incorporate general relativity and therefore the two models do not represent a unified theory of physics.

Nevertheless, there is sufficient common ground to incorporate both into the Lambda CDM version which after all endeavours to present the best available match between scientific theory and observation covering all areas of physics.

The standard model of particle physics is a theory concerning the electromagnetic, weak and strong nuclear interactions which orchestrate the dynamics of the known sub atomic particles. It was developed in the latter half of the 20th century as a collaborative effort of scientists around the world.

The current formulation rests on the discovery of quarks in the 1970s culminating in the further discoveries of the top quark, the tau neutrino and more recently evidence in 2012 supporting theoretical predictions of the Higgs boson. It includes in particular the theory of Quantum Chromodynamics (QCD) which describes the strong interactions.

It has significant limitations which will be referred to below but it nevertheless successfully incorporates in one theory a unification of particle interactions other than gravity and identifies a set of seventeen particles (ignoring anti particles and colour charges) which obey prediction, have all been observed and appear to be fundamental with no indication of a deeper underlying structure. This does not preclude the existence of other particles which extend the existing structure in a non-hierarchical sense such as hypothetical super symmetric particles.

It is therefore used as a basis for building more exotic models that incorporate such hypothetical particles and extra dimensions in an attempt to explain results at variance with the standard model such as dark matter and neutrino oscillations.

The standard model splits particles considered elementary into two main classes matter particles called fermions of spin ½ which obey Pauli's exclusion principle and force carriers called gauge bosons of spin 1 which do not. Pauli's exclusion principle requires that the wave function be symmetric for bosons and anti-symmetric for fermions.

All known fermions have distinct anti-particles. They form the main constituents of matter. Fermions are classified according to the charges they carry. They consist of six quarks (up, down, charm, strange, top, bottom) and six leptons (electron, electron neutrino, muon, muon neutrino, tau, tau neutrino).

Pairs from each classification are grouped together to form a generation, with corresponding particles having similar physical behaviour.

Quarks interact via the strong force giving them a colour charge and are the only known carriers of fractional electric charge but because they combine in groups, only integer electric charges are observed in nature.

Their respective anti-particles are the anti-quarks which are identical except that they carry the opposite electric charge.

The phenomenon known as colour confinement results in quarks being perpetually bound to one another forming colour neutral composite particles (hadrons) containing either a quark and an anti-quark (mesons) or three quarks (baryons).

Hadrons form the nucleus of atoms namely protons and neutrons which have quarks as their primary constituents, the down quark (d) and the up quark (u) triplets of which bind together to form protons (u-u-d) and neutrons (u-d-d).

Quarks interact with other fermions both electromagnetically and via the weak interaction. The six remaining fermions called leptons do not interact through the strong force and hence do not carry a colour charge. The three neutrinos do not carry electric charge either so they are only influenced by the weak force.

Each succeeding generation has greater mass than the corresponding particles of a lower generation. First generation particles are not observed to decay so all baryonic matter is made of them.

Second and third generation particles do decay with very short half lives and are observed only in very high energy experiments. Neutrinos of all generations also do not decay but rarely interact with baryonic matter.

Turning now to the gauge bosons the standard model defines gauge bosons as force carriers and are said to mediate the strong, weak and electromagnetic interactions.

Interactions are the means by which particles influence other particles. In particular electromagnetism allows particles to interact with one another via electric and magnetic fields and gravitation allows particles with mass to attract one another in accordance general relativity. The standard model explains these forces as resulting from matter particles exchanging other particles known as force mediating particles. When a force mediating particle is exchanged, at a macroscopic level the effect is equivalent to a force influencing both and the particle is said to have mediated or been the agent of that force.

The gauge bosons all have spin 1 and do not follow Pauli's exclusion principle, therefore they do not have a theoretical limit on their spatial density.

There are five elementary bosons in the standard model, ignoring colours and anti-particles.

Firstly the massless photons which mediate the electromagnetic force between electrically charged particles described by QED.

Then the massive W^+, W^- and Z gauge bosons which mediate the weak interactions between quarks and leptons.

The W^+ and W^- particles carry an electric charge of +1 and -1 respectively and couple to the electromagnetic interaction. The three bosons and the photon collectively mediate the electroweak interaction.

There are in addition gluons eight in number which mediate the strong interaction between colour charged particles i.e. the quarks. Gluons like photons are massless.

Their multiplicity is labelled by combinations of colour and anti-colour charge and because of the charge they interact among themselves, as described by QCD.

Finally the Higgs boson is a further massive scalar elementary particle originally theorised in the 1960s as a key element of the standard model. It has no intrinsic spin but is classified as a boson (like the gauge bosons which have integer spin). It explains why the photon and gluon are massless while the other elementary particles are massive.

In electroweak theory, the Higgs boson generates the masses of the leptons and quarks. As the Higgs boson is massive it interacts with itself and decays almost immediately when created.

Experimental observations on the Large Hadron Collider (LHC) at CERN in 2012 seem to have confirmed its existence.

Decay rates for particles generally are theoretically estimated to have a wide range varying from a half life of some 10^{34} years in the case of protons to 10^{-25} seconds in the case of bosons for example.

Counting particles by a rule which distinguishes between particles and antiparticles and including the many colour states of quarks and gluons gives a total of 61 elementary particles.

The atomic nuclei consist of protons and neutrons, each nucleus containing a specific numbers of protons which defines its atomic number called a nuclide and a specific number of neutrons which may vary for a given element called isotopes.

Nuclear reactions can change one nuclide in to another. Atoms are the smallest neutral particles in to which matter can be divided by chemical reactions and can be pictured as consisting of a small heavy nucleus surrounded by a relatively large cloud of electrons.

Each type of atom corresponds to a specific chemical element, some 118 having been discovered to date all arranged in accordance the periodic table.

Molecules are the smallest particles in to which a non-elemental substance can be divided while maintaining the physical properties of the substance. Each type of molecule which is a composite of two or more atoms corresponds to a specific chemical compound.

The standard model as presently formulated is, despite its successes, a long way short of being a complete theory of fundamental interactions because it makes certain simplifying assumptions. It does not incorporate a theory of gravitation as described by general relativity nor does it predict the influence of dark energy or contain any viable dark matter particle nor does it account for the behaviour of neutrinos and their non-zero masses.

Observations show that matter predominates over antimatter in the visible universe to the extent that the latter is confined primarily to high energy particle collisions, as demonstrated in particle accelerators.

QFT has difficulty in explaining this and as a consequence provides modern physics with a major dilemma.

The observed imbalance is called baryon asymmetry, a necessary condition for which is the general violation of CP symmetry, not just that observed experimentally in weak interactions.

So physicists have endeavoured to extend the standard model by assigning a super partner to every particle species. These particles have not been observed to date and little is known about their masses, but their interactions are constrained by this super symmetry in turn placing ever greater constraints on its predictions.

The Minimal Supersymmetric Standard model as it is termed involves a significant menagerie of hypothetical particles where to put it bluntly Ockham's razor has lost its edge.

Zero point energy

In connection with the Standard Model of particle physics, it is important to explain the principle of zero point energy.

Zero point energy is the ground state or lowest possible energy that a quantum mechanical physical system can have. All such systems undergo fluctuations as a consequence of their wave like motion, as demonstrated by liquid helium which does not freeze under atmospheric pressure at any temperature down to absolute zero.

Vacuum energy is the zero point energy of all fields in space and therefore all fields in the standard model. It is the energy of the vacuum, which in quantum field theory is defined not as empty space but as the ground state of those fields. In cosmology the vacuum energy is one possible explanation for the cosmological constant.

In 1900 Max Planck derived the formula for the energy of a vibrating atomic unit:

$\epsilon = hv$ / (exp (hv/kT) – 1), where h is Planck's constant, v is the frequency, k is Boltzmann's constant and T is the absolute Temperature.

Then in 1913 using this formula Einstein and Stern suggested that all oscillators have a residual energy at absolute zero. By analysis of the specific heat of hydrogen at low temperature they concluded that the vibrational energy was given by:

$\epsilon = hv$ / (exp (hv/kT) – 1) + $hv/2$. So an atomic system at absolute zero has energy $hv/2$.

Zero point energy correlates with Heisenberg's uncertainty principle which states a particles position and momentum cannot be defined simultaneously at any given quantum state. There is no state at which a system sits at the bottom of its potential well for then its position and momentum would be precisely determined. The energy of the ground state must always be greater than the minimum of the potential well.

The standard model chronological sequence

Based on the standard cosmological model and the standard model of particle physics it is useful to consider the possible chronology of events in the universe from the big bang onwards up to the present time together with an indication of what the future holds. Such chronologies are of course approximate and always in process of modification as experimental and theoretical physics develop.

Descriptions of these can be found on the internet and refer to a succession of epochs or eras representing periods of time, a simplified version being given below:-

Planck (10^{-43} seconds after big bang).

Fine-tuned cosmological models assume unification of the four fundamental forces, gravity, electromagnetic, weak and strong nuclear forces at this time.

Grand (Standard Model) Unification (10^{-43} to 10^{-36} seconds after big bang).

Fine-tuned cosmological models presume gravity separates from the other three forces which form the basis of the Standard Model.

Magnetic monopoles should have been produced but have not been observed.

Inflation theory gives an explanation for this but does not involve this stage.

Electroweak (10^{-36} seconds to 10^{-12} seconds after big bang).

Strong force separates from the electroweak force as temperature falls.

In the case of inflationary cosmology inflation begins produced by a field termed an inflation similar to a Higgs Field. Inflation implies this is the reason for the homogeneous state observed today. Inflation then ends and the inflation decays in to ordinary particle. A quark gluon plasma dominates the universe.

Quark (10^{-12} to 10^{-6} seconds after big bang).

As temperature falls further the Higgs field breaks electroweak symmetry with electromagnetic force and provides elementary particles with mass.

Hadron (10^{-6} seconds to 1 second after big bang).

The quark gluon plasma cools allowing hadrons and baryons to form.

Lepton (1 second to 10 seconds after the big bang).

Hadrons and anti-hadrons annihilate each other, leaving leptons and anti-leptons to dominate. A similar process then annihilates most leptons.

Photon (10 seconds to 380,000 years after the big bang).

Photons dominate the universe interacting with protons and electrons.

Temperature falls to the point where nucleosynthesis occurs and atomic nuclei form, a process of nuclear fusion.

70,000 years after the big bang small structures begin to form. Cold dark matter dominates and gravitational collapse increases density in some regions compared with others.

Approximately 380,000 years after the big bang, hydrogen and helium atoms form in an ionised state. Photons interact with electrons and protons and the universe is opaque.

Electrons then get captured by the ionised material as the temperature falls forming atoms which are electrically neutral, a process referred to as recombination.

Photons can now travel freely and the universe becomes thermally diverse, a process referred to as decoupling now pictured by the Cosmic Wave background.

Dark ages and structure formation (380,000 years to 10^9 years after big bang).

Although initially there was no light, large perturbations arose from a largely homogeneous background and the first stars and quasars formed providing the intense radiation which reionised the surrounding universe.

The first stars to form referred to as population III stars had little metallic element apart from lithium. Subsequent population II stars had low levels of metallic content but they ultimately produced population I stars with higher metallic content.

These stars caused large volumes of matter to condense to form galaxies some 300 million years after the big bang. Such light sources ended the dark ages of the universe.

Galaxy cluster formation (10^9 years to present day, 13.8×10^9 years after big bang).

Formation of clusters of galaxies and super clusters.

Formation of our solar system some 9.2×10^9 years after the big bang.

Acceleration of expansion of the big bang at approximately the same time termed dark energy.

Present day to say 10^{100} years from now.

In the short term the biosphere of the earth will become unstable eliminating life on earth and in 5×10^9 years our sun will in turn become unstable, enter a red giant phase engulfing the earth before collapsing to a dwarf star and ultimately ceasing to shine.

The universe will expand, existing stars burn out and black holes consume matter and eventually evaporate. Particles may well decay leaving the universe in a state of very high entropy in thermal equilibrium.

The problems of quantum gravity

The primary endeavour of theoretical physicists since the time of Einstein and Bohr has been to resolve the dichotomy between relativity and quantum theory and in so doing create a Grand Unification Theory of physics from which the existing standard cosmological and particle models described earlier could be seen in some sense as limited special cases. To this end they have proceeded to develop two main lines of enquiry, namely string theory and loop quantum gravity.

Theories of quantum gravity generally are based on fundamental units of measurement namely the Planck units of energy, time and space (Planck mass, Planck time and Planck length). These units are defined in terms of four universal physical constants (the speed of light, gravitational constant, Planck constant and Boltzmann constant) details of which can be found on the internet.

String theory is based primarily on particle physics but incorporates modifications involving extra dimensions and transforms the idea of a particle in to a string of minute length close to the Planck length.

In the case of string theories the idea of extra dimensions arose from the work of the physicists Kaluza and Klein who had proposed a fifth dimension to cater for electromagnetism additional to the four dimensions of Einstein's theory.

The angle of the fifth dimension mimicked the electromagnetic field and its circumference the relative strength of that field to the gravitational force field.

This work which was of great interest to Einstein was resurrected in the late 20th century and incorporated in to string theory in which the laws of physics are controlled by the shape and size of additional microscopic dimensions.

Thus from a theory of gravity alone in five dimensions there is obtained a theory of both gravity and electromagnetism in four dimensions.

In string theory as described by Bousso and Polchinski (Ref 7.4.10), particles are one dimensional objects, small vibrating strings or loops of size of the order of the Planck length which look like points at larger scales. Under high magnification the string appears as a tube so each zero-dimensional point is a one-dimensional circle on the tube.

String theory in addition predicts membrane like objects called branes of various dimensionalities, with strings having end points on the brane or closed loops not on the brane. A line of space is actually a tube with a one-dimensional brane running through it, populated by strings which loop round the circumference of the tube.

For the theory to be mathematically consistent the string has to vibrate in ten space-time dimensions, six of which are too small to be detected. String theory also contains fluxes or forces represented by field lines much as forces are represented in classical electromagnetism.

In general relativity matter tells space-time how to curve and space-time tells matter how to move, but in string theory there are several extra dimensions and hence many more adjustable parameters. The extra dimensions can take many different shapes or topologies, spheres, doughnuts etc, each doughnut loop has a handle with a length and a circumference resulting in a huge number of geometries. In addition to the handles further parameters correspond to the location of branes and the different amounts of flux which are wound round each loop. Each configuration has a potential energy called vacuum energy when the four large dimensions are devoid of matter or fields.

The geometry of the small dimensions tries to adjust to minimise this energy.

Curves can be plotted showing how the vacuum energy changes with say the overall size of the hidden space. Such curves have peaks and valleys. How the hidden space behaves depends on the initial conditions i.e. where a ball on the curve (representing the hidden space) is initially positioned.

If the ball rolls off to infinity the space will expand and no longer be hidden, otherwise it will settle in a valley where the resulting vacuum energy is either positive, negative or zero. In a universe like ours the size of the hidden dimensions is not changing with time, thus in this scenario it is sitting at a minimum vacuum energy which seems to be slightly positive. There are however many parameters resulting in a multidimensional landscape of hills and valleys the low points of which represent stable conditions or vacua.

It is significant that the theory implies vast numbers of these states with some 10^{500} different configurations. So the question is which stable vacua represents the physical world of the common experience. The world may not be stuck in a particular configuration as quantum processes may allow it to jump from one to another and since the theory endeavours to incorporate general relativity allowing configurations to expand rapidly like bubbles so they may exist side by side each unaware of the others.

It is considered that the universe may experience all possible sequences resulting in a hierarchy of sub universes each having its own laws of physics our big bang being one, similar to the eternal inflation theory proposed by Guth and the chaotic inflation theory of Linde referred to earlier.

The theory attempts to explain the vacuum energy problem which arises when quantum field theory predicts a value for that energy (Λ^{p}) of one Planck mass per cubic Planck length, whereas observations give a value as little as 10^{-120} of that figure.

Using the 10^{500} configurations they will have vacuum energies ranging from say $+ \Lambda^{\mathrm{p}}$ to $- \Lambda^{\mathrm{p}}$ so some values will be nearly zero, as in our visible universe, all in accordance the anthropic principle.

The alternative proposal termed loop quantum gravity as described by Smolin (Ref 7.4.11) is based on general relativity and predicts that space and time are made of discrete pieces, instead of being endlessly smooth and continuous, at scales corresponding to the Planck length involving small loops marked out in space-time.

Two key principles of general relativity are retained in loop quantum gravity. Firstly background independence which says the geometry of space-time is not fixed but instead is an evolving dynamical quantity. To find the geometry the relevant equations including the effects of matter and energy have to be solved. It is to be noted that string theory as currently formulated is not background independent, the equations describing the string being in a predetermined classical space-time.

Secondly the closely related principle of diffeomorphism invariance, namely that one is free to choose any set of coordinates to map space-time and express the equations.

A point in space-time is not defined by its location according to some special set of coordinates but by what physically happens there.

Using these principles it is considered that space is quantised involving small loops marked out in space-time.

The theory predicts that there is no length less than 10^{-33}cms and therefore no area less than 10^{-66} cm^2 or volume less than 10^{-99} cm^3.

Diagrams called spin networks are used to represent quantum scales of space, for example where a cube consists of a volume enclosed in six square faces the corresponding spin network has a node representing the volume and six lines representing the faces. In this case the volume is 8 Planck lengths and each face 4 Planck lengths.

So one quantum of area is represented by a single line and a quantum of volume by one node, many lines and nodes are required for larger volumes. Diagrams can be drawn representing quantum states of volume and area called graphs and these can also represent curved space. Since the distortion of space produces gravity the diagrams or graphs form a quantum theory of gravity. The lines and nodes now express space and continuous space does not exist as a separate entity. The quantum state of our universe can in principle be expressed as a complex spin network of many orders of magnitude of nodes. Particles are represented by adding labels to nodes and fields by additional labels on lines, so matter exists at the nodes of the spin network.

The geometry of space also changes with time which proceeds by discrete ticks of Planck time. To add time to the spin network the lines of the spin network become planes and the nodes become lines forming what is termed spin foam.

Loop quantum gravity correlates with Bekenstein and Hawking's work on black holes.

It also has no difficulty in incorporating the positive cosmological constant or dark energy. The progress it can make with regard to unification theory remains to be seen.

7.4.7. *The role of black holes.*

Black holes are regions of space-time where scientists believe gravity is so powerful that it prevents anything including light from escaping its clutches.

They are represented by the Schwarzchild metric; a spherically symmetric static solution to the Einstein field equations referred to in Section 7.4.4 and can be considered homogenous in all dimensions apart from radius.

The recent observations of gravitational waves produced by the collision of two black holes promises great advances in astronomy and cosmology, including the ability to investigate the early radiation dominant epochs which will make visible, in a sense, that time where there is ironically no light!

Predicted by the theory of General Relativity such holes are formed by the deformation of space-time and have been assumed to have around them a surface defined mathematically, termed the event horizon, that is a point of no return for material falling on to that surface from outside. Just as with a perfect black body in thermodynamics the total absorption of light involved gives rise to the term 'black hole'. General relativity predicts the centre of a black hole to be a singularity of zero volume and infinite density at which point the theory is assumed to breakdown and give way to quantum effects due to the high density and the particle interactions associated with quantum field theory. To date it has not been possible to unify relativity and quantum theory in these circumstances, i.e. a theory of quantum gravity, although great efforts have been made to do so by means of loop quantum gravity an approach using the theory of relativity as a base and by means of string theory using quantum theory and particle physics as a base.

Black holes are an ideal subject on which to test such attempts at unification as they involve conditions where the concepts of relativity and the quantum state are both in some sense relevant, but in apparent conflict.

In the late 1960s and 1970s work by Hawking and Penrose (Ref 7.4.12), Bekenstein (Ref 7.4.13) and Hawking (Ref 7.4.14 and 7.4.15) led to the development of theories concerning the thermodynamics of black holes.

They showed that the characteristics of black holes were analogous to the laws of thermodynamics by relating mass to energy, area to entropy and surface gravity to temperature.

In particular Bekenstein and Hawking showed that spherically symmetrical black holes have a well-defined entropy proportional to the square of its mass and that the natural assumption that the entropy of black holes is proportional to its volume is incorrect and is instead proportional to the surface area A of its event horizon.

The entropy $S = kc^3 A / (4G\hbar)$, where k = Boltzmann's constant, c = the speed of light,

G = the gravitational constant and $\hbar$ = Planck's Constant/2π.

The Planck area is approximately 10^{-66} cm ^{2} and S is proportional to the number of such squares that will fit on its surface. The entropies of black holes are therefore enormous. Large black holes are believed to occupy the centres of galaxies and are likely to be associated with the formation of the latter.

Hawking also showed that black holes radiate like a black body with a temperature proportional to the surface gravity of the hole, which results in such objects having very low temperatures, a hole of solar mass being only about 10^{-7} deg Kelvin.

Therefore black holes should slowly radiate away energy now termed Hawking radiation.

Now in respect of black holes physicists believe they have only three distinguishing properties all distinguishable from outside the hole, its mass which determines the radius of its event horizon, its electric charge and its angular momentum, sometimes referred to as the no hair theorem.

From this theorem it would be expected that the Hawking radiation would be completely independent of material entering the black hole.

However if such material in a quantum state were transformed in to the mixed state of Hawking radiation the original quantum state would be destroyed. It implies physical information could disappear in to a black hole and be lost forever defying the scientific principle that complete information about a physical system should determine its state at any other time, violates a fundamental theorem of physics namely Liouville's theorem and the principle of the conservation of energy.

It is a fundamental aspect of quantum mechanics that complete information about a system is encoded in its wave function whose evolution is determined by what is referred to as unitarity, which in turn implies that information is conserved. The argument over whether or not information is lost in black holes is referred to as the black hole information paradox. The paradox has given rise to something of a split in the scientific community. In defence of the principle of unitarity and quantum mechanics G. t'Hooft (Ref 7.4.16) and L. Susskind (Ref 7.4.17), both string theorists, invoked what is now referred to as the holographic principle. In the manner of a hologram they suggested that as the maximum entropy a region of space can contain is related to the area of its surface not its volume as shown in respect of black holes, correspondingly the event horizon acts as a hologram capturing all the information content of the black holes interior.

An extremely informative background suitable for the layman on the subject of string theory and its relation to the hologram is given by Greene (Ref 7.4.18 and 7.4.19).

As Greene describes, t'Hooft and Susskind generalise this to the whole universe by suggesting that everything that occurs in the interior of the universe is a reflection of that defined on a distant bounding surface necessarily of lower dimensionality than that interior. The physics of string theory at least in certain cases was shown to embody holographic characteristics by the work of Maldacena (Ref 7.4.20) and Bekenstein (Ref 7.4.21).

This embodiment involves a hypothetical universe, technically a five-dimensional anti de Sitter space-time, with four large dimensions and one time dimension that has uniform negative curvature. A universe with a boundary that (like all boundaries) has one dimension less than the shape it bounds: three space dimensions and one time dimension, as indeed it appears to the human observer.

Although the five-dimensional anti de Sitter model being empty is hardly like our universe filled as it is with matter and radiation, the holographic model provides a clue to a better understanding of the relation between relativity and the quantum state and the four-dimensional universe we observe. So a particular quantum theory without gravity is indistinguishable from another quantum theory that includes gravity and has one more space dimension. Both descriptions are regarded as equivalent, a form of duality it might be said. Interestingly physicists use the term 'brane' to describe the boundary.

Arising from all this, the predominant belief among scientists is that if a full quantum gravitational treatment of the problem were available, information and unitarity would be preserved. Hawking contrary to his original opinion that information is inevitably lost, changed his view and suggested that quantum perturbations of the event horizon could allow information to escape from a black hole resolving the paradox.

However Penrose claims loss of unitarity in quantum systems is not a problem as quantum measurements (corresponding to what is termed the collapse of the wave function) are by themselves already non unitary. He claims that quantum systems will in fact no longer evolve unitarily as gravitation comes into play, precisely as in black holes. This being all in accordance Conformal Cyclic Cosmology advocated by him, which is critically dependent on the condition that information is lost in black holes, as discussed in Section 7.4.4.

7.4.8. The basis for interpretations of quantum mechanics that have arisen from the dichotomy between it and the common experience.

The competing theories.

Self-generating fractal universes, parallel universes and the concept of the Multiverse.

Do an infinite variety of physics exist?

A physicist's view of the quantum state as opposed to the classical level inevitably fragments in to many different camps, as there are several issues which are the subject of debate.

A list of the primary issues include:-

Determinism v indeterminism.
Locality v non-locality.
Is there a unique history?
Counterfactual definiteness, yes or no?

Is the wave function real? Does it collapse? Is there a universal wave function?
Are there hidden variables?
What is the role of the observer?

Before proceeding to examine the competing theories evoked to explain the quantum dilemma it is necessary to consider the nature of interpretation in this context, to form the common background to all of them.

They all interpret a formal set of equations and principles to generate predictions from initial conditions and a set of observations including those obtained by formal scientific research and from human experience of what is termed the real world.

They are also in varying degrees epistemological or ontological in character, all as described in Section 7.1.3.

There are various challenges to the familiar processes of interpretation associated with the real world including the following:-

First; the abstract nature of quantum theory such as Hilbert space as described in Sections 7.3.7 and 7.3.8, involving complex mathematical equations.

Second; evidence of apparently indeterministic irreversible processes as opposed to those deterministic as discussed in Section 7.4.5.

Third; the role of the observer is unclear. The Copenhagen interpretation holds that the myriad probabilities associated with a quantum field are unreal yet the act of observation and measurement collapses the wave function and fastens realism to a single possibility. Quantum decoherence grants that all the possibilities can be real but that they quickly interact with the surroundings, equivalent to an unstoppable loss of information which decoheres to the ordinary classical state.

Fourth; there are correlations between objects far apart, termed quantum entanglement which does not comply with principles of locality as discussed in Section 7.2.4.

Fifth; there is the question of complementarity as exemplified by Heisenberg's uncertainty principle also referred to in Section 7.4.5, which holds that no set of physical concepts in the classical sense can simultaneously refer to all properties of a quantum system. In the Copenhagen interpretation for example the wave description and the particle description can each describe the quantum system but not simultaneously.

Sixth; as the size of a quantum system increases its complexity rises exponentially to the point where classical approximations become impossible.

The next aspect to be considered is the description that relates mathematical procedures to experimental practice and prediction fundamental to a scientific theory. Such a description in classical physics unifies the mathematics, the experimental apparatus including measuring instruments and the measured results. The results are each single valued in principle if not in practice i.e. individual experiments do not involve probability distributions. All of this is carried out by human observers who are assumed to have no influence on this process. Indeed every effort is made to ensure this by double blind procedures and by repeated experiments to corroborate original findings.

In the case of quantum mechanics however, measurements by instruments are necessarily performed many times starting with the same initial conditions results in a well-defined probability distribution; furthermore quantum mechanics provides a means of determining the statistical properties of this distribution, including the expectation value. The basic mathematical description is summarised in Section 7.3.7, but the crucial question is whether that description is real in a physical sense and accordingly an identifiable structure. The mathematical description is no longer capable of providing an interpretation since that sense of physical reality is lost.

So the next step is to consider the basis for interpretations before describing each. An interpretation in the context of this section is a meaningful explanation of the formal mathematics of quantum mechanics i.e. to form a relationship with the results measured by the instrumentation. For example Einstein considered such an interpretation should be related to matters already referred to earlier in this book namely realism, completeness (which accounts for every aspect of that reality), determinism and local realism. He and others proposed the

EPR paradox in 1935, when they characterised the element of reality as a quantity whose value can be predicted with certainty before measuring or otherwise disturbing it, while they defined a complete physical theory as one in which every element of physical reality is accounted for by the theory.

An interpretation is then complete if every element of the identifiable structure is to be found in the mathematical description.

Elements of the mathematics are real if they correspond to something in the interpreting structure. Interpretations do not necessarily provide this correspondence.

Determinism requires that states of the system in the future are a function of the state at the present time. Again interpretations are not necessarily deterministic.

Local realism requires firstly that measured values are elements of reality and secondly that the effects of measurement have a propagation speed not exceeding the speed of light so the interpreting structure must be based on locality.

Now as described earlier, Bell's theorem combined with experimental testing restricts the kind of properties a quantum theory can have, with the primary implication that quantum mechanics cannot satisfy both the principle of locality and counterfactual definiteness. Counterfactual definiteness has already been referred to in Section 7.2.4. It is a basic assumption associated with the real world described by classical physics.

In the quantum context such basic assumptions are either overturned or not according to the interpretation made, since counterfactual definiteness is about whether it is meaningful to refer to the results of measurements which have not been carried out. An immediate problem arises from Heisenberg's uncertainty principle which removes the ability to measure particle position and momentum simultaneously. The measurement of the one eliminates any information about the other, so the question arises, is it permissible in that case to maintain such definiteness.

Even the no conspiracy principle also referred to in Section 7.2.4, is not immune to this argument, as it implies nature may hide the 'true facts' from the physicist when she chooses to do so.

Against the above background an assessment of the various interpretations can be conducted.

The competing theories

Of the competing theories evoked to explain the quantum dilemma as it has been called, a number have received more support than others or introduce interesting ideas while not necessarily being popular.

These competing theories which do not arise in classical physics may be termed interpretations though some are not specifically interpretations but emphasise important aspects which should be incorporated in the debate.

They all take in one way or another, different stances to the issues given above and include but are not limited to:-

The Copenhagen interpretation.
The Many Worlds and its variant the Many Minds interpretation.
Decoherence theory.
Consistent histories.
The de Broglie – Bohm interpretation.
Objective collapse theories.
Relational interpretation.
Transactional Interpretation.
Time symmetric theories.
The shut up and calculate approach.

Descriptions of the above can be found on the internet and some with graphical illustrations by Al-Khalili (Ref 7.4.8), which are very beneficial to the layman.

The Copenhagen interpretation advocated so strongly by Bohr was the first on the scene in this debate in the late 1920s and still has many adherents.

The quantum state $|\psi\rangle$ in this version does not represent reality but merely allows the computation of probabilities that those alternative realities might occur.

Accordingly Bohr viewed the quantum state of a system between measurements as having no physical reality but merely a summary of the observer's overall knowledge in the circumstances, which would increase when the measurement is made. This knowledge should take in to account that the system and the measuring instrument and the observer formed one indivisible whole this being taken to correspond with classical objective reality. Whether the level of the classical world was the measuring instrument or the level of the observer's consciousness was an open question.

In other words Bohr regarded the classical world as having an objective reality but that reality did not include the underlying quantum level states.

The state vector reduction process is a consequence of measurement but not regarded as physical reality and relegated to being 'all in the mind'.

This interpretation denies counterfactual definiteness as it considers only measurements performed and ignores those not.

The Many Worlds interpretation as proposed by Everett is diametrically opposed to the above, namely that the state vector $|\psi\rangle$ does represent the reality and the state vector reduction does not take place at all. So when a measurement is made all the alternative outcomes actually co-exist in reality in a grand quantum linear superposition of alternative universes described by a wave function $|\psi\rangle$ for the entire superposition sometimes referred to as the multiverse.

This overall reality is not perceived by the observer as he and his mind also co-exists in this quantum superposition which split attaching to each of the different probabilities of the measurement being performed, one of the alternatives corresponding to the classical level observation.

These separate copies of the observer are however unable to communicate each thinking only one result has occurred.

Again the state vector reduction process is not regarded as having a physical reality, another case of being 'all in the mind'.

The many worlds interpretation also rejects counterfactual definiteness but for a reason exactly opposite to that of Bohr. Instead of ignoring unmeasured values it provides infinitely many in an endlessly branching reality. We cannot perform measurements in these other worlds and measure their values, only the single value in our own world.

The many minds approach extends the many worlds interpretation by proposing that the distinction between worlds should be made at the level of the mind of the conscious observer, though what is meant by mind in this context is not clear.

The above interpretations use the principle of decoherence to their advantage.

Decoherence theory is not an interpretation as such but assumes that quantum interactions with the surrounding environment have an effect on the system in a state of superposition. There is an inevitable loss of information as a result which decoheres that quantum superposition, reducing it to the classical state, with increasing rapidity the larger and more complex the system. The Copenhagen approach regards such a loss of information as a good reason to support the view that it is best to throw a curtain over the quantum process and pay attention only to the results on the screen which exhibits the results of the experiments. The Many Worlds adherents use decoherence as the explanation as to how parallel worlds form and why the various copies of the observer cannot see each other.

The consistent histories interpretation assumes the overall history of a system to be described so that each possible history obeys classical probability, the alternatives having relative probability.

The particle explores all possible paths simultaneously cancelling each other out with the exception of the one physical path taken by the particle.

This interpretation also rejects counterfactual definiteness as though it allows single values to unperformed measurements they are in effect hidden. It then rejects the combination of incompatible measurements whether factual or counterfactual as they do not match with performed compatible measurements.

It therefore insists on single values being obtained as opposed to the ignorance of values or the multiplicity of values in the two previous interpretations.

The de Broglie Bohm interpretation is based on a theory originally formulated by de Broglie in the 1920s and extended by Bohm in the 1950s.

Apart from the books written by Bohm already mentioned, David Z. Albert (Ref 7.4.22) and A. Whitaker (Ref 7.4.23) give accounts for the lay reader, while a comprehensive mathematical account of this theory is given by P.R. Holland (Ref 7.4.24).

Particles always have definite positions and are guided by a wave function (the quantum potential) which is real and universal and evolves according to the Schrödinger wave equation but never collapses. The theory is inherently non local and deterministic. It therefore assumes counterfactual definiteness.

The simultaneous determination of a particle's position and velocity is subject to the usual constraint imposed by the uncertainty principle. The theory is considered to be a hidden variables theory but since it is non local it satisfies Bell's inequality. The measurement problem is resolved since hidden variables or beables exist representing the actual positions of particles at given moments in time.

The reduction process does not involve collapse of the wave function, which is considered phenomenological, i.e. the subjective experience of the observer.

The phenomenological view is that reality consists of objects and events as perceived by human consciousness, rather than anything independent of that consciousness.

The quantum wave properties of particles are related to and guided by the non local form dependent quantum potential or pilot wave. The effect of the quantum potential is regarded as independent of its strength but only on its form which Bohm compares to a ship (or plane say) on automatic pilot being guided by radio waves. The ship moves with its own energy but the radio waves direct the much greater energy of the vessel. Furthermore the effect of the wave does not necessarily fall off with distance so that even remote influences can affect movement. This of course is the opposite to the classical example of a cork bobbing about in the sea when only waves adjacent to the cork have an effect on its motion.

The influence of the quantum potential and the non-local nature of the theory imply that particles interact with deeper structures associated with the overall wholeness of existence.

Non-locality is the coupling of separate systems in circumstances which according to relativity no physical interaction should be possible. It defies the edict that all influences can propagate at or less than the speed of light.

It is accepted that signalling involving the transfer of information at speeds greater than light is not possible. Even though the interpretation gives rise to the statistical results of quantum theory that are known to accord with relativity, at the fundamental level of the individual processes the spirit of relativity is violated.

Superluminal influences are allowed and contrary to relativity there exists a preferred frame of reference in which the influences are instantaneous.

Alternative proposals embedding non-locality in a relativistic interpretation without the need for a preferred reference frame are work in progress.

Albert (Ref 7.4.22) gives a mathematical formulation of Bohm's theory relating it to the conventional Schrödinger equation, consisting of three main elements.

Firstly the deterministic Schrödinger equation; secondly a deterministic law of the motion of the particles; thirdly a statistical rule analogous to one used in classical statistical mechanics. This rule uses probability to reduce ignorance of the exact state of physical systems, with the facility to update knowledge of that state during experimental measurement using a mathematical procedure termed conditionalization.

Objective collapse theories regard both the wave function and the process of collapse as objective, just the opposite of Bohr. Collapse may occur randomly irrespective of observers. They are therefore realistic, indeterministic, no hidden variables theories. Examples include the Ghirardi Rimini Weber theory which modify the Schrödinger equation and the Penrose interpretation which proposes that gravity is responsible for the reduction process related to the Planck mass, details of which can be found on the internet. Penrose and the anaesthesiologist Hameroff further propose the reduction process plays out at the level of neurons and the microtubule structures in the human brain. Such theories can be regarded as entirely explicit in character so the implicit plays no role.

Relational quantum mechanics argues that different observers may give different accounts of the same series of events, as applies in special relativity.

For one observer a system may be in a single collapsed state, while to another at the same time it may be in a superposition of two or more states. So a relationship applies between the observed system and the observer who has degrees of freedom. The physics therefore is about the relation between objects and not the objects themselves.

The transactional interpretation describes a quantum interaction in terms of a standing wave formed by the sum of a forward in time wave and a backward in time wave, between the emitter or source of a particle and subsequent detection.

It is like the de Broglie Bohm interpretation blatantly non-local, but in this case in terms of both space and time.

Time symmetric theories add retro causality to the muddy waters of interpretation apparent to the reader. Retro causality implies future events affect events in the past in addition to our familiar assumption that the past affects events in the future. The equations of quantum mechanics are modified accordingly.

Two measurements at different times are required rather than a single measurement to determine the state of the system. The collapse of the wave function is therefore not a physical change, just a change in knowledge of it due to the additional time measurement. Hence quantum entanglement is regarded as an illusion, associated with our refusal to accept the retro causal concept.

The shut up and calculate approach is to carry out experiments, ignoring the philosophical implications but instead note the results of practical interest and apply them to advancing technology, which in a reductionist sense is all that is needed.

Finally it may be said to all boil down to whether the wave function is real or just information; some thinking yes and being driven to a many worlds theory or Bohmian mechanics, alternatively thinking no, it's just information and the collapse is natural, like flipping a coin. Before you look at the coin its state is 50 per cent heads 50 per cent tails, after you look it changes to 100 per cent heads 0 per cent tails say or vice versa.

The above interpretations take varying positions with respect to determinism, locality, the wave function, history, counterfactual definiteness, hidden variables and the role of the observer as listed at the beginning of this section.

In one sense each interpretation can be related to differing attitudes to animacy and consciousness; some considering these characteristics to be merely an accident or an anthropic by product of existence, others regard those characteristics as fundamental to it and some take a neutral stance.

Those in the first category are typified by the many worlds/many minds/anthropic/decoherence interpretations held by many physicists, for example Tegmark (Ref 7.3.3).

Supporters of this general thesis are happy to accept the bizarre consequences of endlessly splitting universes.

The first level of Tegmark's thesis involves a vast number of parallel universes which are different in detail, but ultimately start to produce exact replications based on calculations involving the possible arrangements of particles with size based on the Planck length and the diameter of the observable universe.

The second level relates to the bubbles of space assumed to arise in relation to chaotic inflation described in earlier sections. A third level relates to the universes presumed to arise in correspondence with quantum super positioning i.e. many worlds also described earlier. Finally a fourth level is deemed to be the so-called multiverse in which differing mathematical structures apply involving endlessly different physics.

This and the Many Worlds interpretation correspond to a state where consciousness is merely a rare sideshow and even then one where that consciousness is totally fragmented.

Physicists who hold these views are also keen on tying the occurrence of consciousness to the weak or strong anthropic principle.

Those physicists who take a neutral stance are typified by the shut up and calculate approach which essentially avoids asking questions other than those arising from the abstract nature of the mathematics. Included in this neutral category are physicists who support the Copenhagen interpretation excluding perhaps Bohr's adherence to the wholeness of that being observed with the experimental apparatus and the observer.

In addition those physicists who believe in the reality of collapsing wave functions can also be regarded as taking a neutral position. These parties seek a theory of quantum gravity compatible with the common human experience.

Finally there are those physicists limited in number who believe that consciousness is a fundamental characteristic of existence, comprehend the principle of wholeness and regard science and the concept of order associated with that principle as totally compatible. Bohm and his associates essentially fall in to this category.

This does not mean that Bohmian mechanics and the modified Schrödinger equation incorporating the quantum potential is correct as it stands, but it is at least a major step in the direction of introducing the concept of order into the debate, while removing the bridle on non-locality.

Accordingly in this interpretation determinism, locality, uniqueness of history and counterfactual definiteness, while apparent to the human observer, can be considered a limiting case in which for practical purposes those characteristics exist independently of the observer.

A further contemporary discussion of this and other interpretations of the quantum theory are given by L. Smolin (Ref 7.4.25).

7.4.9. The extraordinary preciseness of the big bang for which the original phase space volume would have to be of the order of one part in 10 to the power 10^{123} of that compatible with the second law of thermodynamics and the universe now observed.

The concept of fine tuning and the associated cosmic coincidences apparently necessary to explain life in the observable universe.

It was explained in Section 7.4.4 that the original phase space volume would have to have been of the order of one part in 10 to the power 10^{123} of that compatible with the second law of thermodynamics and the universe now observed. Penrose (Ref 7.4.3) increases that power to one part in 10 to the power 10^{124} to allow for dark matter.

This extraordinary figure appears a coincidence par excellence which is crying out for an explanation.

Physicists regard the universe apparent to us as fine-tuned on the basis of the anthropic principle and the proposition that the conditions that allow life in the universe can only occur when certain fundamental constants lie within a very narrow range, so that if these constants were only slightly different the universe would not be conducive to the establishment and development of matter, astronomical structures, the elements or life (at least carbon-based life) as it is understood.

It is not evident though that the anthropic principle distinguishes in any case between universes in which life is minimally present and those in which life proliferates and grows with increasing complexity.

It has been noted earlier in Section 7.4.3 that the ratio of the actual density of the universe to the critical density for the universe to have zero curvature is the incredibly low value of 10^{-60}.

Other coincidences include that if the strong nuclear force were 2 per cent stronger, hydrogen would fuse so easily into di protons (two protons) instead of deuterium (proton plus neutron) and helium, that it would be consumed minutes after the big bang. Approximately 74 per cent of the mass of the universe is hydrogen and 24 per cent helium, and all the other elements less than 1 per cent. The synthesis of heavier elements including carbon is critically dependent on the interplay between the strong and weak nuclear forces, such that if the strong force were to be weaker or stronger by 1 per cent, no carbon or heavier elements would have been produced.

If the weak force had been slightly stronger primordial neutrons would have decayed faster and less helium produced, so as carbon is critically dependent on helium for formation there would be little or any carbon. If the weak force had been slightly weaker this would have lowered the proton to neutron ratio reducing the amount of hydrogen, starving stars of fuel for nuclear energy.

The electromagnetic to gravitational strength ratio is approximately 10^{40}. If gravity were slightly stronger all stars would be radiative instead of convective and planets might not form. If gravity were slightly weaker then all stars would be convective and supernovas might not happen, meaning elements from carbon up would not be synthesised in such stellar explosions. The proton to electron mass ratio is such that the neutron's mass is slightly more than the combined mass of the proton, an electron and a neutrino. As a result free neutrons spontaneously decay. If the neutron mass were slightly lower by 1 per cent isolated protons would decay instead and few atoms heavier than lithium would form.

The cosmic wave background is virtually uniform but has a slight anisotropy, roughly one part in 10^5, just enough to form stars and galaxies. If that anisotropy were smaller the early universe would have been too smooth for such formations; alternatively if it were larger, galaxies would have been denser resulting in many stellar collisions and stable stars with planetary systems would have been rare.

One of the most startling coincidences concerns the cosmological constant.

Based on the principles of quantum mechanics the zero point mass density or vacuum energy density implies empty space weighs 10^{93} grams per cc.

The actual average is 10^{-28} grams per cc. With the discovery that the cosmological constant has a very small positive value, this left physicists with need to explain the fact that the positive and negative contributions to the constant cancel to 120-digit accuracy but not to the 121st digit. Curiously the physicist Stephen Weinberg predicted in 1987 from basic principles that the cosmological constant must be zero to within roughly one part in 10^{120} as otherwise the universe would have dispersed too fast for stars and galaxies to form or else would have collapsed on itself long ago.

The recent discovery of what appears to be the Higgs boson at CERN in 2012 gives it a mass of 126 GeV, but calculations of interactions with other particles gives a value of 10^{19} GeV. This means the rest mass of the Higgs boson must be almost exactly the negative of this enormous number. Supersymmetry might solve this problem if the existence of super partners applied, but there is no experimental evidence for their existence to date.

It is believed by most physicists that all these coincidences can be explained by the anthropic principle together with the chaotic inflation/ many worlds/multiverse scenarios referred to earlier in this section.

Otherwise they believe that the diversity of view is likely to go away when existing observations/calculations/models are found to be in error.

7.4.10. Science and the Reference Model.

The mathematical physics referred to in Sections 7.4.2 thru 7.4.9 is that considered by the author to be most relevant to the Reference Model put forward in this book, for reasons given in this Section 7.4.10. With regard to the scientific method described in Section 7.4.1 it is clearly a means of establishing reality at the mean level of order (related to the local real axis) as opposed to religious belief based on imagination (related to the non-local imaginary axis). Another way of putting it is to say science is a bottom up view of the Ordering Centre as opposed to the top down religious view, the one being the inverse of the other.

7.4.10.1. *The Lambda CDM Model, dark matter and dark energy.*

The Reference Model regards the Theory of General Relativity, the concept of the Big Bang and the Standard Cosmological Model described in Sections 7.4.2 and 7.4.3 as being the explicit description unfolded from the mean level of order in an Ordering Centre all as described in Section 4.0. It is therefore unnecessary to repeat that description here.

Section 7.4.3 goes on to describe the scientific ramifications of the discovery of dark matter and dark energy and it is now necessary to interpret these phenomena in terms of the Reference Model. It does so in respect of dark energy by including the effect of those levels of order below that of 'our' apparent big bang which must increase disorder indefinitely to the point where the Ordering Centre decoheres completely, returning the system to the random state.

Dark energy appears uniformly distributed as if our 3D world is but a point in the multidimensional ensemble of lower ordering as shown in Fig 4.5.1.

The cosmological constant related to it must be tuned to a value consistent with the mean level of order associated with our level of consciousness. If it was significantly different, conscious entities such as Homo sapiens would never have appeared.

Dark energy can therefore be associated with the concept of particles and their ever-increasing diffusion at lower levels of order in the Ordering Centre, apparent to us only through its global (non-local) effect on the four-dimensional unfoldment housing our inanimate core, an effect that repels gravity.

Almost the same explanation applies to dark matter also discussed previously in this section except that those levels of order above that of 'our' apparent big bang, are responsible for the gravitational effects of it apparent to the physicist.

Dark matter is not uniformly distributed as if the multidimensional ensemble of higher ordering appears but a point in our 3D world again as shown in Fig 4.5.1.

Again there are strict limits on the strength of dark matter compatible with the mean level of order of 'our' apparent big bang. If it were stronger the universe would have quickly collapsed and the structures observed in the universe would never have developed. Likewise if it were weaker the expansion of space would have prevented the aggregation of matter which has made those structures possible.

Dark matter is evidently related to higher levels of order in the Ordering Centre again apparent to us through its local gravitational effect on the four-dimensional unfoldment housing our inanimate core, which accounts for its resistance to scientific enquiry dominated by the concept of particles.

The mean level of order gives rise to the unfolding of a four-dimensional gravitational field. Therefore the Reference Model has no difficulty in accommodating General Relativity while at the same time forming a relationship with other levels, in accordance the description given in Section 4.0.

Section 7.4.4 engages with the conflict between inflation and fine-tuning versions of cosmology in the physics community.

Chaotic inflation is clearly a reductionist thesis (opposed to the holistic view) which implies inflationary cosmology applies to low levels of order when chaotic conditions prevail leading ultimately to the breakup of the Ordering Centre concept.

In support of this it is difficult to reconcile chaotic inflation with increasing order, it is instead related to the breakup of order into explicit disorder.

The conformal cyclical cosmology as proposed is equally clearly related to the mean level of order, its cyclical nature associated with the principle of fine tuning being supported by the Model.

It however implies a closed loop without the ordering process described in Section 4.0 which supports the growth of consciousness and human imagination. It is scientific humanism on a grand scale omitting the rich picture based on the concepts of Wholeness and the Implicate Order proposed by Bohm (Ref 2.5.1) which provide the background from which the principle of the Ordering Centre is developed in this book.

The rich picture corresponds to higher levels in the Ordering Centre where unlimited fine tuning and the present state dominate as evidenced by the low entropy at the Big Bang, giving rise to the forms of life with which we are familiar. The relation between the reduction process and consciousness is further discussed in Section 7.4.10.7.

7.4.10.2. The quantum state.

Section 7.4.5 describes the quantum state which is incorporated into the Reference Model from the outset. While the observed (real) classical state is associated with the horizontal grid in two dimensions the unobserved (imaginary) unitary quantum state is associated with the corresponding vertical grid, just as real and imaginary axes are portrayed on an Argand diagram. The standing wave system symmetrically disposed about the present state (corresponding to the time independent Schrödinger equation), takes the role of the probability distribution.

The fractal unfolding and enfolding processes which give rise to the expansion and contraction of the present state is exactly what the doctor ordered to explain quantum super positioning and the non-local character of the quantum state.

It is emphasised that the imaginary present state orthogonal to the real distributed state, together with probability and self-similarity all evident features of the quantum state are built in to the Reference Model from the start, so those very features which remain inexplicable to the scientist come as no surprise to the Reference modeller.

The scientific community takes no account of imagination other than in a mathematical sense. Imagination enables us to comprehend that the quantum state, at higher levels of ordering than that corresponding to our explicit world, can be considered as enfolded to a point in our space-time as portrayed in Fig 4.5.1, which is why particles appear to have zero dimension and why they can be in two positions at once before conversion to the unfolded four dimensional space-time we observe and use for measurement.

It is the contention of the author that the Reference Model provides the essential background against which the features of the quantum state as described in Section 7.4.5 become comprehensible.

These features are ultimately a manifestation of the Wholeness of the Implicate Order. It is a state in which four dimensional space-time becomes less meaningful as non-locality takes over. The intersections of the grid structure give rise to the wave particle interaction and the associated duality so readily apparent. On measurement the particles appear on the distributed state corresponding to the mean level in the Ordering Centre where locality applies. The particle orientated component of this reduction process to that level is responsible for our sense of reality and the wave orientated component for our sense of imagination.

7.4.10.3. The inevitable conflict between experiment and theory.

Section 7.4.6 refers to the difficulties in relating experimental observations to theory in terms of both the Standard Cosmological Model and the Standard Particle Model. The Reference Model offers important clues in an effort to resolve these matters and has further implications with regard to the question of Grand Unification Theory and also the role of gravity in the overall scheme of things.

It is suggested that the many problems apparent in trying to formulate such a theory and the discrepancies between the experimental results supporting the Standard Cosmological Model as opposed to the Standard Particle Model arise from ignoring the Implicate Order and the processes involved in unfolding and enfolding order as illustrated in Fig 4.7.1.

The peak of the explicit big bang ordering envelope emphasises attractive gravity, mass diffusion, particle aggregation and a holoarchical structure. The dominant holoarchical structure centred on the present (inanimate) state emphasises symmetry with energy density increasing in the ordering direction, together with short wavelength high frequency wave motion.

Decreasing order and energy density towards the base of the ordering envelope implies repulsive gravity, mass aggregation, particle diffusion and a hierarchical structure.

In this case the dominant hierarchical structure centred on the distributed (inanimate) state emphasises anti symmetry with energy density decreasing, together with long wavelength low frequency wave motion, reducing towards that of the scalar field.

While the Ordering Centre holds together, a general degree of symmetry and anti-symmetry applies but chaotic inflation and the related concepts of many universes associated with string theory, endless vacuum states and the consequent collapse of the Ordering Centre will ultimately follow, any notion of overall symmetry or unification of any kind collapsing with it, all governed by the quantum fluctuations of the random state.

It is for these reasons that the scientist is pushed towards the anthropic principle to explain life.

It follows that the present situation with regard to the scientific standard models can be explained in terms of the intermediate or mean level of order in which we find ourselves, where both attractive and repulsive gravity is weak relative to the other three forces giving rise to the particles evident to us using particle accelerators.

All such experiments are clearly associated with and tied to the mean level of order and the explicit four-dimensional world which unfolds from it, where measurements forming the basis for both the Standard Cosmological Model and the Standard Particle Model apply.

The Special and General theories of Relativity are also essentially tied to the explicit world witness Einstein's reluctance to accept the 'spooky world' of quantum super positioning.

The unification of the quantum state with gravity therefore depends on allowing for the non-local implicit order.

The situation in respect of the Standard Particle Model is more complicated but it is apparent that it attempts to eliminate unwanted infinities in an effort to define explicit reality while trying simultaneously to incorporate quantum theory, which is why physicists feel themselves forced in to renormalisation procedures, while ensuring important physical constants are tied to observations rather than the theory.

One particular conflict in this connection is evident since the Standard Particle Model is totally at variance with observation in respect of the cosmological constant as opposed to the zero point energy or vacuum state, commonly assumed to be equivalent.

If the universe is described by an effective local quantum field theory down to the Planck scale then a corresponding cosmological constant would exceed the observed value by variously some 60 to 120 orders of magnitude. It cannot be so; therefore an alternative approach must be found to the present one which evidently, is equivalent to throwing the baby out with the bath water.

The Reference Model implies that Fig 4.5.4 provides the key to understanding that the quantum state of infinities and endless vacuum states is governed by Hilbert space and the endless infinite particle state by phase space, all in turn related to the explicit level of order with limited dimension we call the real world.

Infinities and vacuum states multiply along the imaginary axis because Feynman diagrams give integrals which diverge when momentum values get indefinitely large and distances indefinitely small.

The Standard Particle Model in a sense therefore correctly represents the collapse of the above infinities and the unfoldment of explicit particles, as the explicit world unfolds.

The same argument applies to the vacuum energy density which arises as a result of the quantum fluctuations constituting energy, which would add to the cosmological constant. Since, as with the infinities, the profusion of vacuum states collapse as the explicit world unfolds, so the quantum fluctuations no longer add to the cosmological constant over and above its minute observed value.

7.4.10.4. *Measurement and the independence of consciousness.*

Both the scientific cosmological and particle models are based not only on the assumption that explicit measurements made are independent of consciousness, but also on local realism.

The Reference Model defines the explicit world as being a region with a balanced vertical and horizontal grid structure supporting not only wave motion but the formation of particles while confirming that at the mean level of order scientific measurements should be effectively independent of consciousness.

Otherwise there is a significant gap between the objective and subjective philosophies which this book seeks to bridge.

Unification of our concepts of reality with imagination is dependent on regarding the present state on the imaginary axis and the mean level of order on the real axis as being in a state of motion all as described in Section 4.0, as opposed to being fixed in an absolute sense as both science and religion in effect demand.

The Standard Cosmological Model and the Standard Particle Model based on General and Special relativity and quantum field theory are clearly related to the fixed real axis and the principle of locality. The cosmological constant as observed is also related to the fixed real axis as the ease with which it is incorporated in to Einstein's field equations demonstrates.

On the other hand the quantum state is also clearly related to the fixed imaginary axis.

When non locality invades the scene a rich picture involving unfolding and enfolding axes is required. Consciousness and awareness involving the non-local theory proposed in this book can then in turn be related to the asymmetric motion about those axes without interfering with scientific observations.

It therefore should not be a surprise that uniting the quantum state with General Relativity is a difficult enterprise especially when the scientific community in general is hell bent on emphasizing locality and de-emphasizing non-locality in their basic assumptions.

It is the combination of the local theory with non-locality that is necessary to explain consciousness together with our observations and our sense of reality in balance with imagination.

7.4.10.5. *Matter and antimatter.*

The Reference Model gives a clue perhaps to the imbalance between matter and antimatter. It is suggested that they will be in a state of balance along the present state imaginary axis, an inanimate state at the core of the Ordering Centre.

Figures 4.4.1 and 4.5.4 are most relevant to the argument here. If the full lines in Fig 4.4.1 tend to represent matter separated from the dotted lines correspondingly representing antimatter, then increasing decoherence corresponds to increasing imbalance between matter and antimatter.

Increasing coherence gives rise to the structure implied by Fig 4.5.4, and the principle of the Ordering Centre. Increasing decoherence is associated in the Ordering Centre with the dominance of unfolding explicit processes involving explicit particles apparent along the real explicit axis and increasing coherence with the dominance of enfolding implicit processes involving anti particles.

This proposal is supported by Feynman diagrams associated with the CPT Theorem which uses the time reversal aspect to distinguish particles moving forward in time (unfolding) from anti particles moving backwards in time (enfolding).

As these are symmetrical processes there is no need to introduce consciousness into the argument as explained in the previous Section 7.4.10.4.

Unfoldment gives rise to the observations in accordance with General Relativity and the observable elements of quantum field theory based on locality while enfoldment gives rise to the non-local aspect of the quantum state, hence the dichotomy between the two and all the confusion only to fully apparent.

Ultimately in the Whole Ordering Centre matter and antimatter can presumably be regarded as one whole stuff.

7.4.10.6. Black holes.

Section 7.4.7 concerns the phenomena of black holes so it is necessary to consider the relation between black holes and the Reference Model. Black holes have important beneficial effects in forming and sustaining galaxies yet at the same time are clearly associated with disordering processes involving states of very high entropy.

This corresponds to the requirement of the Reference Model for order in the form of information gain to be rising and disorder in the form of information loss to be falling in the overall scheme of things.

Black holes may in some sense lose information but it turns up elsewhere, a process involving the Whole Ordering Centre, as Section 4.0 of this book describes.

Just as apparent singularities become evident when the levels of order applicable to 'our' big bang are exceeded, so the same scenario applies in the opposite sense when the corresponding levels of disorder are exceeded; this time in the form of black holes.

Once again as Fig 4.7.1 implies, there are no absolute singularities, so the information would reappear in a less ordered form in lesser big bangs applicable to the bottom part of the figure.

Ultimately this leads to the chaotic inflation scenario and the further collapse even of that, followed by the reinstatement of the random state previously referred to.

The central present state in the Whole Ordering Centre ensures that the apparently lost information is ultimately recirculated from the random state to the central ordering column. So the whole system can therefore be considered an order recycling process, but where there is no absolute ordered state, chaotically disordered state or random state, so the cycling loop never closes.

7.4.10.7.   *Interpretations of quantum mechanics and its relation to consciousness.*

Section 7.4.8 enquires in to the many varied interpretations associated with quantum mechanics. It is therefore essential to examine these from the perspective of the Reference Model.

The many worlds interpretation is found to relate equally to the present state and distributed state axes. In this case the wave function does not collapse and indeed this condition applies to the present state since it includes all orderings and disorderings but enfolded as opposed to unfolded. Nor does in effect the wave function collapse along the totality of distributed unfolding states since wave motion then encompasses the range of high frequency short wavelength waves to those of low frequency long wavelength. Where the wave function can be said to objectively collapse is when the implicit mean level of order converts to the explicit mean level as described by the Reference Model.

The many worlds concept regards such events as decoherence processes which give rise to the endless splitting into alternative worlds all mutually unobservable, each of which can be regarded as local.

The Reference Model confirms these conditions apply towards the base levels of the Ordering Centre.

The majority of scientists adopt the neutral shut up and calculate approach assuming objective collapse of the wave function corresponding to the mean level of explicit order, with a minimal allowance to cater for the quantum state, which corresponds to some degree of enfoldment (or in their terminology entanglement) in the mean level of implicit order. Conformal cyclic cosmology is a significant extension to the neutral stance, its cyclic nature and the explicit unfolding of space-time being features supported by the Reference Model.

However as stated in Section 7.4.10.2 it implies a closed loop which fails to support the growth of consciousness and imagination as described in Section 4.0 of this book.

The further relation between the reduction process and microtubule structures in the brain proposed by Penrose and Hameroff referred to in Section 7.4.8 is also supported by the Model, provided it is regarded as representing the basic explicit description enfolded in the mean level of order. It should not therefore be associated with consciousness, merely the organic and biological substrate from which consciousness unfolds again as described in Section 4.0.

In this sense it corresponds to the frog's eye view of the Ordering Centre as opposed to the bird's eye view and the vast range of higher and lower levels of consciousness which arise from the latter.

The Reference Model therefore provides a robust explanation for the reduction process all supported by cosmological observations and simultaneously an explanation for consciousness.

The Penrose interpretation may be a limited view of the reduction process at the level of the Planck mass but it certainly does not provide an explanation for consciousness.

Only a limited number of scientists support the rich picture as it emphasises implicit properties and non-locality, which other than that revealed by Bell's Theorem (see Section 7.4.5), are not accessible to the real world discipline of scientific measurement.

In particular the horizontal grid representing phase space in Fig 4.5.4 is related to reality, the scientific process of measurement corresponding to the mean level of order when the principle of real atomistic particles dominates.

Hence the evidence provided by the two slit experiment where the particles are observed using detectors going through a particular slit.

In turn the vertical grid representing Hilbert space in the figure is related to the imaginary, the process of non-measurement corresponding to the present state when the principle of imaginary wave motion dominates.

Hence the evidence of wave motion in the two slit experiment where the unobserved particles with detectors switched off behave as waves going through both slits simultaneously.

Since the central axes of the Ordering Centre grid are inanimate conditions we can go from measurement in phase space to non-measurement in Hilbert space and back again without involving consciousness in our deliberations.

It is only when interference between these conditions and Implicate space, with the effects of consciousness all too apparent, that arguments over the relative merits of realism and anti realism arise.

In particular as shown in Fig 4.7.1 the infinite extent of the Whole Ordering Centre with lesser Ordering Centres enfolded within it, implies a range of mean levels of order each representing reality local to that level.

It is clear that those physicists obsessed with the reality of particles are looking at conditions close to any given mean level of order while accepting the principle of non-locality applicable to the present state. By avoiding the involvement of consciousness as above, the dichotomy between locality and non-locality then disappears as far as they are concerned.

It is only then it becomes apparent that realism and anti realism should be associated with the inanimate framework and the implicit should be associated with animate motion relative to that framework.

Consciousness in the general population then expands such arguments in to that between the objective and subjective, science and religion etc with consequences familiar to all.

Bohr was opposed to this emphasis on realism, namely that there is no quantum world only an abstract quantum physics description; that it is wrong to think that the task of physics is to find out how nature is, that it is only concerned with what we can say about nature. Although most modern physicists in practice follow this argument they prefer to keep non-locality in the background and remain entranced by particle physics, wanting to build ever more powerful accelerators with which to smash any new particle they can find to smithereens in an effort to get closer to reality.

The Reference Model overturns all these limitations looking at a much wider picture including consciousness by the introduction of Implicate space which forms in the Ordering Centre, i.e. the space where animacy and consciousness reigns as opposed to inanimate Hilbert space and phase space related to the imaginary and real axes respectively. It is when the wave function involves implicit motion as opposed to explicit motion that animacy and consciousness unfold from the inanimate.

So the many worlds decoherence and chaotic inflation approaches correlate with the conditions which apply towards the base of an Ordering Centre while the Implicate order approach correlates with more ordered conditions towards the peak. Neutral or intermediate interpretations correlate with the mean level of order.

It is particularly significant that both the de Broglie Bohm and the many world interpretations do not support the collapse of the wave function, while the neutral interpretations insist on it.

The Reference Model offers explanations for these and the other interpretations in terms of the Ordering Centre principle described in this book, while providing a means of relating the micro world of the quantum state and the macro world of classical physics, by eliminating the concept of absolute boundaries and absolute singularities.

With regard to the speed of light, the Reference Model implies this is relative as opposed to absolute. The apparent absolute value associated with Special and General Relativity, which are holistic theories, are basic assumptions of the Standard Cosmological and Particle Models as described earlier.

These experimental results are however tied to the mean level of order and the Reference Model confirms experiments at this level will produce no other result.

However this and other mean levels or harmonics in the Whole Ordering Centre have a preferred reference frame, incorporating the present state imaginary axis, along which space-time effectively evaporates.

Along each mean level of order, the corresponding real axis there will, again based on holistic considerations, be a maximum finite speed for the transmission of information. Hence the speed of light should be regarded as relative when taking a bird's eye view of the whole, as opposed to the apparent absolute value seen from the frog's eye view presented by a given mean level.

7.4.10.8. *Life and the coincidences associated with it.*

Section 7.4.9 considers the extraordinary, fine-tuned preciseness of the Big Bang and the many extraordinary coincidences apparently necessary to explain life.

The Reference Model gives a much more subtle explanation than the scientific models, where the ordering process operates in a cycle which never closes between unlimited order and unlimited disorder and vice versa, human consciousness being merely one level in this process.

In particular the Model explains how and why the Ordering Centre forms the differences and similarities of which order is composed, the peak of the Ordering Centre representing an ever more accurate present time.

The Reference Model relates the preciseness of the big bang to the exact symmetry with respect to space-time that must apply when viewing the Ordering Centre in its inanimate state and setting the mean level of order as merely one level in the ever-greater set of orderings which extend to infinity in the Whole Ordering Centre.

The coincidences described above follow from the growth of animacy and consciousness from an inanimate base to a peak of all knowing wholeness, with the mean level of order representing the human level of comprehension.

All of the opinions and beliefs expressed can be interpreted in terms of the unclosed cycle referred to above, the range and diversity of philosophies being due to the ordering process in which similarities collect together and differences spread apart.

The continued failure to understand this is in part due to ignorance but above all the absence of a working model of existence in the human mind.

It is therefore also absent from everyday conversation and the written word, a void which this book seeks to fill.

7.4.10.9.   *Unification of local and non-local theses.*

It is contended that the description of the Ordering Centre provides a basis for the unification of the Standard Scientific Models.

As stated in Section 4.7 the peak of the Ordering Centre ultimately explains the source of energy, attractive gravity, dark matter and processes relating to mass diffusion and particle aggregation. The distribution towards the base of that structure ultimately explains the dissipation of energy, repulsive gravity, the source of dark energy and processes relating to mass aggregation and particle diffusion. The breakup of the distribution leading to the random state corresponds to the collapse of order to a state of minimum quantum fluctuations, where the notions of particle, wave, space and time have no meaning.

The above scenario rules out the absolute singularity and absolute beginnings or endings in time, replacing them with ever increasing wholeness ultimately in accordance with the concept of the Whole Ordering Centre which forms the core thesis of the Reference Model.

The Implicate order implies that increasing energy density in the Ordering Centre is consistent with Supersymmetry but not with a profusion of explicit particles, more likely a one diffuse mass, an implicit unified particle state, all discrete wave (short wavelength high frequency) scenario. So, at this high level of order the atomistic concept in general may have little validity.

In this case attractive gravity is likely to be more powerful than the very weak value relative to the other three forces (the strong force, the weak force and the electromagnetic force), at the mean level of order with which we are familiar. The further implication is that these forces and attractive gravity become comparable until a state of equilibrium is achieved at high energy densities involving one unified force, assuming some degree of commonality of physics with that applicable at our level of order.

The inanimate grid associated with an Ordering Centre in the direction of increasing order is vertical rather than horizontal, as shown in the figures illustrated in Section 4.0. Hence the phenomena of dark matter evident to us, but no evidence of associated particles, since it is enfolded or implicit relative to our level of order.

When the system converts to the Whole Ordering Centre with one consciousness, a Supersymmetric Grand Unification Model should apply along the inanimate imaginary central axis, about which the Holomovement oscillates.

The enfolded implicit order then explains the observed absence of the hypothetical particles assumed in the existing Minimal Supersymmetric extension of the Standard Particle Model, which like the Standard Particle Model ignores non locality.

It follows that decreasing energy density in the Ordering Centre will promote the atomistic concept, a diffusion of explicit long half-life particles, indiscrete (long wave length low frequency) waves and repulsive gravity all tending to a state of equilibrium.

Relating the above to consciousness, it is evident that it appears to the conscious observer that the non-local quantum state collapses to the local explicit reality as described in Section 4.0. The processes described therein confirm that the conscious explicit side of our nature (the realist) tunes in to the real four-dimensional distributed state while the aware implicit side (the dreamer) tunes in to the multidimensional imaginary present state.

The scientist tries to relate everything to reality and measurement while the dreamer ignores these and tries to relate everything to the imaginary.

Section 4.0 therefore also explains the inevitable confusion that arises, since order made up of reality and imagination cannot be formed without sequences made up of both these elements using Implicate space. Hence the inability of the two parties to agree.

Having considered the above there is a further complication arising from the nature of enfolded order as portrayed in Fig 4.7.1, and the figures in Section 4.8 dealing with consciousness. There is little likelihood of our being able to describe mathematically the relation between reality and imagination; reality only yes, imagination only in the form of the unitary quantum state yes, but combinations of both, almost certainly no.

Hence the difficulties of relating general relativity to quantum theory even though they both deal with the inanimate only. When consciousness is added calculation appears impossible.

Furthermore while these more ordered structures seem to be but a point in our four-dimensional world, when unfolded the disordered stuff returning to the base of the Ordering Centre involves distributed dimension in which our four-dimensional world is in turn reduced to a point.

So it is best to concentrate on finding a measure of understanding by fusing our comprehension of reality and imagination in the form of a unified metaphysical model of the type described in this book, a process which must not be beyond us.

The structure of the Ordering Centre and in turn the Whole Ordering Centre, offer a way out of this dilemma.

The concept of the Whole Ordering Centre in the Reference Model implies a progression towards a unified physics from a potentially infinite variety of physics in the opposite sense.

Crucially the Model does not imply all statements are true only that they all exist relative to an ordering framework governed by the present and distributed states which oppose collapse to the random state.

However, the question remains how at high levels of order the unification of reality and imagination, particle and wave, locality and non-locality is achieved.

The Reference Model therefore implies that absolute unification is not possible, forcing the Whole Ordering Centre into a state of never-ending motion.

Hence the formation of mean levels of order relative to which consciousness forms, in turn giving rise to the evident conflict between the objective and subjective.

7.4.11. Time and the Reference Model.

What is time, past present and future?

Time asymmetry i.e. the forward flow of time.

The frozen past. We can reverse a film reel, why not time?

To what extent is existence symmetric and to what extent asymmetric?

Existence has no limits nor is it stationary; it is in a state of motion.

The Ordering Centre and levels within it (our level being one), involve cycling between order and disorder, but a cycle that never closes. That part of the cycle involving increasing chaotic disorder from a given level, gives rise to the unfoldment of dimension (space-time) from what is termed the present time and therefore the passage of time from that present to the future.

This passage of time due to the unfoldment of dimension in an explicit sense is to be associated with our comprehension of the first and second laws of thermodynamics, giving rise on average to increasing entropy and the inevitable unfolding of disorder in that sense.

Inanimacy is a property which increases with this explicit disordering process and can be regarded as symmetric with respect to the present time.

Now in the opposite sense to the above, that part of the cycle involving increasing order logically enfolds dimension from the present time to the past. This process by definition gives rise on average to decreasing entropy and an inevitable unfolding of order in an implicit sense.

The inanimate state as seen by human observers is the condition which applies at the current level of explicit order looking outward in the direction of unfolding dimension and variance from what is termed the present time.

From the viewpoint of the inanimate state the present time at the big bang was less than 10^{-40} seconds, is currently approximately 13.8×10^9 years and will ultimately unfold to a value far in excess of 10^{100} years. Accordingly the variation in the inanimate present time can be related to the explicit ordering envelope as shown in Fig 4.3.3.

For a given level in the Ordering Centre the inanimate grid fixes the explicit present time and provides the reason why the equations of physics in theory run backwards and forwards in time with equal facility.

The Reference Model implies the mean level of explicit order occurred roughly 8.8×10^9 years after the Big Bang, unfolded from the mean value of implicit order as shown in Fig 4.8.1, which relates the explicit and implicit states.

Human comprehension of the present time is far more accurate than 13.8×10^9 years which the Model implies is the inanimate figure corresponding to our appearance on the scene; the former is related to the short period not precisely determined but of the order of a second taken by an individual to be conscious of an external stimulus. We are therefore capable of measurement of the present time to an accuracy similar to the explicit value that applied close to the time of the big bang, because the level of order of the human brain is close to the level of explicit order which applied then.

In fact the brain is physically capable of response to such a stimulus within a tenth of a second.

This evidence of a time delay in consciousness arose from observations made following EEG experiments on the brain originally by Deeke, Grotzinger and Kornhuber (Ref 7.4.26) and further experiments by Libet et al (Ref 7.4.27).

In particular that there is a build-up of electrical potential in the brain over a second before the individual is conscious of responding to the stimulus, implying a form of a priori knowledge is involved. Relevant details can be found on the internet.

A possible explanation for all this according to the Reference Model, is illustrated by the figures shown in Section 4.0.

The present time is a fractal state which expands and contracts in an implicit sense as shown in Fig 4.4.5, as opposed to the inanimate explicit state which is always unfolding. The apparent reduction process shown in Fig 4.5.3 is opposed by the expansion of consciousness using Implicate space as per Fig 4.5.4.

Human observation of the present time is then that conscious perception corresponding to any given level of implicit order reached by an animate/conscious entity on the present state (imaginary axis), as opposed to the explicit level of the real world, all as per Fig 4.8.1, 4.8.2 and 4.8.3.

The global 'now' as perceived by Homo sapiens corresponds to the level of implicit order of that species, while 'now' measured by the real world is that corresponding to the relevant level of explicit order.

The level of explicit order it should be noted has been falling below the mean explicit level for the last 5×10^9 years, counterbalanced by the rise in implicit order from the real base level to that of ourselves over this period.

It is the grid which is responsible for the inanimate core of the conscious observer being in correspondence with the level of explicit order, from which he derives the common physical (measured) experience that there is only forward motion of time.

It takes consciousness to physically create a film reel and set it in forward or reverse motion, mimicking the motion involved in increasing or decreasing implicit order. For example the observation that a glass of water falling from a table and crashing to the floor, shattering and spreading the contents increases the entropy i.e. reduces the order of the system, while the reverse is never seen.

Measured time therefore only expands from the relevant explicit level while conscious time can expand and contract relative to the relevant implicit level.

The enfoldment of the present time is a process which can increase or decrease order, whereas unfoldment can only increase disorder on average.

Such enfoldment is to be associated with increasing wave motion in the belief and imaginary worlds as opposed to the restricted motion in the real world.

Animacy and consciousness are primarily properties of belief and imagination which can be associated with both the ordering and disordering processes all related to the mean level of implicit order. Furthermore they unfold and enfold the present time, otherwise fixed by the inanimate, according to the direction of the animate vectors relative to the Ordering Centre grid. Such vectors as previously described emerge from wave motion asymmetric to that grid.

It is the asymmetric wave motion that enables the observer to remember the past and make plans for the future, because it is his conscious element which unfolds or enfolds the present time increasing or decreasing his level of implicit order relative to the mean level. It is this wave motion which drives consciousness in the direction of greater wholeness (opposing the otherwise inevitable decoherence of existence to a totally inanimate disordered state) and this same motion which consciousness can unfortunately, use to create unwholeness in the opposite sense.

The future to a conscious entity is always that which is unfolded and the past always that which is enfolded. The enfoldment is the source of the frozen past, as Omar Khayyam said, 'the moving finger writes and having writ moves on, nor all thy piety and wit, shall lure it back to cancel half a line, nor all thy tears wash out a word of it'.

7.5   Human belief and Religion

7.5.1.   Atheism.

Atheists split into those associated with state control as in communist China for example and those who claim to be 'free thinkers', some egocentric in character who regard secularism as a superior stance to religion. Their grounds for this being that reductionist analysis aimed at defining reality is better than the holistic process of unfolding imagination as a means of viewing the world.

As has been said this results in most scientists being atheistic in outlook, to the point where some of them believe life may be exceedingly rare in the universe, hence the necessity of developing artificial intelligence.

Such intelligence would in their view allow 'life' under their definition to be exported across the galaxy and given time further still, when human life is long gone.

Science is struggling to define, even comprehend, what consciousness is, but assumes artificial life is merely another form or extension of the human version.

It has limited comprehension of good and bad since reality remains neutral in this regard. However they are quick to use these terms to distinguish good science namely that which stands the test of time with respect to the repeatability of experimental results (all in accordance the principles referred to in Section 7.4.1), as opposed to bad science which has a slovenly attitude to those principles.

7.5.2   Agnosticism.

The spiritual but not religious.

Agnostics are individuals who take the view that nothing can be known about God or anything not involving material things but stop short of the severe attitude of atheists who rule the possibility of the deity out completely. In short agnosticism is a state of suspended belief. It may be this book will clarify things for them so that they do not convert to the extreme views taken by absolutists whether religious or scientific.

Like agnosticism believers in the spiritual are individuals or small groups of individuals. Just as agnostics diverge from atheism so the spiritual individual diverges from theism and the mainstream religions. Inevitably this gives rise to more variation not only in character as they refer to themselves as spiritual but not religious, but also more changeable with time.

7.5.3. Theism.

The cosmological argument and its validity.

The ontological argument and its validity.

The argument from design and its validity

The debate between intelligent design, and the theory of natural selection.

Intelligent design as opposed to so called creation science.

As with science, confusion rules the religious roost, if not more so, at least there is only one science to contend with, but with religion comes a profusion of beliefs and rampant imagination, out of control of course.

Nevertheless philosophic discipline has some effect and theologians generally condense their basis for religious belief in a deity to the cosmological, ontological and design arguments, descriptions of which can all be found on the internet.

The cosmological argument, which has its roots in Greek philosophy (particularly Plato and Aristotle), is one in which the existence of a unique being, referred to generally as God, is deduced or inferred from facts alleged or otherwise from causation, motion, contingency etc in respect of the universe as a whole or processes within it. It is commonly called the argument from first cause or the causal argument.

The basic cosmological argument runs along the following lines:- that which begins to exist has a cause, the universe began to exist, therefore the universe had a cause, the cause is God, the cause exists, therefore God exists.

The Islamic philosopher Avicenna (Ibn Sina) in the early 11th century enquired in to the question of being, arguing that existence (Wujud) could not be accounted for by the essence (Mahiat) or intrinsic nature of existing things and was therefore due to a cause that adds existence to essence. In other words cause and effect must coexist.

Aquinas in the 13th century used Aristotle and Avicenna as a guide and formed the argument from contingency to explain why the universe exists.

It could conceivably not exist (contingency), so its existence must have a cause, namely God, which must exist by necessity in order for anything to exist.

In effect this equates to existence being meaningful and non-existence meaningless.

The philosopher Liebniz asked why there is something rather than nothing, there must be a substance bearing the reason for its existence within itself.

Aquinas' argument is termed in esse (essence) and in fiere (becoming) a theistic view, while Liebniz' argument is in fiere, a deistic view.

Objections to these cosmological arguments include what caused the first cause?

The philosopher and empiricist David Hume also argued that the premise of causality has been arrived at via a posteriori (inductive reasoning) which is dependent on experience and not a priori (deductive reasoning). In other words it is unsatisfactory to exclude either form of reasoning from the debate.

In addition the argument does not necessarily identify the first cause with a God; it could merely be for example the big bang singularity.

Further objections include the possibility of causal loops, where a powerful entity travels backwards in time to a point before its own existence creating itself and everything which follows. Finally an infinite causal chain could be its own cause.

Scientists originally argued that the big bang represented the start of all things including space and time and that asking what happened before was like asking what is north of the North Pole. Now however new concepts in physics challenge this limitation.

It may be noted that others consider the main question is not how the universe started but what keeps it going.

The ontological argument is associated with arguments about the state of being or existing, of an a priori nature.

Anselm in the 11th century proposed a version of the ontological argument, which has the following short form basis:- by definition a being greater than God cannot be imagined.

A being that exists in reality is greater than a being that does not exist. Thus, by definition, if God exists as an idea in the mind then God necessarily exists in reality.

Descartes argued that God is a supremely perfect being (it is assumed here existence is a predicate of perfection). If the notion of God does not include existence, it would not be supremely perfect, therefore God must exist. Liebniz considered that this assertion lacked coherence as perfection is impossible to analyse but reasoned that all perfections could exist together in a single entity and therefore Descartes argument was still valid.

Mulla Sadra, the Islamic philosopher, put forward an alternative version, the argument of the righteous.

Kurt Gödel (see G. Oppy Ref 7.5.1) proposed a mathematical proof of the ontological argument based on modal logic, which includes possibility as well as necessity.

For this and other versions of the ontological argument the reader is also referred to the internet sources previously mentioned.

Gaunilo of Marmoutiers a contemporary of Anselm in criticism said that if the argument is sound so are others of the same logical form, which cannot be accepted.

In addition that God by nature cannot be conceived i.e. comprehended therefore the argument does not work.

Anselm replied that the argument only applied to concepts with necessary existence.

Aquinas objected to Anselm's argument saying that people could not conceive God, that only God can know his essence, so only He could use the argument.

Other criticisms include the devil corollary and other forms of it which propose that a being than which nothing worse can be conceived exists in the understanding. If it exists in the understanding, a worse being would be one that exists in reality, therefore such a being exists.

David Hume argued nothing could be proven to exist using only a priori reasoning. Nothing is demonstrable unless the contrary implies a contradiction.

Whatever we conceive as existent, we can also conceive as non-existent. There is no being therefore whose non-existence implies a contradiction. Consequently there is no being whose existence is demonstrable.

It follows that it is therefore not a contradiction to deny that God exists.

The philosopher Immanuel Kant in his Critique of Pure Reason (1787) put forward a further influential criticism of the ontological argument aimed at Descartes and Leibniz. He distinguishes between analytic and synthetic propositions. In an analytic/synthetic proposition the predicate concept is/is not contained in its subject concept.

He questions the concept of a necessary being. If the proposition p exists, it follows that if p exists, it exists necessarily, without meaning it exists in reality.

Contradictions only arise when the subject and the predicate are maintained and therefore, a judgement of non-existence cannot be a contradiction, as it denies the predicate. Kant says that the statement 'God exists' must be analytic or synthetic.

If the proposition is analytic, as the ontological argument takes it to be, then the statement is a tautology which says nothing about reality. If on the other hand the statement is synthetic, the ontological argument does not work, as the existence of God is not contained within the definition of God.

Existence merely indicates its occurrence in reality; it is not a predicate or quality, because it does not add to the essence of a being. Finally Kant claims that the concept of God is an object of pure thought outside of the realm of experience and nature, it is therefore impossible to verify his existence in contrast to material things evident to the senses.

The third argument for the existence of God is the argument from design, more recently termed intelligent design, otherwise the teleological argument, which is the doctrine of final causes, the view that developments are due to the purpose or design that is served by them.

The earliest recorded versions of these arguments are again associated with the Greek philosophers including in particular the Stoics.

Early Islamic philosophy played an important role in developing the understanding of God among Christian and Jewish thinkers in the Middle Ages.

Al Farabi took the emanationist approach (issuing from a source) that whereby nature is rationally ordered, God is not a craftsman who literally manages the world; phenomena follow natural laws that God (the source) has created.

Avicenna was also convinced of this and also proposed a cosmological argument for the existence of God as explained earlier.

The Islamic philosopher Averroes (Ibn Rusd), a defender of Aristotelian philosophy, also accepted this argument and even Al Ghazali, the Muslim theologian.

Averroes proposed that order and continual motion in the world is caused by God's intellect, essentially in line with Al Farabi.

The principal demonstrative proof is Aristotle's proof that from motion in the universe there must be a first mover, which causes everything else to move. This physical proof rather than a metaphysical one was in opposition to Avicenna, and this influenced Thomas Aquinas who took a position closer to the latter.

Aquinas considers that the principle of cause followed by effect, can only be explained by an appeal to intelligence. He reasons that as natural bodies do not possess intelligence, there must exist a being that directs causes at every moment, that being we call God.

Newton and Leibniz also accepted the argument, though the former believed God constantly intervenes to keep His design adjusted, while the latter thought the universe was created in such a way that God would not need to intervene at all.

The empiricist John Locke developed the Aristotelian idea that, excluding geometry, all science must obtain its knowledge a posteriori, through sensual experience.

George Berkeley, in response, advanced a form of idealism in which things only continue to exist because they are perceived. When human beings do not perceive objects, they continue to exist because God perceives them, therefore God exists omnipresently.

Derham in the early 18th century published a number of books on the teleological theme which influenced the later work of William Paley with his watchmaker analogy.

According to the theologian Alister McGrath, Paley argued that the same complexity and utility evident in the design and functioning of the watch can also be discerned in the natural world. Each feature of a biological organism, like that of a watch, shows evidence of being designed in such a way as to adapt the organism to survival within its environment.

There are probabilistic supporters of the design argument. These include the version that while natural selection may account for each of the individual biological adaptations, it cannot explain their totality since the whole exceeds the sum of its parts.

Another Bayesian approach is based on the probability of God increasing with the growth of order in the world.

Other variations on the probability theme are built on the concept of the fine-tuned universe, in particular the amazing precision of nature's physical constants, all as discussed in Section 7.4.9. The fact that life as we know it could not have been possible if the delicate balance of conditions observed did not apply, an argument supported by the philosopher Antony Flew.

Unfortunately so-called creation science theories have become confusingly mixed with intelligent design arguments. Unlike Paley's openness to deistic design through God given laws, proponents of creation science seek scientific confirmation of repeated miraculous interventions in the history of life and argue that their theistic science be taught in science classrooms.

Virtually all literature promoting creation science presents the design argument modified in this way, thus rebranding it as a scientific theory rather than a religious one. These creation scientists assert modern science is atheistic, which indeed is generally true, and contrast it with their preferred approach of a natural philosophy which welcomes supernatural explanations for natural phenomena and supports 'theistic science'. This ignores the distinction between science and religion, established in ancient Greece, in which science cannot use supernatural explanations.

David Hume, using his character Philo a religious sceptic, argued that the design argument is based upon a faulty analogy as unlike manmade objects, we have not witnessed the design of a universe, so we do not know whether the universe was the result of design. Moreover the size of the universe makes the analogy problematic: although our experience of the universe is of order, there may be chaos in other parts of the universe. The order in nature may be due to nature alone. If nature contains a principle of order within it, the need for a designer is removed. Even if the universe is designed, it is unreasonable to justify the conclusion that the designer must be an omnipotent, omniscient, benevolent God, the God of classical theism.

Kant accepted Hume's criticisms in general and wrote that the argument, 'proves at most intelligence only in the arrangement of matter in the universe, hence the existence not of a supreme being, but an architect'.

Voltaire observed that, even if the argument from design could prove the existence of a powerful intelligent designer, it would not prove this designer is God.

Kierkegaard rejected all rational arguments for the existence of God on the grounds that reason is inevitably accompanied by doubt. The argument from design does not take into account future events which may serve to undermine the proof of God's existence.

As a theology student Charles Darwin was influenced by Paley, though of course he later proposed the alternative theory of evolution in his book *On the Origin of Species*. He had a problem with understanding suffering in life generally and this may have formed the basis for his turning away from theism to an atheistic view.

Another criticism of the teleological argument is based on the human perspective of needing the universe to seem special when it is no more so than the infinite alternatives of form it can take.

R. Dawkins is severely critical of theology and intelligent design in his books, *The Blind Watchmaker* (Ref 7.5.2) and *The God Delusion* (Ref 7.5.3) and others in the same vein.

He contends that intelligent design can provide no explanation for biology because it does not explain the origin of the designer. Furthermore it requires that an intelligent designer be far more complex than anything designed by that intelligence.

Nevertheless Dawkins believes the anthropic principle implies the chance of life arising on a planet like the earth has a very low order of probability. However having occurred that very improbable event is followed by a process of natural selection which is not based on random chance. From this it follows in his view that the existence of God is also very improbable.

With regard to the perception of purpose in biology, Dawkins rejects the claim that biology serves any designed function, but only mimics such purpose.

He states animals are the most complex things in the known universe and that natural selection should suffice as an explanation of such biological complexity without the need for divine authority.

Professional biologists support the modern evolutionary synthesis not merely as an alternative explanation for the complexity of life but a better explanation with more supporting evidence. Living organisms obey the same physical laws as inanimate objects. Over long periods of time self-replicating structures arose and later formed DNA.

George H. Smith in his book, *Atheism: The Case Against God* (Ref 7.5.4) opposes the design argument as in principle it appears to remove the basis whereby we distinguish between natural objects and human artefacts.

Finally the geneticist James A. Shapiro has suggested 'a third way' proposal emanating from advances in genetics and molecular biology which have implications for the teleological argument. This involves cells having molecular networks processing information with respect to internal operations and the external environment and making decisions regarding growth and motion etc.

This might be said to represent a half-way house between intelligent design and Darwinism.

7.5.4. *The organised religious belief philosophies.*

Smart (Ref 3.2.1) describes religion in terms of seven main categories which can be found on the internet and summarised below:-

(a) The practical and ritual dimension.
This refers to the traditional practices to which the religion adheres, such as regular worship, preaching, prayers and other rituals such as pilgrimages etc.

(b) The experiential and emotional dimension.
The emotions and experiences on which religion feeds concerning its great central figures together with numinous experiences of individual believers, such as feelings of awe, peace, hope, love, visions etc.

(c) The narrative or mythic dimension.
The story telling side of religion, including tales about the great figures, referred to in (b). Stories considered in some cases to have divine origin and others in the form of parables.

(d) The doctrinal and philosophical dimension.
The intellectual component of religion involving arguments which endeavour to explain the nature of God and doctrines laid down as fundamental to a particular religion.

(e) The ethical and legal dimension.
The Laws, which a tradition incorporates in to the fabric of a religion.

(f) The social and institutional dimension.
That which relates to the sociology of religion, whereby the religion spreads and maintains its teachings through the relevant system of leadership to the lay adherents by social means, involving birth, coming of age, marriage and death.
The institutional aspect of the religion i.e. the system of governance and organisation of leaders and teachers.

(g) The material dimension.
Religion in its material form as buildings, works of art, icons, shrines etc.
Other sacred natural landmarks and human creations.

Some general comments on organised religion can be made.

While science operates effectively as one unit, in this sense more holistic than religion; the latter breaks up in to endless variations, the very opposite of holism.

Apart from Judaism (which is included because it was the precursor to the formation of other much larger religions), they have adherents numbered in the hundreds of millions, whether superficially so or devoted to the point of dogmatism, but all have had great influence on large numbers of people in historical terms up to the present day.

Primitive religion stretches back at least 35 millennia, in a historical sense some five millennia, and have developed to their present forms in the second half of that period. Ignoring technology, which in its primitive forms has a much longer past, this is in stark contrast to science which realistically has only some 500 years to its credit, from the time of Leonardo de Vinci say rather than Roger Bacon.

The general explanation for this is down to the corresponding state of the science of materials, use of material mixtures involving metals, cement, paper and glassware etc.

Nevertheless it is a remarkable difference and the advances of science to this point in time provide the clue as to why this book can be written now and not earlier.

The main religions of influence are specifically, those which have arisen in the near and middle east namely Judaism, Christianity, and Islam and those related to the far east, namely Hinduism and Buddhism.

There are significant variations within these religions and other religious belief systems which differ from those listed above. It is not considered however that reference to these alternative systems would significantly alter the conclusions drawn in this book.

All taken together those who claim adherence to one or other of the main religions, together with those individuals who regard themselves as spiritual but not religious, form a significant majority of the world population.

Non-religious systems namely atheism and agnosticism, again for practical purposes can be considered to apply to the remainder, but they have been considered in Sections 7.5.1 and 7.5.2 and do not fall under the heading of this section. Nevertheless criticisms they make of religion in general are included in answer to questions in this section 7.5.

Religion emphasises goodness as opposed to evil and whatever that involves can be labelled good religion as opposed to bad religion.

Of course many people of all persuasions are at times or even most of the time behaving in a manner which may be termed bad religion.

To the extent they do so, they cannot be regarded as true believers in what may be termed good religion and the proportion of the population who believe in that sense must fall accordingly. The numbers of true adherents cannot therefore be established as it must be subject to the swings of behaviour of groups large and small in the total population; consequently the majority referred to above is of doubtful provenance.

The differences and similarities of each of the main religions are described by Smart (Ref 3.2.1) using key terms and the seven categories of dimension referred to above.

The established religions referred to above only are considered in this section concentrating on the elements which they have in common and the differences between them in respect of those elements.

The purpose being to identify as far as possible a coherent theme to represent a standard model of religion, as has been done in earlier sections dealing with science.

7.5.5. The differences and similarities of the main religions.

The concepts of heaven, hell, reincarnation and the afterlife/beforelife in these religions. The concepts of goodness and evil.

The importance of prayer.

The belief in virgin birth and miracles generally.

The basic concept of heaven is that it is a transcendent place where heavenly beings such as gods, angels, saints and venerated beings are said to live. It is commonly believed that heavenly beings can descend to earth or incarnate and that earthly beings having lived holy lives on earth can ascend to heaven. It is described as a holy, higher place or paradise in contrast to the unholy or lower place referred to as hell or the underworld.

It is considered accessible to the good soul of earthly beings variously on a universal basis or on conditions according to relevant standards of divinity, goodness, faith, will of God, etc.

Others believe in a heaven on earth in a future world.

In any case heaven is the desired destination or afterlife following death, and is generally considered to last for eternity.

Hell is a place of torment or punishment sometimes described as hell on earth but generally located in the underworld or another dimension. It is commonly regarded as an afterlife destination of the bad soul following an unholy earthly life.

Hell is sometimes portrayed as populated with demons or devils in various forms which torment those dwelling there.

In any case hell is the least desired destination or afterlife following death and is also commonly but not always considered to last for eternity.

Other afterlife destinations include purgatory or limbo, generally as an intermediate station prior to entering heaven or hell based on whether the individual carries out penance or not for his/her sins committed during earthly life.

Some religions which do not consider the afterlife as a place of punishment or reward merely describe hell or the underworld as an abode of the dead, the grave or a neutral place.

In religion generally, the afterlife is a place, whether physical or spiritual, in which an essential part of an individual's identity or consciousness termed the soul or spirit continues to exist after the death of the individual's body.

Some belief systems hold that the dead go to levels of existence determined by a higher being, dependent on their actions and beliefs good or bad during life.

In Hinduism and Buddhism however the spiritual soul is subject to a potentially endless cycle of birth and rebirth or reincarnation in different living forms according to its karma, unless broken by the soul achieving Moksha or Nirvana by meditation, goodness, etc. The nature of the continued existence is then determined directly by the actions and beliefs of the individual in the ended life, rather than the decision of a higher being. Reincarnation can be described as a beforelife/afterlife cycle.

Belief in an afterlife is in contrast to the belief in oblivion after death.

However, regardless of the above beliefs religion is concerned at least equally with earthly matters otherwise it would lose adherents.

In this respect, as already stated, what may be labelled good religion emphasises the concepts of goodness, conscience and altruism and opposes the concepts of evil and wickedness and this is the main message delivered from pulpits.

In practice of course religion commonly assumes with ease the very opposite stance, i.e. bad religion.

All the religions believe in the importance of prayer as a means to overcoming the difficulties of life, achieving enlightenment, spiritual sustenance and entrance to heaven or at least higher levels of existence. In association with these beliefs the religions practice rituals of worship, prayer and meditation.

In addition various degrees of belief in specific miracles apply, particularly in relation to religions associated with the near and middle east.

These beliefs and practices though widespread in human society are in direct opposition to the principle of the natural sciences which is analytical in character and distances itself from the emotional aspects of life.

The attitude of the main religions to the above is correlated below. Taking firstly the concept of heaven, Judaism has less to say about it than Christianity or Islam; it is less focused on getting in to heaven but on life itself and how to live it.

The Torah has little to say on the subject of survival after death, it regards the future as inscrutable with no clear guidance about what is to come.

The Kabbalah, the source that describes Jewish mysticism is less reticent and refers to seven heavens in order from the lowest to the highest and some Jewish writings refer to a new earth as the abode of mankind after the resurrection of the dead.

So these ideas are held by some but not by others in Jewish thought, presumably influenced in some degree by other religious philosophies.

A similar situation applies to the concept of hell, though there are references to Gehenna originally meaning a grave but later a sort of purgatory or waiting room where one is judged on one's deeds. Hell is not necessarily physical, rather an intense feeling of shame, with the door open to a return from this shame if the person realigns his will to that of God.

Mainstream Christianity believes in the Nicene Creed; it is concerned with death, heaven, hell, the Second Coming of Christ, the resurrection of the dead, the end of the world, the last judgement, a new heaven and a new earth and the consummation of God's purposes. In all this the concept of the afterlife is not straightforward and a spectrum of beliefs applies depending on which sect of Christianity is examined.

Christianity teaches that heaven is the throne of God in the company of holy angels.

It is considered a state of existence, the supreme fulfilment of theistic vision.

Heaven is generally regarded as the abode of the righteous in the afterlife, prior to the resurrection of the dead. The resurrected Jesus is said to have ascended to heaven, where he now sits at the right hand of God and will return to earth at the Second Coming. Others said to have entered heaven while still alive, apart from Jesus include Mary, the mother of Jesus.

The word heaven in Christian writings is applied to the sky and in phrases relating to astronomical bodies; it is also used metaphorically as a living personal relationship with the Holy Trinity.

The Christian doctrine of hell derives from the Greek Hades or the Hebrew word Gehenna, the latter referring to the Valley of Hinnom, which was a garbage dump outside Jerusalem. In the Roman Catholic Church and many other Christian churches, hell is taught as the final destiny of those who have not been found worthy after the general resurrection and last judgement, where they will be eternally punished for their sins and separated from God. The Catholic Church defines hell not as punishment imposed on the sinner but as the sinner's self-exclusion from God, also that some go to purgatory where they undergo purification to enter heaven.

The Orthodox Church as with Judaism is in contrast more reticent on the afterlife, acknowledging the mystery of things that have not yet occurred.

The concept of eternal damnation in punishment for sin is a tenet of belief in the afterlife for some Christians, other more liberal Christians believe in Universal Reconciliation and that all souls will be eventually reconciled with God and admitted to heaven, with a range of views in between.

The concept of heaven in Islam differs in many respects to that of Judaism and Christianity. Heaven is described primarily in physical terms as a place where every wish is immediately fulfilled when asked. If one's good deeds outweigh one's sins then one gains entrance to heaven, conversely one is sent to hell.

Heaven has seven levels, the highest being the seventh heaven.

A similar set of levels apply to hell. Enemies of Islam are sent to hell upon their death.

On judgement day all souls pass over a bridge which those destined for hell will find too narrow and fall to their new abode, hypocrites and polytheists in particular. Not all are agreed on whether this fate is permanent however.

The Islamic belief in the afterlife as stated in the Qu'ran is descriptive. The level of comfort in the grave depends wholly on the level of Iman or faith of the individual in God or Allah. The individual must practice righteous deeds or else his level of Iman can wither away. Islam teaches that the sole purpose of life is to worship Allah including being kind to other human beings and animal life. The life we lead on earth is nothing but a test to determine each individual's ultimate abode be it eternal punishment or paradise. Islam teaches the continued existence of the soul and a transformed physical existence after death, deceased souls remaining in their graves until the Last Day of judgement comes when those souls will be divided between one fate and the other.

Hindu doctrine involves the personal God who takes many forms and the non-personal God, Brahman, these many forms being manifestations of one divine being.

Heaven in Hinduism is not a final destination because it is ephemeral and there are several heavenly levels. According to Hindu cosmology there are other planes above the earthly one, and since heavenly abodes are also tied to the cycle of birth and death any dweller of heaven or hell will again be recycled to a different plane and in a different form as per the karma. This cycle is broken only by self-realisation termed Moksha. Moksha stands for liberation from the cycle of birth and death and final communion with Brahman.

Early Vedic religion does not have a concept of hell but in later Hindu literature, the Puranas for example; a realm similar to hell is mentioned called Naraka which is generally a place of punishment for sins. Individuals who finish their quota of punishments are reborn in accordance with their balance of karma.

In Hinduism the belief is that the body is but a shell, the soul inside is immutable and indestructible and takes on different lives in the cycle of birth and death.

The soul called Atman leaves the body and reincarnates itself according to the deeds or karma performed in the latest life. The five elements of the body namely earth, water, fire, air and sky die, none of them harming or influencing the soul.

So if an individual performs good deeds they have good karma and will be rewarded in the afterlife and vice versa for bad deeds. Effectively this amounts to a system of beforelife and afterlife.

Buddhism emphasises the doctrine of impermanence, so the concept of an absolute God is accordingly replaced by spiritual levels.

In Buddhism there are therefore again several heavens, all of which can be related to samsara (illusionary reality). Those who accumulate good karma may be reborn in one of them, but their stay in heaven is not eternal, they will use up their good karma and revert to humans, animals or other beings. Because heaven is temporary and part of samsara, Buddhists focus on escaping the cycle of rebirth by reaching for enlightenment namely Nirvana which is not a heaven but a mental state.

According to Buddhist cosmology the universe is impermanent and beings migrate through a number of existential planes in which the human world is only one realm or path. A sort of vertical continuum it

351

can be said with the heavens existing above the human realm and the realm of animals and hellish beings below.

With respect to hell, Buddhism teaches that there are several realms of rebirth which can be divided into further degrees of agony, of these the lowest realm is the hell realm or Naraka, this being split in to further levels, the worst involving great suffering. Again none of these realms are permanent and even the worst individual can be reborn again.

Buddhist eschatology maintains that rebirth takes place without the soul passing from one form to another, the type of rebirth being conditioned by the moral tone of the person's actions or karma. Yet the mechanism of rebirth is not deterministic, it depends on the level of karma.

The overall similarity between Buddhism and Hinduism in all of the above is not surprising since they have similar origins in Asia, with Hinduism eventually replacing Buddhism in India.

The narrative and mythic dimension of virtually all religions is littered with stories of virgin births of great figures, these stories, generally apocryphal in nature, are nevertheless commonly taken seriously to the point where they are essential features of the main theme of that religion, Christianity for example. The same applies to the belief in miracles in general, the parting of the waters when the Jews escaped from the clutches of Pharaoh, the feeding of the five thousand, the raising of Lazarus from the dead, similarly Jesus and so on.

The concept of virgin birth in the Christian religion was further entangled with the concept of original sin, so to some extent it became obligatory for Jesus to be born that way.

The essential point being that the story of the religion has to be accepted by its adherents in a holistic sense, the one whole in fact. Analysis in any scientific sense is frowned upon, effectively ignored and replaced by imagination.

Now while there is in the animal kingdom the phenomenon of parthenogenesis, a natural form of asexual reproduction in which growth and development of embryos occur without fertilisation, it is not to be related in any meaningful way to the religious concept which after all is concerned with deep philosophical ideas.

Now it is obviously necessary to consider criticism of the above beliefs, essentially the atheistic view.

It has been said that consciously or unconsciously, most atheists see in gods and devils, heaven and hell, reward and punishment, a means of brainwashing the people in to a state of obedience and contentment.

George Orwell's use of Sugarcandy Mountain in his book *Animal Farm* is considered to be a literary expression of this view.

Others have argued that a belief in a reward after death is poor motivation for moral behaviour while alive. The linkage between religion and morality gives bad reasons for helping other human beings when good reasons for doing so to the overall benefit of mankind are available. Such benefits apply to life in an earthly sense, forms of life not entirely explained by reality being considered illusory.

Many people who come close to death or have near death experiences report meeting relatives or entering what is termed 'the Light' in an otherworldly dimension which shares similarities with the religious concept of heaven. 'The Light' is reported as being an immensely intense feeling of love, peace and joy beyond human comprehension. Others though report distressing experiences and negative feelings more associated with the concept of hell.

Neuroscientific explanations are put forward by some to explain these experiences, accepting the precepts as real neural activations with subjective precepts involving altered states of consciousness.

The scientific community is basically sceptical about the religious belief in life or consciousness of any kind after death and generally dismisses it as an illusion.

Regarding the mind-body problem most neuroscientists take a physicalist position according to which consciousness derives from and/or is reducible to physical phenomena such as neuronal activity occurring in the brain.

The implication of this premise is that once the brain ceases functioning at brain death, consciousness cannot survive and ceases to exist.

Furthermore worship, prayer and meditation have accordingly little or no meaning unless they have some benefit in real terms.

A similar reductionist position is applied to the religious belief in the virgin birth and miracles generally.

7.5.6   *Human belief, religion and the Reference Model.*

7.5.6.1. *Atheism, agnosticism and theism.*

It is necessary now to relate human belief as discussed in Sections 7.5.1 thru 7.5.5 to the Reference Model.

The religious dimensions (a) thru (f) defined by Smart (Ref 3.2.1) and referred to in Section 7.5.4 relate primarily to belief and imagination and the imaginary axis as portrayed in the figures shown and discussed in Section 4.0 and further discussed in this Section 7.5.6.

Dimension (g) is clearly the material substrate of (a) thru (f) and relates to the real axis as portrayed in the figures shown and discussed in Section 4.0.

The spiritual but not religious and indeed agnostics and atheists share with the mainstream religions their own versions of some elements of these dimensions.

The relation between the various groups is therefore depicted as shown in Fig 4.8.11.

The Reference Model associates atheism and to a lesser extent agnosticism with the mean level of order as described in earlier sections and with the unfoldment of the real elements of the differing worlds (referred to in Section 4.8) on to that real axis and the enfoldment of the imaginary elements of those worlds on to the imaginary axis (see Fig 4.8.2). In effect the vertical wave motion centred on the present state and the horizontal wave motion centred on the distributed state appears to collapse on to those axes just as the unitary quantum state when observed appears to collapse to the four-dimensional real world via the reduction process, which the Model views as real and objective at the mean level of order but otherwise phenomenological.

The scientific humanist nevertheless endeavours to equate his own humanity with science but in effect does what the Model says he must do, namely extend his comprehension to include for the more ordered region above the mean level which includes the concept of goodness and likewise the less ordered region below the mean level which includes the concept of evil; as indeed do other humanists who do not base their belief on scientific principles.

The Reference Model identifies deism as being associated with the imaginary present state axis and theism with the Holomovement in perpetual motion about it.

There is nothing much to doubt about the existence of that motion as without it there would not be any order, disorder in the opposite sense and individual conscious entities in between, with propensities for both.

Before proceeding with the confused state of religious belief concerning the idea of a deity or the associated belief in 'nirvana' which is effectively a replacement for God, but for all intents and purposes means the same, it is worth considering the scientific attitude to deism etc.

Scientists have no difficulty in accepting the highly ordered state referred to as the big bang as the moment of creation and its extension to include other physics. They further accept the apparent inevitability of the laws of thermodynamics and the conclusion that ultimately in time the system decays to totally distributed disorder. This is all incorporated in the Reference Model together with the random state where order and distribution have no meaning. Science in general accepts that in addition to material being increasingly disordered there is the opposite motion towards increasing order which they attribute to natural selection working as it were from any given state upwards. They are unable to comprehend that there are higher levels of order in existence at a more accurate present time acting as a great attractor in the same way that the disordered state is a great attractor. The absence in the Model of any first cause should remove confusion as there is no boundary to existence and no first present time, since there is no limit to the Holomovement; so the idea of infinity or infinite regression is preferred in the Model to stipulating absolute boundaries.

Whereas the disordered state is about unwholeness the ordered state is about wholeness and both relativity and quantum theory are about precisely that.

The wholeness or oneness state is ultimately responsible for belief in good religion and the unwhole state for bad religion and people are in the main able to distinguish between right and wrong, with regrettably many exceptions which nevertheless confirm the rule. So increasingly, conscious stuff moves up the ordering ladder and increasingly unconscious stuff downwards in the generally accepted convention.

The inanimate present state is then responsible for the non-personal God and the Holomovement for a personal God both referred to in religious philosophy. Correspondingly disorder is to be associated either with chaos (explicit disorder) or with evil (implicit disorder). The further notion of the devil is demonstrably weak as consciousness reduces with increasing disorder, so the scientist does not have to worry about him!

Now having explained the relation between theism and the Reference Model and leaving the scientist to his deliberations on the above, the religious basis for belief in deism (non-personal God) and theism (personal God) will now be considered.

The Reference Model finds the absolute nature of the cosmological or first cause argument unsatisfactory and replaces it with the processes of ever-increasing order and ever better definition of the present state as described in Section 4.0 and in this section.

These processes replace notions of first cause, causal loops, causal chains and the absolute singularity with a far more subtle description of ever greater complexity without limit, cyclical in general terms but where the loops never close; that is what keeps it going.

The ontological argument involving absolute perfection like the first cause is incompatible with the requirement for ever increasing order as implied by the Reference Model described in Section 4.0 and again later in this section.

The Model clearly supports the proposition that the probability of God increases with the growth of order, though opponents object to any kind of probabilistic reasoning in respect of theistic belief. Criticism of the design argument based on the human perspective of needing the universe to be seen special effectively assumes life is limited to a particular form or level of order as opposed to the Reference Model where life grows with the complexity of ordering sequence.

In respect of the points made by R. Dawkins the Reference Model answers them immediately since it shows there is no absolute origin, only a recycling process from ever more disordered states to ever more ordered states and the reverse of that process associated with wave motion. The motion towards ever greater ordering is a tuning process inducing an ever more precise present time or state of wholeness and consequently an ever-increasing complexity far beyond that comprehensible to the human race, referred to as the Whole Ordering Centre.

The Reference Model relates the anthropic principle to lower levels in the Whole Ordering Centre where the probability of life is indeed minimal, but the opposite applies at higher levels of order in what is an integrated structure with no starting or end point in time. The Model does however confirm that life regardless of source is incompatible with the random state as explained in Section 4.0.

The Reference Model finds that life is entirely purposeful in the direction of increasing order and the opposite as order decreases, all as further stated in Section 7.5.8.

Natural selection and the increasing complexity of biology is directly related in the Reference Model to ordering processes which oppose the laws of thermodynamics, processes which have their source in the imaginary (implicit) present state axis.

The Reference Model has no problem with the principle of natural selection but it is derived from the limited perspective of science (which is related to the mean level of order which houses the real [explicit] distributed state axis) as opposed to the range of human belief (which is related to all levels of order).

With regard to the criticism made by George H. Smith it is clear from the Reference Model that natural objects are associated with the mean level of order and the material (spatiotemporal) substrate of human artefacts likewise; it is the non-spatiotemporal aspect of life in general in motion about the real axis and to a greater extent about the imaginary axis of the Ordering Centre which distinguishes that which is human and indeed alive from the inanimate natural state aligned with those axes.

The simple assumption that either the argument or the opposite applies is superseded by the complex situation emphasising the need for coherent thinking described by the Model.

Returning to what the Reference Model has to say about all the arguments for God and against, given above. It has indeed important points to make.

Firstly that it is mistaken to pursue the absolute either way, absolute God or absolute non-God or absolute anti God for that matter. The Model is based on the assumption that existence has no upper bound to order or any lower bound to disorder whether chaotic or random in nature. While this cannot be proven the alternative is the endless debate as demonstrated above in which for every argument in one direction there is another in the opposite sense. This should have been a clue to both parties. It is clear that God has no interest in being absolute regardless of human stubbornness to make him so.

In addition the distributed state can be enfolded in to the present state, just as it can be unfolded from the latter. The random state of disorder can also be unfolded from the peak ordered state and vice versa therefore those states are directly related.

Therefore the insistence on an absolute God is as meaningless as the equally dogmatic insistence that God does not exist. In fact technically for this reason it is more correct to say that God is either nearly certain to exist in the ordering direction, or nearly certain to not exist in the direction of disorder, being replaced by non-God at any mean level of order or anti God in the direction of implicit disorder.

The reason for the dogmatism is the misapplication of boundaries as has been said previously, arising from the profusion of boundaries imposed by the four-dimensional real world.

Secondly account must be taken of the mean level of order and the associated four-dimensional reality as seen from both the multidimensional Hilbert space state and the multidimensional phase space state, both of which for all practical purposes may be regarded as infinitely dimensional.

The Reference Model shows that scientists and many theologians of all religious beliefs are incredibly naive, as they are unable to contemplate that both science and religion are relevant to understanding what is meant by the Implicate Order, only giving the appearance of being mutually exclusive.

Science limits existence in effect to the explicit real world. However the Model explains that the extension of existence in to the animate, belief and imaginary worlds removes this limitation and provides the basis for comprehending religious belief in all its complexity, all in accordance the more detailed description given in Section 4.8.

The empathetic religious individual sees higher degrees of order stretching upwards using his sense of imagination from the mean level of order associated with motion towards the great attractor, namely the holoarchical central present State (the deistic non personal God) and the Holomovement swirling about it (the personal God).

This is the region where oneness and consciousness is greatest. The atheist is attracted towards the mean level of order associated with the hierarchical distributed state (non-God).

The empathetic humanistic atheist and agnostic are influenced by both attractors, hence their humanism.

It is important to also note that the anti-God element, the crooks and charlatans of this world, see ever lower degrees of order (higher degrees of disorder) stretching downwards from the mean level of order and are attracted towards the distributed chaotic base state or the random undistributed base state, hence their lack of empathy, evil in other words. Many who claim to be religious, but practice bad religion also fall in to this category, a region where empathy evaporates and a state of unwholeness applies.

Now with regard to the design argument the same states of mentality are apparent.

The empathetic religious individual is influenced by the present state and the Holomovement as before, hence their faith and ability to explain why they believe in intelligent design and correspondingly their total inability to explain how.

The relation between the Reference Model and Darwin's theory of natural selection has already been described in Section 4.4 and Section 7.2.2 but the following general comments apply.

The atheist sees the increasing degree of order extending from the base state where inanimacy applies and natural selection is absent, to the point where natural selection begins to take over until the mean level is reached where this manner of evolution (seen in terms of reality) meets the belief in evolution by intelligent design (seen in terms of imagination). Again we see the atheist is able to explain how natural selection applies but totally unable to explain why and is forced to use the anthropic principle.

The Reference Model explains the probability problem because it covers the immense range of states that applies to the Whole Ordering Centre.

Firstly the base state, where consciousness and the probability of consciousness is effectively nil, where reality predominates, approaching a state of totally unfolded disorder, centred round the random state.

Then the range of states where natural selection is most clearly evident, up to the mean level where reality and imagination are in balance, where order and disorder seem of equal status, the mean level of consciousness applies and where the debate between the probabilities of God, non-God and indeed anti God thrive.

Finally the states of order above the mean level, where consciousness derived through oneness is most clearly evident, where imagination predominates, ultimately approaching a state of totally unfolded order. This is the region where the non-personal deistic God and the animate personal God in the form of love apply with ever increasing probability, without the need for absolutism.

Just as all the scientific elements discussed in Section 7.4 were related to the Ordering Centre grid, so religion can be related to the motion across that grid.

It should be remembered that all the physical four-dimensional levels in space-time unfold from the mean implicit level in the Ordering Centre.

It is relatively easy to identify the common threads applying to religion in general.

The concept of absolute God corresponds to the great attractor of the 'peak' of the Whole Ordering Centre, bearing in mind there is no absolute peak. So the cosmological and ontological arguments, in insisting on the absolute, are at fault in this respect.

In particular Buddhism for example replaces absolutism with impermanence involving many levels of spirituality.

The further concept of the personal God corresponds to the Holomovement where consciousness of the Whole, Wholeness and Oneness reigns supreme.

Consider now just that level in the Whole Ordering Centre represented by what has been termed the Ordering Centre, also multidimensional in character.

The peak of the Ordering Centre and its highly ordered state gives rise to the belief in the absolute deity and heaven. The process of rising upwards in the ordering system towards that state corresponds to the religious concept of goodness.

The mean level of order (the earthly level) in that Ordering Centre in which four-dimensional space-time is enfolded, corresponds to the non-God philosophy of the atheist and levels unfolding from that mean in a multidimensional sense to concepts of artificial intelligence.

Levels of order below the mean level either take the form of chaotic explicit disorder, giving rise to confusion and error, or implicit disorder involving the deliberate intention to deceive and harm, which corresponds to the religious concept of evil, corresponding ironically to bad religion as opposed to good.

As implicit disorder is a conscious act ultimately psychopathic or sociopathic in nature and as destructive consciousness degrades to inanimacy eventually, the concept of evil spirits and the notion of the devil become meaningless.

States of implicit disorder correspond to the religious concept of hell decaying ultimately to the random state, while states of chaotic disorder decay to the inanimate scalar field base state of the Ordering Centre.

Implicit disorder is equivalent to enfolded chaotic disorder. Evil therefore equates to being disorder isolated in a prison from which it struggles to escape, the ultimate prison being the random state. Implicit order is the process of escaping from that prison.

More importantly in the direction of increasing implicit order consciousness and therefore intelligence increases.

It is the growth in order in explicit terms in Fig 4.3.4 which gives rise to the belief in natural selection, but that atheistic concept is in implicit terms the process of movement towards the present state of the Ordering Centre. Convergence towards the present state is the process of intelligent design. Intelligent design and natural selection are therefore the same process seen from the imaginary and real perspectives respectively.

The argument against the concept of God based on the enormous improbability of His existence is simply negated.

The structure of the Ordering Centre confirms the enormous probability that order is enfolded in it and disorder unfolds from it and inversely the enormous probability that order unfolds as disorder enfolds, the one being a reflection of the other. The whole process being cyclical in nature as Fig 4.2.4 implies but where the loops never close, as there is no absolute state.

7.5.6.2. Religious concepts in terms of ordering processes.

The Reference Model interprets science in terms of ordering processes and the same applies to religion. Accordingly the belief in the 'afterlife' in the Western religions associated with resurrection and the belief in the 'afterlife' and 'beforelife' associated with reincarnation and Eastern religions arise from failure to comprehend multidimensional animate wave motion relative to the mean level and the present state, as portrayed in the figures in Section 4.8.

The use of the above misleading terms gives a false picture as in fact this motion applies at all times arising from the contraction of the present state giving rise to differing levels and worlds, all as related in this book to Implicate space.

Figures 4.5.1, 4.5.2 and 4.5.3 show that concepts of before and after relate to the explicit world and are at cross purposes with the profusion of potential orderings and disorderings applicable to Implicate space.

It is clear from Section 4.0 that the real state of the conscious entity is totally dominated by locality, the intermediate state of belief relatively equally by locality and non-locality while the imaginary state is correspondingly dominated by non-locality.

Locality imposes the reduction process, creates boundaries, expands space-time and generates concepts of before/after, either/or and the overall common-sense view of reality. Non-locality contracts the present time dissipating the above effects imposed by locality.

So a localised conscious entity with associated boundaries then loses validity being superseded in the direction of increasing implicit order relative to the mean level, by an unbounded global consciousness governed by the concept of Wholeness. Locality explains the life/birth/death cycle of conscious entities while non-locality explains the dissipation of such cycles when global consciousness prevails. The individual consciousness is conceived and ultimately dies in accordance the effects of locality while global consciousness is to be associated with the enfoldment of space-time dimension as the present state contracts.

This dissipation of boundaries and growth of global consciousness is to be related to the degree of Wholeness achieved by the individual, for which that individual is responsible.

Alternatively in the direction of increasing implicit disorder relative to the mean level of order the localised conscious entity again loses validity as a result of the unbounded decay of global consciousness governed by the concept of unwholeness, for which again that individual is responsible.

By Wholeness/unwholeness here is meant the degree to which an individual increases/decreases implicit order/disorder with respect to the mean level of implicit order housing that individual's inanimate core.

All this is opposed to the idea that existence pushes the individual around, so that he/she is not so responsible but is the victim of it.

The consciousness that the above entities possess whether constructive or destructive is therefore transferred by wave motion towards a more central present time, their individual consciousness being converted to the Whole/unwhole implicitly ordered/disordered version as applicable.

The reader can visualise such motion and the processes involved by studying the principles of wave reinforcement and cancellation referred to in previous sections, from which Figures 4.7.1 and 4.7.2 are derived.

As an analogy the use of a wave machine for example to generate wave motion in a tank of water sends a wave from one end A of the tank to the other end B, but the water itself remains in the same position with respect to the ends of the tank as demonstrated by a block of wood at A which does not move horizontally. The block of wood represents locality and the wave non locality.

The approach of an individual to both locality (reality) represented by the mean level of order and to non-locality (imagination) represented by the present state are tuning processes as illustrated in Figures 4.6.1 and 4.6.2.

Therefore an atheist who favours reality and objective beliefs tunes in to a fixed mean level of order and a corresponding fixed present state at the expense of variable levels while the theist who favours imagination and subjective beliefs tunes in to variable levels of order and present states at the expense of the fixed axes.

In accordance the above the Reference Model shows that the explicit reality of birth and death in the unfolded four dimensional world has progressively less meaning for wave motion in the higher dimensional worlds as the relevant vector motion becomes increasingly divergent from the mean level of order i.e. at significant angles to that level; to the point where birth and death relatively speaking have no meaning for vectors close to but not aligned with the present state, as such boundaries dissipate under the influence of the Whole Ordering Centre.

So motion in the implicit worlds is related to the explicit world, the latter being the form which the higher worlds can in a sense refer to, so as to establish the level above which things are increasingly ordered and below which things are increasingly disordered. This of course giving rise to our sense of right and wrong, encapsulated in our concept of conscience.

Vector motion in the implicit worlds above the mean level becomes increasingly more whole to the extent that individual consciousness is superseded by the Whole global consciousness. This should be thought of not as a series of more lives in the ordering direction but rather one more Whole Life and increasingly so the further one rises in the system, rendering the terms 'afterlife' and 'before life' meaningless.

The world of belief associated with the senses corresponds to Popper's world of suffering where the endless battle rages between good and evil giving rise to both human virtue and sin.

Regarding the human virtues, they are essentially common to all religions, including altruism, the unselfish regard for others as a principle of action, unconcerned with reward in this life as it were. The belief that such people are rewarded in heaven then derives from the vector motion described above.

Motion in the opposite sense below the mean level is increasingly less whole in a chaotic sense causing or ultimately leading to explicit death; or alternatively more related to motion close to the present state but in the disordered direction, where the entity becomes increasingly unwhole under the influence of that disordered motion associated with evil, a state which religion describes as hell.

Nevertheless the Model confirms that, as noted earlier, the belief commonly held that the wicked are punished for their sins by existence is an error which can be regarded as bad religion.

Rather such individuals choose their own vector motion by free will made available by the multi-dimensional Implicate space referred to earlier and therefore are themselves responsible for their own fate, as implied by the figures in Section 4.8.

Summarising, for practical purposes the transition from the bounded explicit state of an individual entity to the ultimately unbounded implicit state should be seen as a process. The distribution of the entity on the imaginary axis whether one of greater wholeness or unwholeness grows as levels of consciousness and awareness increase, that distribution being related to explicit time, as illustrated in Fig 4.8.1.

So the Reference Model provides the reason for the existence of the world, based on the principle of no absolute boundaries all in accordance with the ordering process.

It further explains the complementary nature of the inanimate and the animate, with the structure of present states and levels representing the former and the motion about that structure the latter. It explains the growth in consciousness as order unfolds and vice versa the collapse of consciousness as disorder unfolds. It explains the belief in God and the associated belief in goodness, the belief in non-God and the prevalence of both chaotic disorder and evil, anti-God in other words.

It remains to explain why religion believes in events which seem miraculous and from the scientific viewpoint inexplicable and therefore unbelievable.

The Reference Model associates levels of order in the Ordering Centre with reality and the present state with imagination. As imagination and consciousness grows so reality reduces and vice versa as reality grows so imagination and consciousness reduces. Imagination is concerned with Wholeness and Oneness as represented by the peak of the Ordering Centre, while reality is ultimately reduced to the inanimate levels at the base of the Ordering Centre. At the mean level imagination and reality are in balance and this is the earthly level where atheists view the mean level as absolute, or nearly so in the case of scientists who discovered and then needed to explain the quantum state.

Previous sections have described the relation between the multidimensional scenarios of classical and quantum states which give rise to consciousness set in Implicate space.

The relation between that non-local consciousness and local explicit reality has also been described. The conscious state allows a whole range of scenarios between the inanimate present state and the equally inanimate mean level, as the figures in Section 4.0 show. Is it surprising that religion and science seem in conflict, that analysis is in conflict with emotion, goodness with evil, and any one viewpoint concerning philosophy, politics, economics, religion, etc with another? No it is not.

The conclusion to be drawn concerning religious miracles is that they have no place in reality at the mean level or even above that level and arise from confusion regarding the nature of imagination and ignorance of the processes describing the formation of order. The formation of reality at other levels in the Whole Ordering Centre should be related to harmonics, in the same manner as a sense of pitch and the range of notes is fundamental to musical appreciation. So the Holomovement consists of a series of points of reality at these other levels, as imagination and consciousness increase.

It is therefore unwise to insist that miracles are absolutely true in any real sense; instead they should be looked on as being related to the construction of order which has both real and imaginary states.

It is accordingly nonsense to try to prove or indeed disprove religious concepts relating to miracles, virgin birth or otherwise, in terms of scientific proof; they are in any case to be associated with imaginary stuff, not real stuff. They are concerned with the process of things becoming more whole as described earlier.

Since we are more concerned with processes here, which are considered as ordered sets, it is best to regard the use of the term miracles as a distraction which inhibits understanding of the Implicate Order.

Virgin birth would ultimately seem related to the concept of the Holomovement, the state where the concept of individuality and distinction from others has little meaning, the Christian doctrine of original sin associated with it, emanating from confusion between the ordering processes, as opposed to disordering relative to the real world.

The same applies to other miracles such as rising from the dead, since the states of birth, followed by the one life, followed by death have progressively less meaning as the present state becomes dominant. Extrapolating to the level of the Holomovement the concept of birth and death translates in other words to a state approaching all life.

In the opposite sense, an increasingly disordered scenario towards the base of the Ordering Centre is in other words, a state approaching all death.

7.5.6.3. *Faith as opposed to scientific proof.*

The spiritual as opposed to the physical.

The belief in the soul.

The belief in love.

Science is ultimately concerned with the real processes of measurement and measurement only. In principle though, great scientists have to use imagination when developing theories of how nature works, witness Newton, Einstein, Darwin and so on. Indeed as figures in section 4.0 show scientists must use the belief and imaginary worlds to achieve such remarkable insights, but it is nevertheless assumed that everything can be reduced to the explicit world. So belief and its wilder form imagination, play little part in day-to-day science which is primarily concerned with experimentation and the analysis of results, using the principle of reduction; the latter meaning the reduction of complex systems in to simpler isolated parts. At all times attention is to be paid to the necessity for mathematical and scientific proof as described in Sections 7.3 and 7.4.

Religion on the other hand does exactly the opposite, all things being looked at as representing aspects of a whole thing, on the basis that such holistic entities cannot be reduced to the sum of their parts.

The belief world has at least a relationship with the explicit real world, whereas religion accepts the imaginary world as paramount and happily accepts ideas on this basis which fit in to the particular holistic edifice it supports.

It is not considered necessary to prove anything in accordance with the scientific and mathematical principles referred to above. Scientific proof is replaced by faith, which is simply acceptance of the collection of beliefs and imaginings which comprise the religion concerned, without criticism. In practice such believers are often assailed by doubts to a greater or lesser degree, not entirely dissimilar to the way in which scientists have doubts about their theories and are always checking their validity against experiments carried out since the original formulation.

The holistic emotional side of the imaginary world contrasts with the analytical reductionist side as shown in figures included in Section 4.0. However as religion treats that world as paramount so science treats the explicit real world as paramount, so religious belief is drawn to the spiritual region close to the present state and science to the explicit world which unfolds from the mean level in the Ordering Centre.

The physical self unfolding in four-dimensional space-time is the explicit form of the individual, while the ego, the self and the soul are clearly associated with implicit properties applicable in Implicate space.

Clearly these concepts can be related ultimately to the whole ego, the whole self and the whole soul that can be considered to apply to the Holomovement.

Similarly the feelings of love that the individual experiences, can be considered to again apply in a totally holistic sense to the Holomovement.

The belief that God is love can be equated to that relationship, in the unconditional sense, where the concept of individuality and distinction from others is meaningless.

The above conclusions apply in the direction of increasing order; in the opposite sense love becomes destructive and turns to the contrary feeling of hatred to the point where the system falls below the animate level when of course such feelings disappear.

Such destructive feelings primarily harm the perpetrator rather than the victim even if the contrary appears the case in the explicit world.

7.5.6.4. The belief in prayer.

Prayer, like love at the mean level in the Ordering Centre has no meaning since this is an inanimate state, but above that level increases without limit in the form of asymmetric motion about the present state.

As far as science is concerned prayer has no proven role in human affairs since there is no scientific evidence for it doing so. The Reference Model confirms this to be the case as far as the explicit real world is concerned.

It is therefore to be associated with the belief world, which grows to the point where complex ideas take over in the imaginary world.

Prayer is entirely consistent with the concept of asymmetric wave motion in these higher worlds, and has all the relevance accorded to it by spiritual and religious conviction.

However to take this to extremes, as religious believers sometimes do and try to enforce prayer rituals as if the deity demands it, amounts to prayer by rote contrary to its very purpose.

Again as with love, belief in prayer turns destructive below the mean level of order, when taken to extremes it being supposed that curses or incantations, suitably phrased, will cause harm to the intended recipient. Harm is indeed the result but this fate falls again not on the victim but the perpetrator, even if this is not apparent in the explicit world.

7.5.6.5. Is existence purposeful or blind to purpose?

If existence is purposeful what is that purpose?

Viewed in the direction of increasing order existence is entirely purposeful. The purpose is to seek without limit perfection as conveyed by a sense of increasing wholeness or peace, a state which it never achieves in an absolute sense as existence implies endless motion.

In the case of implicit disorder, namely evil, there is indeed a purpose, but one that isolates and ultimately destroys the evil doer.

In the direction of increasing explicit disorder the purpose just referred to is destroyed by chaos. That chaos is not purposeful in any sense, that's what chaos is.

The collapse of the Ordering Centre to the core state of its reference frame leads to the collapse of consciousness, any sense of purpose and ultimately to the random state.

7.6   Human characteristics and activities.

7.6.1   *The physical and mental attributes of the human brain.*

A good start to understanding these attributes is to utilise the efforts of Popper and Eccles (Ref 2.7.4) as referred to in Sections 2.7 and 4.8.

Penrose (Ref 7.2.6 and Ref 7.3.1) also examines these subjects in an effort to identify a scientific explanation for consciousness, in which Popper's differing worlds are connected on the basis of Platonic realism avoiding subjective considerations.

A further reference which explains the conventional scientific view of the brain to a lay audience is that by Greenfield (Ref 7.6.1).

Only the essential details of this subject will be considered here, in sufficient detail to form a basis against which the arguments put forward in this book can be judged.

The reader is invited to read the references and internet sources to obtain further background material.

Firstly there is much to be learnt from the single celled organism, the Paramecium.

This creature, roughly of the order of 10^{-4} metres in length and a fifth of that in diameter, swims around ponds using hair like legs called cilia, all supported by a structure referred to as a cytoskeleton, which not only holds the cell in shape, but is also the equivalent of skeleton, muscles, legs, circulation and nerves all in one.

The cytoskeleton is an important component of the animal cell and consists of protein like molecules arranged in various types of structure, neurons, dendrites, axons, synapses and cytoskeletal filaments.

Neurons consist of dendrite structures connected together by axons which split in to separate strands which terminate in junctions to other dendrites called synapses.

Axons are distinguished from dendrites by their shape being long and of constant radius, the dendrites being tapered and restricted to small regions in the cell body.

Dendrites receive electrical signals from other neurons by means of these axons via the soma or body of the neuron from which the dendrites project. So the soma forms the trunk of the tree and the dendrites the branches.

Axons are another type of projection from the neuron, this time typically transmitting signals to other neurons as opposed to dendrites which receive signals.

Synapses are structures that permit neurons or nerve cells to pass an electrical or chemical signal to another neuron. The junction from which the signal emanates is called the pre-synaptic neuron (or bouton) and that receiving the signal the post synaptic neuron.

Cells contain three main kinds of filament, microfilaments of polymers of the protein actin, microtubules of the protein tubulin and intermediate filaments of various proteins.

The cytoskeleton includes a central structure known as a centrosome, which acts as an organising centre or hub for the microtubules and consists in turn of two orthogonally arranged cylinders of nine triplets of microtubules termed centrioles.

Microtubules come in bundles with a fan like or Catherine wheel cross section of which the Paramecium cilia are an example. Each microtubule is like a hollow tube normally consisting of 13 columns of tubulin dimers in a spiral arrangement.

Each microtubule has two principal geometrical configurations, alpha and beta. For reference the dimers have dimensions of the order of 10^{-9} metres.

Before DNA replication, cells contain two centrioles, the older mother and the younger daughter attached by interconnecting fibres.

During the cell division cycle a new centriole grows from each side of the mother.

After duplication the two centriole pairs remain attached to each other orthogonally until cell division (mitosis) occurs, referred to previously in Section 7.2.

At that point mother and daughter separate, each dragging a bundle of microtubules with it. Since the microtubules connect the centriole to the separate DNA strands in the nucleus of the cell, these strands separate completing the process of cell division.

Each daughter cell formed after cell division inherits one of the pairs of centrioles.

The human brain consists of many neurons comprising dendrites linked together by axons. Dendritic spines may number some 200,000 spines per cell and dendritic branching can provide as many as 100,000 inputs to a single neuron.

It is estimated there are some 10^{11} neurons, which interestingly is of the same order as the number of stars in the galaxy and the number of galaxies in the visible universe.

If the ability of the human brain is based on the number of neurons, the number of operations it might achieve per second per neuron may be many orders of magnitude greater, certainly sufficient to rival that of computers.

Having looked at the relevant biology at small scales it is necessary now to view the brain at human scales.

The human brain has much in common with all vertebrate brains including three basic elements, the forebrain, midbrain and hindbrain together with fluid filled cavities and brain structures. In addition it has further features common to mammals including the cerebral cortex and other structures.

The hindbrain is the oldest in evolutionary terms resembling the reptile brain.

The hindbrain's stem connects to the body receiving information from sense receptors and issuing commands to muscle and body via the spinal cord.

The cerebellum, part of the hindbrain, stores motor commands and produces smooth movement.

The midbrain is the next evolutionary development and the forebrain termed the cerebrum, the most recent and by far the most complicated, including a limbic system, thalamus, basal ganglia and two large cerebral hemispheres. The limbic system is a complex system of nerves and networks in the brain concerned with instinct and mood. It is central to the basic emotions and drives hunger, sex, dominance and care of offspring. The limbic system's pea sized hypothalamus performs many of these functions using electrical commands, while the thalamus and basal ganglia control conscious state and initiate movement.

The cerebral cortex is the convoluted part of the brain which is far more developed in humans than other animals. Especially so the frontal lobes which are associated with executive functions such as self-control, planning, reasoning and abstract thought. That area of the cortex devoted to vision is also greatly enlarged in humans.

The cortex is divided in to four lobes, the frontal, parietal, temporal and occipital lobes.

The main functions of the frontal lobe are physical reactions, planning, abstract thinking and problem solving, behaviour and personality.

The parietal lobe is involved in somatosensation (touch, temperature and pain), and spatial cognition.

The temporal lobe controls language, hearing and speech and associated memories.

The occipital lobe is the smallest with functions such as visual reception and processing, spatial movement and colour recognition.

Basically the occipital lobe sees, the parietal feels, the temporal hears and the frontal thinks and commands.

The lobes are divided into left and right hemispheres, similar in shape, with most cortical areas replicated on both sides. There is however a degree of lateralisation with the left-hand side in most people dominant in language and analytical ability while the right-hand side is dominant in visual and spatial ability.

The cerebral hemispheres termed the cerebrum form the largest part of the human brain being covered by the cortical layer (the cerebral cortex) which is pinkish in colour but referred to as grey matter, while the cerebrum is referred to as white matter.

The two cerebral hemispheres are connected by a very large nerve bundle called the corpus callosum. It is the main avenue of communication between the two hemispheres.

Underneath the cerebrum lies the brainstem resembling a stalk, to which the cerebrum is attached. At the rear of the brain and behind the brainstem is the cerebellum a structure with a furrowed surface, the cerebellar cortex.

There are two areas on the left side of the brain connected by nerve bundles associated with language processing, Broca's area (the formation of sentences) and Wernicke's area (the comprehension of sentences) and associated regions of sound processing and speech.

The left and right auditory cortex in the temporal lobe deal with sound the left and right olfactory cortex in the frontal lobe with smell.

The somatosensory cortex in the parietal lobe, deals with the sensations of touch.

The left and right motor cortex in the frontal lobe activates the movement of various parts of the body. However there is a complication.

While each hemisphere of the brain interacts primarily with one half of the body as might be expected, the connections are crossed so that the left side of the brain interacts with the right side of the body and vice versa. Motor connections from the brain to the spinal cord and sensory connections in the opposite direction cross over at the level of the brainstem.

Visual input follows an even more complex rule; the optic nerves from the two eyes come together at a point called the optic chiasm and half of the fibres from each eye split off to join the other. Connections from the left half of the retina, in both eyes, go to the left side of the brain, whereas connections from the right half of the retina go to the right side of the brain. Consequently visual input from the left side of the world goes to the right side of the brain and vice versa. Thus the right side of the brain receives somatosensory input from the left side of the body and visual input from the left side of the visual field.

Each of the four lobes consists of three cortical areas, primary, secondary and tertiary.

The regions of the cerebral cortex dealing with visual, auditory, olfactory, somatosensory and motor functions are called primary and represent the input and output functions of the brain. Next to these primary regions are the secondary regions which receive and process information from the primary areas and provide more specific directions for actual muscle movement by the primary motor cortex.

The tertiary regions in turn adjacent the secondary receive information from the latter and carry out the most abstract and sophisticated activity of the brain.

The tertiary motor region sends refined instructions back to the secondary regions which in turn deliver instructions to the primary motor region from which specific instructions for movement are issued. Detailed maps of areas of the brain involved in these processes called Brodmann maps have been produced.

So the brain has input and output rather like a computer with an immensely complicated processing system in between, which provides us with the immensely complicated phenomena we call consciousness.

It appears the cerebrum is responsible for learning basic skills with precision control being handed over to the cerebellum acting as an auto pilot, when such skills have been mastered.

The hippocampus plays a vital role in laying down long term memories, the actual memories being stored in the cerebral cortex. The brain can hold memories on a short-term basis but damage to the hippocampus causes these memories to be lost so that no new memories are retained once they have left the subject's attention, with unfortunate consequences for the individual concerned.

To an artificial intelligence enthusiast this seems the perfect description of an algorithmic Turing Machine as discussed in Section 7.3, awareness arising when the degree of sophistication reaches a sufficient level.

Attempts have been made to evaluate whether consciousness can be related to any particular areas in the brain rather than others.

The hippocampus must play a part in consciousness since it is crucial to the laying down of long-term memories.

The cerebellum seems for example more of an automaton than the cerebrum, so the most likely area to be associated with consciousness is therefore the cerebral cortex.

However it is evident that attempts to locate consciousness in a particular part of the brain are fraught with difficulties.

Of particular interest is the effect on the brain of severing the connection between the brain hemispheres by cutting through the corpus callosum which normally provides that linkage. Such operations have been performed in an effort to cure patients with extremely severe epileptic seizures.

Subsequently such patients were subjected to psychological tests to ascertain the consequences of the operations.

The differences between the two halves of the brain in these circumstances as deciphered by Popper and Eccles (Ref 2.7.4) and Gazzaniga (Ref 2.7.5) have already been introduced in Section 2.7.

After the left and right brain are separated (excluding the hypothalamus) as referred to above, each hemisphere appears to have its own separate personality including how it sees the world together with associated impulses and actions arising.

If a conflict arises one generally overrides the other.

Now when split brain patients are shown an image only in their left visual field, they cannot put a name to that image. The image is sent only to the right side of the brain but the speech control centre is generally on the left side of the brain and since the usual means of communication has been disrupted the patient is unable say what the right brain is seeing. When touching an object with the left hand without the benefit of cues in the right visual field, the patient cannot put a name to that which the right side of the brain is aware of, as each hemisphere only communicates with the opposite side of the body.

Patients shown differing images in each visual field are then unable to meaningfully associate one image with the other.

Eccles conclusion is that the left hemisphere is primarily objective and specialises in linguistic, analytic, calculation and ideational tasks, while the right hemisphere is primarily subjective and while nonverbal specialises in synthetic, holistic, visual and spatial tasks.

However this is not a view that brain scientists who are also reductionists would support. Their view is that the left and right brain while serving opposite sides of the body in a physical sense, are purely explicit in nature and that brain scans indicate no difference in activity levels for either hemisphere.

It is clear that the corpus callosum facilitates communication between the hemispheres.

Split brain patients can still live a coordinated and consistent life and are able to compensate so as to be largely indistinguishable from normal adults.

So long as the hypothalamus is intact there is still limited communication between left and right sides of the brain.

However while this information is of great benefit to an understanding of the brain, it is of course the whole brain and its activities we are primarily interested in.

It is in this respect that Popper and Eccles (Ref 2.7.4) is most valuable as it discusses the modes of interaction between the hemispheres, while defining their main constituents.

This holoarchical model of the brain is further supported by the work of Karl Pribam (Ref 2.7.6), who developed the holonomic model of brain processing.

In addition to the circuitry associated with large fibre tracts in the brain, processing also occurs in webs of fine fibre branches such as the dendrites and in the dynamic electrical fields surrounding these dendritic trees. This processing can influence trees associated with nearby neurons whose dendrites are entangled but not in direct contact, a form of non-local contact. This type of processing is related to the Fourier Transform processes which are the basis of holography as invented by Dennis Gabor in the late 1940s.

Holograms can store prodigious amounts of information, the inverse transform returning the results of correlation into spatial and temporal patterns which are the guides used by the brain to navigate.

7.6.2. Intelligence.

A useful background to this subject is contained in the following references:- Eysenck (Ref 7.6.2); Eysenck and Kamin (Ref 7.6.3).

Intelligence is a term used to refer to a person's cognitive powers and intellectual abilities. Some believe intelligence to be entirely measurable, others not; some believe intelligence is largely inherited genetically from parents (nature), others that at least in part it is more related to environmental factors (nurture).

There is no doubt the whole issue is a complex matter subject to intense political and social agendas that obscure and distort even the most established scientific findings as Gottfredson (Ref 7.6.4) says.

The concept of intelligence has been known from the times of Plato and Aristotle, who distinguished between the cognitive aspects of human nature i.e. those concerned with thinking, problem solving, reasoning etc and those concerned with emotions, feelings, passions etc. This can be considered the foundation for the split between those who regard intelligence as explicitly measurable and therefore subject to scientific experiments and those who to some degree regard it as not measurable or implicit.

This dichotomy is evidently related to the debate between nature and nurture.

Considering firstly intelligence viewed as measurable by the scientific community, the measured index is the Intelligence Quotient or IQ given by the mental age of the individual divided by the chronological age, multiplied by 100.

The primary tests involved in determining this parameter are generally:- verbal ability and verbal fluency, numerical ability, spatial ability, perceptive ability, inductive reasoning and memory, (see Eysenck Ref 7.6.2).

The questions have only one correct answer, a matter which itself assumes intelligence is innately measurable.

The subject is set a list of questions involving the above and given a set time to complete the exercise. The tests can be standardised and when given to a large number of individuals a bell curve can be drawn indicating the median or mean IQ which is set at 100, other scores being higher or lower relative to that mean.

Such tests have been systematically applied as shown by Gottfredson (Ref 7.6.4) for example, to a wide range of chances in life from high to low risk, which can be related to the style of training involved for the relevant work content from simple to complex, and the career potential (or that actually achieved) all measured against various set outcomes in life. By life chances is meant the risk of having difficulties in meeting life's challenges. By style of training is meant the range from slow, simple and supervised to gathering and inferring information of significant complexity on one's own. The outcomes typically relate to the likelihood of school dropout, unemployment, need for welfare, divorce, poverty, prison etc.

The records show the increasing likelihood of such problems as the IQ score falls.

The correlation of such tests implies there is some global or general factor at work, sometimes called g by psychologists.

Science treats intelligence as a measurable quantity which has been described as being in the form of a hierarchy of neural functions, in which a basic type of activity develops in to higher and more specialised forms.

Although there is a consensus among scientists that there is a measurable function, there is only limited agreement on whether this is due to genetic factors i.e. hereditary or due to environmental factors whether they be, social, economic, political or cultural conflict etc. Even where it is agreed that both play a part, the argument continues over what proportion is due to each. The pro genetic sympathizers favour 80 per cent genetic and 20 per cent environment and their opponents the opposite or even that it is not possible to quantify such a ratio.

For these two views the argument in the 1980s between Eysenck and Kamin (Ref 7.6.3) defines the respective hereditary and environmental divisions very well.

Eysenck distinguishes between fluid intelligence (g_f) i.e. tests which are culture free or culture fair minimally dependent on knowledge, education or cultural factors and crystallised intelligence (g_c) which draw on knowledge and information more likely to have been acquired by intelligent persons than dull ones. It is in any case possible to devise tests which allow for these effects.

Notwithstanding this, a not insignificant disparity in g is displayed by different cultures and races, which inevitably provoke outrage among those who regard themselves as it were, marked down by the results. This disparity is of course regarded as hereditary by one side and as due to one or more environmental effects by the other.

Even the sexes are not immune to differentiation with regard to their respective abilities at one thing or another, though it is then more difficult to involve the nature nurture argument.

Eysenck also refers to the regression of the mean applying to sexual reproduction and referring to the tendency for parents with extremes of a characteristic to produce less extreme offspring. Arising from this, class divisions in society may be inevitable, but regression makes for social mobility and inhibits attempts to impose a caste system (which is the bane of Indian society for example). However if a totally classless system applied, everybody would be reduced to the mean with no sense of order.

Although he demonstrates test results revealing differences to good effect, Eysenck is at pains to point out that hereditability is not God given but applies to a given population and that results at any one time may not apply if conditions alter, including environmental ones. IQ tests do not create inequality as social problems have always existed, it is not the measurement that creates the problem; it merely clarifies them.

All racial and cultural groups that have been tested show great overlap in their abilities and each individual must be treated on his merits. So trying to create quotas favouring one group as opposed to another to even things out, are counterproductive.

Another error is to exaggerate the importance of intelligence; there are many other qualities which may be more important such as faith, hope and charity, honesty, hard work, kindliness, a passion for justice and many more.

Hitler and Stalin for example may have been of above average intelligence but were the greatest mass murderers in history, the one out to prove racial inequality the other to enforce equality. Eysenck supports the truth finding principle and refers to geniuses like Newton, Faraday and George Washington Carver who all rose from poor or appalling backgrounds to achieve greatness.

Kamin opposes hereditary views vociferously. Belief in heredity led to sterilization programmes, immigration quotas and consignment for some to the educational scrap heap. Much of his argument is based on elementary flaws in the research of those supporting the hereditary viewpoint, including vagueness about method, manipulation, built in false assumptions, camouflage, unconscious bias, invalid scores, poor standardisation, problems with volunteer subjects etc.

Kamin then goes on to discuss studies of adopted children, since the adoptive parent goes on to provide the adopted child with environment only as opposed to biological parents who provide genes and environment. In some cases the adoptive parent has children of their own which is another positive source of comparison of genes and environment. He says the studies of such cases favour environmental factors.

In adoption cases there is also the effect of selective placement effecting results.

Further studies of identical and non-identical twins, same sex and opposite sex twins lead in his view to the same conclusions.

Kamin believes the significant differences in measured IQ between those of particular ethnicities are clearly attributable to educational, social and economic discrimination.

He criticises slipshod references, misleading claims and fictitious matching by Eysenck. He refers to myths created by IQ tests relating to sexual variance, school attainment, equal environments, that regression to the mean is a genetic consequence etc.

It is evident that the genetic adherents take a purely scientific analytical approach while that of the environmentalists is to a greater extent emotional in character. The latter may be due in part to the lack of emotional content in the tests used to determine the intelligence factor g.

Although the arguments of these protagonists date back to the 1980s, the general debate has continued to roll on with others taking similar vantage points.

A significantly different view of intelligence is taken by those who believe in a multiplicity of intelligences such as Gardner (Ref 7.6.5), who not only expand the range of intelligences considered to apply but do include some which can be described as clearly related to emotion.

It is his view that human intelligence encompasses a far wider, more universal set of competences, some which do not always reveal themselves in standard tests for g.

The range referred to consists of the following:-

1. Linguistic.
2. Logical-mathematical.

3. Musical.
4. Spatial.
5. Bodily-Kinesthetic (controlling and orchestrating body motions and the skillful handling of objects).
6. Intrapersonal intelligence (accurately determining moods, feelings and other mental states in oneself and using it to guide behaviour).
7. Interpersonal intelligence (as 6, but with respect to others).
8. Naturalist (recognising and categorising natural objects).
9. Existential (capturing and considering fundamental questions of existence).

He emphasises firstly that while the concept of intelligence should be expanded, it should not be conflated with other virtues such as creativity, wisdom or morality.

Secondly, that there is a need to get away from standardised short answer proxy instruments to real life demonstrations or virtual simulations.

Thirdly, that the concept of multiple intelligences provides for more effective assessment. It should be possible to look directly at an individual's performance, to see how they argue, look at data, criticise experiments, execute works of art and so on.

The argument between nature and nurture in determining both fluid and crystallised intelligence was further complicated when Flynn, a university researcher working in the 1980s, rediscovered earlier findings that showed worldwide IQ scores had apparently risen significantly since at least the 1930s.

These findings are described in Flynn (Ref 7.6.6) and related background arguments discussed in detail in Neisser (Editor) (Ref 7.6.7). A further short summary for the layman of what is termed Flynn's effect is given by Holloway (Ref 7.6.8).

Holloway refers to Flynn's discovery that certain IQ tests had new and old versions which were sometimes given to the same group of people. Children for example did much better on a 1949 test than a 1974 one. Flynn noticed that groups worldwide performed much more intelligently on older tests.

The greatest improvements were associated with tests on fluid g (g_f) or on the spot problem solving, not educationally or culturally loaded, and in other tests measuring only verbal ability. Less pronounced improvements in crystallized intelligence (g_c) were also found. The effect was particularly pronounced in younger groups, as might be expected since g_f is found to reach maturity at an earlier chronological age than g_c.

These discoveries enhanced environmental arguments as opposed to genetic ones as the rising IQ scores occurred over a few decades, since most, though not all, genetic effects occur over long time scales. They also emphasised the importance of standardising tests over time.

Holloway refers accordingly to the various explanations offered for the Flynn effect, namely, 'better nutrition and parenting, more extensive schooling, improved test taking ability, and the impact of the visual and spatial demands that accompany a television laden and video games rich world'.

However other later studies imply reversals of the Flynn effect and the reader is advised to consider this whole subject as another 'work in progress'.

7.6.3. *Humour.*

Humour arising from life experiences which provoke laughter and provide amusement has deeper more subtle connotations.

People of all ages and cultures create and respond to humour to a greater or lesser extent. Those who seldom smile or laugh are regarded as humourless.

Those who primarily create humour or try to, the stand-up comedian, the witty raconteur, the wisecracker, the best man at a wedding are generally popular, while those who do not see the joke for whatever reason are not.

The individual who creates or receives humour favourably is said to have a sense of humour. Whether an individual responds to humour depends on many things including their character, taste, culture, maturity, education, intelligence and the relevant context. Humour varies from the slapstick version popular with children, such as puppet shows and cartoons, to sophisticated forms such as satire popular with mature audiences, involving comprehension of the relevant context be it social, political or otherwise.

Humour is an important factor in relieving tension in social situations, in the workplace, as a form of evening entertainment, in social media (radio, television, cinema and the internet). In relations between the sexes it ranks second only to physical attraction even among the young.

Furthermore it contributes to both physical and psychological health and well-being.

It can also have however detrimental effects in people on the receiving end of humour, where the source or transmitter regards the relevant material humourous but the recipient not so. This occurs where the humour is aggressive (racial, religious, sarcastic, disparaging for example) or where the humour is self-defeating as in cases of self-disparagement.

While it is useful here to have some knowledge concerning the types of humour, how it is communicated and used in human society, what is also of interest is its nature and how it relates to other human characteristics such as emotions. Indeed it does not appear to be an emotion rather a means of reducing stress induced by them.

It should be noted that some claim humour cannot or should not be explained.

The author E.B. White said that, 'humour can be dissected as a frog can, but the thing dies in the process and the innards are discouraging to any but the scientific mind'.

This is very true, but inevitably one asks why it is that having to explain a joke goes a long way to rendering it barren and lifeless.

There are many theories of humour for example relief theory, superiority theory and incongruity theory. It is likely that all these theories and others briefly mentioned below all play a part in what is a very complex phenomena.

Relief theory maintains that laughter is a mechanism which reduces feelings of tension, releases nervous energy and overrides social and cultural inhibitions.

We and particularly children laugh while being tickled, when the tickling stops it is followed by a build-up in tension prior to the tickler striking again.

Superiority theory can be traced back to the Greek philosophers. The idea being that a person laughs about the misfortunes of others because they feel superior to them, as they regard those misfortunes as being in some sense the fault of those who suffer from them.

Incongruity theory states that humour is perceived at the moment of realisation of incongruity between a concept involved in a certain situation and the real objects thought to be in some relation to the concept. The humour is said to provide resolution to the situation involved.

As Arthur Koestler put it, 'humour of this type results when two different frames of reference are set up and a collision engineered between them.'

These and other theories of interest involving riddles and punch lines can be found on the internet.

One further aspect that needs highlighting is that of timing.

Theories of humour arising from both the study of information systems and of biological systems emphasise the importance of timing. In the former it arises from the need to avoid malfunctions by rapid deletion of false information and in the latter the need to quicken the transmission of processed information, allowing a more efficient use of brain resources. This implies humour has a biological origin, which accords with the apparent possession of a sense of humour by monkeys and even rats. Who can doubt that all these theories apply in complex situations?

7.6.4 *The importance of work and the work ethic in human society.*

Mating, marriage, child rearing and family formation.

The Arts.

Physical sport and mental games.

Work and the work ethic are of fundamental importance in human society. By work is meant self-employment, or work for an employer, or engagement in some meaningful day-to-day activity of benefit in one way or another not only to himself but also to society regardless of occupation. In this context work may mean the arts as discussed below.

In the normal turn of events an individual will provide for himself and any dependants once he has become essentially independent of those who reared him. For this he will have acquired skills to a greater or lesser degree, to enable him to do so through schooling, university and/or other forms of training.

The work ethic is quite independent of the capitalist or anti-capitalist viewpoint.

Where an individual does not or will not work, then to some extent he is at risk of having no function in society and of creating and experiencing an increasing state of disorder. Where an individual cannot work through some disability, then others must, to a greater or lesser extent, look after him, but to the extent he overcomes that disability he also gains from the work ethic.

Another fundamental human activity is that involved in acquiring a mate and providing security for same in the form of marriage with a view to conceiving and rearing children in a family environment.

In relation to this subject it is difficult to find a more apt statement of principle here than that provided by Ferraro and Briody (Ref 7.6.9).

They refer to the imperative of systematic procedures for mating, marriage, child rearing and family formation. Failing to do so imperils the very existence of society as human children are for a long period totally dependent on adults. If children do not survive in to adulthood themselves survival of the species is clearly in jeopardy.

The Arts are the imaginative adventures of the human mind arising from the impulse to create as opposed to scientific analysis. The primary examples include literature (novels, short stories, poetry, epic sagas etc), the performing arts (music, dance and theatre), the visual arts (drawing, painting, pottery and sculpture), the media arts (involving radio, television, photography and cinematography) and the culinary arts (food preparation, cooking, baking and wine making). The purposes from prehistoric to modern times have been to provide a means of storytelling and to pass on beliefs and knowledge to later generations. In general, to provide a relationship between humanity and the world both real and imaginary.

Only one element of the above, namely music, will be considered in a little detail, the idea being to use it to distinguish the creative non-measurement process from that of analytic measurement.

Music is to be associated with wave motion, with harmonics in particular.

Fundamental to music as described for example by Cooper (Ref 7.6.10) are the concepts of pitch, the stave for perceiving same, the clef representing the zone of pitch and the notes ascending from A to G or in the reverse descending. Using this system it is possible to follow the shape of the melody.

The next step is to comprehend the concept of rhythm concerned with the time element of music involving pulse, stress and speed.

Rhythm values relating to the length of note can be simple ranging from the semibreve to the demisemiquaver in steps of two at a time or compound values which add 50 per cent to the length of the note.

Rests are of significance in music as they relate sound to silence.

Barlines divide music in to sections called bars which contain an equal number of pulses or beats. Within the framework of each bar certain pulses will be stronger than others, usually the first beat of the bar will have the strongest stress.

Time signatures indicate how many beats to the bar and which rhythm value applies; they can again be simple or multiple. From all this rhythm patterns are developed.

Tied notes are those with the same pitch. Syncopation is the name given to notes which would normally be weak but are given strong accents. Triplets and doublets divide simple rhythm values in to three parts and complex rhythms in to two parts.

Pitch can be organised into structures, consecutive sounds form melody and simultaneous sounds harmony.

The sonic distance between two different notes is called an interval, those comprising eight notes being an octave. A piano keyboard for example has 7 or 8 octaves, the interval between any note and its neighbour is called a semitone, a tone being two semitones.

In musical notation a key signature is a set of sharp or flat signs placed together on the stave, written immediately after the clef at the beginning of a line of musical notation. A key signature designates notes that are to be played higher or lower than the corresponding natural notes and applied until the next key signature. A sharp sign raises the note one semitone above the natural and a flat lowers the note one semitone.

A natural sign cancels the sharp or flat sign.

A scale is a set of notes consistently rising or falling and contained within an octave, the most common forms being major and minor.

It is apparent that music can be expressed in two dimensions on paper in the same way as the written word.

In general terms these rudiments express the rules whereby music is created (composed), written, read and played whether it is as complex as that of Beethoven, Mozart or Chopin, down to the simplest tune; whether played by a symphony orchestra or beaten on a drum.

It appears the right side of the brain correlates with the musical abilities of most people.

It has the pictorial and pattern sense, the geometrical and spatial sense, the synthesis with time and the holistic characteristic which are to be associated with musical comprehension.

Another important feature of human society is the physical activity of sport and the corresponding essentially non-physical activity of games.

Sport is a physical activity normally competitive in nature which demonstrates relevant ability and skills and at the same time provides entertainment to both participants and spectators. All sports involve equipment of one sort or another.

It is primarily physical and therefore requires an appropriate degree of fitness achieved through training, but with essential mental comprehension incorporated.

Players may be involved professionally in a given sport, receiving remuneration for same, or take part on a purely amateur basis.

The contest is usually between two sides each attempting to better or exceed the other. In some cases ties are allowed, otherwise tie breaking methods are used to ensure one winner and one loser. Tournaments can be arranged to produce a champion.

In certain sports many contestants compete each against all with one winner.

Regular tournaments may be staged between many teams in the form of leagues, over a season, resulting in a top to bottom hierarchy developing according to each team's proficiency over that period. The hierarchy may involve several leagues again with generally diminishing ability from top to bottom.

That hierarchy applies whether to team sports or competing individuals. However the holoarchical principle is equally apparent in the necessity for teams to integrate in a holistic manner for maximum effectiveness or the individual to deploy his/her strengths and disguise weaknesses to the same purpose.

Sport is generally governed by a set of rules or customs designed to ensure fair competition and consistency in adjudication of the winner. It is designed to be competitive emphasising skill and strategy and to reduce the element of luck to a minimum.

A key principle is that the result should not be predetermined and that both sides have in principle an equal chance of winning. Where players cheat by whatever means, by use of drugs or deliberate foul play or by creating a situation where the play is not real but a sham, then if discovered such players are disbarred from further participation by the adjudicating body applicable.

Non-physical mental games follow the same principles but are essentially means of stimulating the brain, as opposed to achieving fitness of body. In general they involve competing individuals and relevant material accessories, but in some cases the competitive element is replaced by an individual playing a game on their own.

A great variety of such games exist, board games such as chess, card games, tile games, pencil and paper games and more recent innovations such as video games.

The general purpose of these activities is for the participants to order their thoughts and actions so as to provide the relevant mental effort required to ensure the greatest margin of victory or successful achievement within the rules applicable.

However it is of little interest if victory or success is thought to be inevitable in advance or likewise defeat, hence the development of hierarchies of skill, where generally sides of roughly equal ability meet in competition, or individuals match themselves to a given level of skill in pencil and paper puzzles for example.

7.6.5   *Crime and the criminal mind.*

Human society endeavours to safeguard its citizens from harm through loss, injury or death by the actions of individuals, in accordance the framework of a set of laws set down by that society. These laws are in principle agreed collectively by society for its own good and should be distinguished from those applied against its will, by state dictatorship for example.

Crime is the deliberate, to the point of systematic, intent to break or defeat those laws by individuals in the pursuit of gain, or other motive arising from human emotions.

For these reasons the definition of crime is in general terms common worldwide i.e. the whole of human society agrees with it. Of course there are exceptions to this which complicate the issue, but we are concerned with the average situation here.

The law is laid down by the state through acts passed in accordance the legislative procedures of that state. A force is set up to police adherence to the law, when that law is considered broken the perpetrators where identified, are arrested by that force and investigations carried out to confirm whether there is a case to answer.

Such cases are heard in law courts set up for this purpose by a judiciary system consisting of magistrates and judges who oversee proceedings and make final judgements on whether the law has been broken. The facts of the case are presented by legal practitioners acting on behalf of both the complainants (either the state or those individuals offended against) and the accused, with evidence given on both sides. Such evidence is heard by the judiciary which may include a jury selected at random from the public. The judiciary as defined, weigh the balance of the arguments presented and find the accused guilty or not guilty. Those found guilty are punished in accordance with statutory regulations laid down for that purpose, those not guilty are freed.

The reader will no doubt be familiar with the above, so why should it be necessary to describe it here.

The reason is to show that crime is an offence against the whole of society and not merely the individuals harmed in a particular case. The perpetrators of crimes are aware of this fact but deliberately pursue their activities with a view to gaining satisfaction for themselves at the expense of the whole.

7.6.6 Insanity including schizophrenia.

The nature of sociopathy and psychopathy not associated with insanity.

Basically a crude version of what is termed insanity is that significantly divergent from the mean behaviour apparent to human society, involving abnormal patterns of behaviour whether mental or physical. Examples are a person becoming a danger to themselves or others, though not all such occurrences are regarded as insanity as in relation to others they may be straightforward acts of criminality. The term is used generally to denote mental instability. The medical profession does not use the term referring instead to specific mental disorders such as the presence of delusions or hallucinations called psychosis, collectively described as psychopathology.

The term insanity is commonly used in human affairs to discredit or criticise individuals for views or actions regarded as extreme.

Whereas it was formerly the practice to incarcerate most of those regarded as mentally ill in asylums, modern practice is to endeavour with the use of drugs or other forms of treatment to keep such individuals as far as possible within the social framework of society.

There are legal systems in place to cover situations where the mentally ill commit offenses against the law, in particular did the defendant know the alleged behaviour was against the law, could he/she distinguish between right and wrong etc.

To this end it is useful to consider the particular form of disorder termed schizophrenia.

This is a mental disorder often characterised by abnormal social behaviour and failure to relate to reality as perceived by the majority of individuals and particularly the mean or average individual. Medical diagnosis is made on the basis of observations of the patient's behaviour and on that individual's description of their experiences. Common symptoms include confused thinking, hallucinations, social disengagement, emotional extremes, high levels of anxiety and depths of depression etc. Social problems complicate the situation including, long term unemployment, poverty, homelessness and drug abuse.

Individuals with schizophrenia may experience hearing voices, delusion, disorganised thinking and speech, sloppiness of dress and hygiene etc. They may exhibit symptoms of paranoia and generally an inability to perceive normal human emotions indicated by a lack of response to others. Many fail to appreciate they are ill and are neglectful of treatment provided. As a result of all these difficulties their life expectancy is inevitably reduced relative to the norm.

Genetic and environmental reasons are put forward to explain the symptoms as is the case with intelligence. The problem is viewed in medical terms as related to the splitting or disordering of mental functions, rather than the public perception of split personalities. Courses of treatment include the use of drugs, counselling, and social rehabilitation.

Many psychological mechanisms have been implicated in the development of schizophrenia. In addition subtle neurological differences in brain structure have been found in some 40–50 per cent of cases including the frontal lobes, hippocampus and temporal lobes. The best that can be said is that medical science combats the illness to some degree to the benefit of the patient.

Whereas schizophrenia is associated with the word insanity, other forms of personality disorder occur in some individuals which cannot be meaningfully equated with that term. Psychopathy and sociopathy are forms of antisocial personality disorder characterised by uninhibited or bold behaviour with a lack of empathy and remorse for their actions.

In some degree psychopaths tend to be born i.e. their condition has genetic origins as opposed to sociopaths who share similar characteristics but are thought to acquire them from their environment.

In any case both types have little regard for the safety and rights of others. While not necessarily violent they are deceitful and manipulative.

They typically flout the law, lie or deceive others, are impulsive, prone to aggressive behaviour, have little or no regard for the feelings or safety of others, are irresponsible and feel no remorse for their actions. In most cases these traits are apparent in the teenager and continue to develop in to adulthood.

Psychopaths are thought to have underdeveloped parts of the brain associated with the regulation of emotion and impulses; in particular neurological studies have indicated structural and functional differences in areas of the brain.

They are manipulative and treat others as pawns replacing deep emotional relationships with those of a shallow or unemotional nature so as to benefit themselves. In doing harm to others they typically exhibit no sense of guilt.

As opposed to sociopaths they may seem charming and trustworthy, hold steady or even executive jobs, marry and have families. However they may be cruel to their children or subordinates while charming to their superiors, engage in wilful or criminal behaviour, carefully disguising such activity with a view to escaping detection. Although psychopaths in the workplace represent only a small percentage of staff they are most common at higher levels in corporate organisations and their detrimental effects (involving bullying, conflict, stress, staff turnover etc) sometimes pervade the entire business.

Sociopathy is thought to be the result of environmental factors such as an upbringing in a dysfunctional household involving physical and emotional abuse or childhood trauma. The actions of sociopaths tend to be correspondingly dysfunctional, impulsive and incoherent with no regard for the consequences. While having difficulty forming relationships, where family breakdown occurs they are able to form an attachment to others who are like minded. So they meet others in the same situation and join street gangs. As they grow older they are generally unable to hold down a job long term or engage in normal family life and instead engage in criminal behaviour.

They become agitated, anger easily and are given to violence which has the sole benefit of increasing the chance of their being apprehended.

Both sociopaths and psychopaths present significant risks to society, the former as they may in a few cases be drawn to terrorism (see next Section 7.6.7) and the latter as they have no sense of emotional involvement and feature highly among the list of predatory serial killers.

7.6.7   The nature of terrorism.

As part of finding the relationships that govern human behaviour, it is necessary to examine the nature of terrorism, which in some measure has taken the place of warfare between nation states as a primary source of upheaval, just as the latter in some measure replaced the tribal warfare of earlier times. This is not to say that traditional forms of warfare will not continue to raise their ugly heads again, but at least they are reasonably well understood. Where terrorism arises from tribal and political conflict, though again equally ugly in nature, it is at least comprehensible to both parties involved and to some extent even to those who are not.

In particular it may arise as a result of communities within larger controlling populations who feel they have limited opportunities for enhancement in life and consequently feel disadvantaged and marginalised.

Terrorism however, in the case of extreme forms involving bad religion and suicide which have come to prominence since the turn of the millennium, is less understood.

The following references:- Horgan (Ref 7.6.11) and Borum (Ref 7.6.12) are excellent sources of explanatory background material on this subject.

Borum (Ref 7.6.12) defines terrorism 'as acts of violence (as opposed to threats or more general coercion) intentionally perpetrated on civilian non-combatants with the goal of furthering some ideological, religious or political objective'.

Studies have shown that terrorists are in general not mentally ill in the medical sense, nor are they generally psychopaths. The reasons for this become apparent when the processes needed to support and maintain terrorism are examined. These processes are incompatible with the outward disorder that the mentally ill display. Psychopathic personalities are incompatible with suicidal terrorism as they are in general too concerned for their own welfare to even consider it. Other problems such as abuse or trauma in formative years likewise are not in general instigators of terrorism, though they may be commonly markers for it. There is no accurate profile of the terrorist which accords with the random nature of their behaviour.

The development of the violent behaviour of terrorists is caused by a complex interaction of environmental, social, and emotional factors. These are then reinforced by ideological objectives inevitably involving many activists and supporters.

The normally powerful inhibitions against human killing can be dissipated by the above reinforcement which grow to dominate the perception of individuals particularly disposed to the belief in and use of violence. Such individuals fall in to two distinct groups the leaders and the followers. The former do not normally carry out the violence but plan for it and exhort the latter to do so.

Effective terrorist recruitment is important and relationships critical since the falling out among thieves situation can otherwise frustrate objectives. Recruiters concentrate on people who feel deprived and dissatisfied with life, emphasising urgency for and importance of the cause concerned.

Terrorist leaders must form organisations which inculcate a belief system, establish strategy, tactics and communications under conditions of secrecy, manipulate followers with incentives real or imaginary, train potential attackers, identify and source suitable materials as weaponry, select targets and instruct actions to be taken against those targets.

Certain themes appear consistently as incentives, perceived injustice, the need for identity and the requirement to have a sense of belonging.

Disadvantaged and marginalised individuals, like the communities referred to earlier, fall easy prey to extremist organisations which cater for these incentives in particular ideas and values (an identity) which merges with the group identity of the terrorist organisation, all invariably adopted without critical examination.

The sense of belonging and affiliation provided by the organisation, offers an easy replacement for what has up to then been a lifetime of rejection.

The deprived easily develop hateful attitudes, but most people who hate don't kill. The trick is to erode the barriers that inhibit killing.

This involves creating justifications for one's actions, dehumanising the intended victims in some sense and casting them as evil to justify their slaughter.

So successful are these methods that suicide terrorism is a growing problem resulting in significant innocent civilian casualties. Terrorist operations are relatively inexpensive and the psychological effects on the civilian population are devastating. The terrorist groups find that it works, with gains for them in terms of publicity out of all proportion to expenditure.

It is evident from studies of such incidents and their perpetrators that there is no particular evidence that the latter suffer from suicidal tendencies associated with mental states of depression, as might apply in the general population. On the contrary most attackers view themselves as martyrs to the cause.

In the case of extreme religious ideology the primary aim is not suicide, because they assume eternal reward in the afterlife.

With regard to such ideology it must first provide a set of beliefs that guide and justify their behaviour. Secondly those beliefs must be inviolable, therefore cannot be questioned or be subject to critical examination; rigid adherence to a simple idea is all that is necessary. Religion can and is learned by rote as far as they are concerned, there is no room or time for debate on philosophical matters.

Thirdly all activity must be directed towards a goal seen as serving some meaningful cause or objective.

Such ideology is therefore absolutist, dividing the world in to black and white, into ideal and evil realms, their own being ideal of course. All aggression is directed against the non-believer as a necessary defence and retribution, indeed a moral imperative, sanctioned by their version of religion.

7.6.8 Human characteristics and activities and the Reference Model.

7.6.8.1. The human brain and intelligence.

In respect of Section 7.6.1 the great advantage of Popper and Eccles (Ref 2.7.4) is that it relates the brain and mind to the worlds of reality, belief and imagination making allowance for the subjective.

Similar worlds are derived in this book from a different perspective, namely that of the universal ordering sequence in turn derived from the totality of human records as referred to in the introduction and illustrated in Fig 3.2.1.

Although much further work has been carried out since this publication and the other references given in Section 7.6.1 they don't affect the conclusions which can be made using the Reference Model.

The lateralisation of the brain referred to in References 2.7.4 and 2.7.5 is clearly related to asymmetry which develops as the complexity of the different worlds increases. The scientific symmetrical description is equally clearly derived from the perspective of the mean level of explicit order.

Obviously the brain acts best when it operates in unison in any case.

This confirms what the Reference Model is at pains to explain, that experimental measurement is confined to the real world, whereas what is described in References 2.7.4 and 2.7.5 and this book incorporates the unfolding animate, belief and imaginary worlds and their increasingly implicit nature. So both the researchers and subjects of the split-brain tests are, unwittingly or not, including the implicit worlds in their deliberations.

The author of this book has put his own slant on all of this, extending the basic ideas referred to above and incorporating the results in to a comprehensive model of the human mind and its relation to the world as a whole in Section 4.0.

The Reference Model has clear answers to statements that the brain and the mind are merely a complex algorithm and to questions concerning the seat of consciousness.

Firstly the idea that the brain and by association the mind are a complex computer are not completely invalid since artificial intelligence has some similarities to human intelligence.

It is however related primarily to motion of a hierarchical description unfolding the present time state in the Ordering Centre and thereby increasing distribution in the direction of increasing chaotic disorder, as opposed to human intelligence which has motion of a holoarchical nature enfolding the present state, all as explained in Section 7.2.8. Enfolding the present state then gives human intelligence the option to either increase or decrease order as shown in Fig 4.8.1.

It is the whole point of artificial intelligence and its advocates that the mind is the brain, with no additions. For them there is no level other than defined by scientific measurement, but this is not correct, there are other levels as Figures 4.4.5 and 4.6.4 show. Indeed like human intelligence the artificial kind has a material substrate each unfolding levels of order enfolded within that substrate, but in opposite senses. In the artificial case this explains the strong AI belief that the material of the thinking device is a basis only for the algorithm it uses which determines its mental attributes.

For these reasons and the further explanations given in Section 7.2, imaginary intelligence diverges from that of the real world such that there is no manifest extra-terrestrial evidence of it to the human observer.

The alternative, that we are the only example of consciousness in the universe, simply does not bear examination as has again been explained in Section 7.2.10.

The brain and on closer examination some particular internal structures can indeed be considered the seat of consciousness in an explicit sense, but these are enfolded implicitly in the asymmetric motion ultimately derived from the Holomovement.

With regard to human intelligence as discussed in section 7.6.2, Eysenck's views correspond to the Reference Model description of the ordering process from a reductionist viewpoint.

Kamin's views correspond to the ordering process seen from the holistic viewpoint. Gardner's description corresponds to the balanced viewpoint taken by the Reference Model.

It is time to relate the above diverse views on intelligence in terms of the Reference Model. The Model confirms that measurable intelligence is the explicit form unfolded from its enfolded implicit form, the former being viewed from the perspective of the latter. Standard intelligence tests are therefore forms of measurement which correspond essentially to explicit life which in turn corresponds to scientific genetic biological models. It is therefore meaningless to try and include emotional aspects into these tests, whether in terms of nurture, effects of environment etc.

Nevertheless the Model keeps analysis and emotion in balance as the figures in Section 4.8 show. It does so, while supporting the concept of g based on standard tests, by using the concept of multiple intelligences to include the emotional. Some of these multiple facets are explicit and measurable and some implicit, largely immeasurable and emotional in character.

The Model further shows that the opposing emotions of love/hate for example are propelled by wave motion upwards/downwards in the Ordering Centre, whereas measurable intelligence is to a considerable degree fixed by genetics at birth. It is how these emotional and analytical intelligences are used that counts.

All this renders the nature versus nurture argument a waste of time, a situation again arising from the failure to build a model of consciousness which corresponds to and explains the range and diversity of human opinion on the subject.

Figures 4.2.4 and 4.4.1 can be used to support the above by keeping difference and similarity in balance where two waves in motion move together to become a reinforced single wave at the centre line or conversely move apart so as to weaken the degree of reinforcement.

If the left-hand wave denotes analytical intelligence (A) and the right-hand wave emotional intelligence (E) then the degree of reinforcement represents what may be called coherent intelligence (C), so $A + E = C$, where A and E are not merely high and roughly equal, but in addition **cohere.** Where there is limited coherence, no matter what the subject there will be an argument, A or E may temporarily win the battle, but coherence ultimately wins the war. High C can be denoted 'ACE' (the winning card!).

In the opposite sense incoherence leads of course to chaotic fragmented arguments with limited or no validity, or in the case of direct disorder to evil, as discussed in the previous section 7.5.

Quite apart from increasing coherence, A in general has in any case been increasing due to the enormous increase in explicit scientific knowledge, particularly in the last century. This alone apart from increasing coherence could explain the Flynn effect.

The Reference Model demonstrates the advantage of providing an overall model of consciousness on which to support the arguments it puts forward, when both science and religion fail to do so, thereby causing endless confusion.

7.6.8.2. Understanding humour.

In respect of humour it remains again to view this subject against the background of the Reference Model which indeed contains the requisite ingredients.

The belief and indeed the imaginary worlds provide the subjective aspect of humorous situations while the apparent (phenomenological) collapse of the wave function from that world to the four-dimensional real world provides the sense of shock posed by reality. The sense of timing necessary for humour is to be associated with that shock.

Fig 4.8.13 reveals that the contrary sets of emotions are in balance about the present state and distributed state axes.

Humour appears to be activated when the emotions collapse towards the centre of the diagram in concert with the enfoldment of the implicit world, all in reaction to the sudden comparison with the explicit world.

In particular, human beings have two distinct moods arising from the ordering/disordering sequence, namely optimism and apprehension which are generally in balance. Humour is of course a major defence against human suffering so evident in the belief world and provides a means of establishing this balance.

Humour subjects optimism to an increasing degree of shock, giving rise to amusement which escalates to laughter and even hysterics, before collapsing again as the situation returns to the normal state of balance.

Simultaneously apprehension is subjected to a similarly increasing degree of shock giving rise to surprise which escalates to amazement, before collapsing again to the balanced state as before.

If the joke is not well delivered i.e. does not create shock, it will not be appreciated and the emotional state rapidly converts to one of weariness and disinterest.

So it is essential the situation being set up in the imagination is carefully ordered so as to be understood so it can stand in shocking contrast with the apparent (phenomenological) collapse to the real world as the joke is literally cracked.

Disorder presents itself when the joke falls flat either because the necessary ordering has not been applied, or the timing is imprecise or the recipient(s) find the humour in one way or another offensive. Humour must use Implicate space as a base from which to operate as both science and religion are in principle absolutist in nature, i.e. attached to the main inanimate axes and are consequently essentially humourless in content. It therefore involves a degree of relativism, a conclusion supported by the Reference Model.

7.6.8.3. *Constructive processes.*

Re Section 7.6.4, the Reference Model implies that the principle of order or disorder is a description that governs the work ethic since the greater the contribution the individual makes to society the greater the degree of order and vice versa in the opposite sense. Patterns relating to mating, marriage, child rearing and family formation are created to increase or at least preserve the degree of ordering in society, avoiding the disintegration and disorder that will otherwise inevitably follow.

The Arts correspond to the emotional and holistic side of human nature as opposed to the analytical and reductionist side..

It is also to be noted that there are levels of complexity in music from the simple drum beat through the many types of popular music generally simple and homogenous in character, to the complexity of classical music which fits with the coarse grained belief world and the fine-grained imaginary world as described in Section 4.8. A similar creative process is involved in the other forms of art referred to earlier.

The principles of physical sport and games provide yet another excellent description of the ordering process forming the theme of this book.

7.6.8.4. *Destructive processes.*

Re Section 7.6.5, the ordering and disordering process is epitomised by the adherence to processes of law as opposed to criminal activities.

In place of saying, as religion would have it, that the criminal descends in to hell, the Reference Model implies the wave motion the criminal creates forms vectors in the direction of decreasing order from the mean or neutral level determined by the whole of society. Such motion is represented by the destructive element of the belief world as illustrated by the figures in Section 4.8. The explicit world (in this case the body of the criminal) stays exactly where it is, namely at the mean level.

In other words while the criminal body is in the explicit world, the criminal mind is in the disordered version of the belief world.

These vectors and the motions they create apply at all times, when the individual dies is irrelevant, religion being under the impression that the sinner goes to hell when he/she dies. Science of course assumes there is no motion from the mean level of any kind, since it tries to define everything in terms of that level.

In respect of Section 7.6.6 it is necessary to relate mental disorder to the Reference Model. All forms of insanity and personality disorder are inevitably related in the Reference Model to the negative portion of the present state as described in Section 4.8 and associated figures. Courses of treatment endeavour to reverse these illnesses so as to enable the patient to recover the ability to order their thoughts and actions in a positive sense.

For those who are not mentally disordered in the medical sense, but are determined to pursue criminal tendencies they will inevitably fall in the ordering sequence.

In cases of extreme evil the Model implies the psychopath is to a limited degree oriented towards the differential objective aspect of destructive nihilism and the sociopath to the integral subjective aspect.

Considering the subject of terrorists referred to in Section 7.6.7 they are clearly orientated towards the integral subjective aspect of destructive nihilism in the figures shown in Section 4.8, similar to the criminal sociopaths who provides the bulk of the followers referred to above.

In all the subjects previously examined the Reference Model has already found excellent correlation between the principle of Implicate order and the objective and subjective descriptions available of those subjects, from whatever source.

In this case the Model shines a light on the inhuman examples of our species described above, which exposes with devastating clarity the utter futility of their thinking and their actions. Their actions are indeed effective in the explicit world. However their inhumanity is directly related to the totally disordered wave motion of the belief world, and sends that disorder unerringly in the totally opposite direction to that they assume, namely total decoherence and oblivion.

All of the figures in Section 4.8 place these wretched people on a downward path directly aligned with the negative present state, as illustrated. They compete well in the rush for nihilism with the most extreme forms of fascist and communist ideologies.

It should be noted, as previously mentioned, that the ordering system does not take revenge on them; instead they bring their fate upon themselves.

As a final point the ordered wave motion of their victims in the belief and imaginary worlds remains operative, since it is wholeness that prevails, irrespective of the reality of the explicit world.

7.6.8.5. The relation between reality and imagination.

The associated concept of different worlds.

The range of human personality and emotions.

The background to these questions is illustrated in Section 4.8, Figures 4.8.1 through 4.8.5, which should be viewed in conjunction.

Fig 4.8.1 shows the use of the mean bell curve approach to order and disorder and the relation between the implicit and explicit.

The relation between reality and imagination and the associated concept of different worlds arising from that relation is illustrated in Fig 4.8.2 and the correlation of these two figures in Fig 4.8.3. Figures 4.8.4 and 4.8.5 further illustrate the increasing complexity and asymmetry of the different worlds relative to the inanimate state of the real world.

The inanimate, the state of having no life, corresponds to the mean level of order in the Ordering Centre within which the explicit world is enfolded, its distribution being orthogonal to the vertical present state axis which is also inanimate. Then the real world which unfolds implicitly from the mean level, termed zone A, is interpreted as comprising the inorganic, organic, biological and autonomic material substrate.

The transition from autonomic to animate characteristics gives rise to zone B the basic animate world prior to the mitosis meiosis transition.

In turn the mitosis meiosis transition gives rise to zone C the more complex animate world which culminates in the animal to conscious transition.

The animal to conscious transition gives rise to zone D and the development of the brain involving the unfolded physical senses as actions, sight, sound etc and the enfolded non-physical senses such as thoughts and memories, feelings and intentions etc associated with the mind. Basic beliefs concerning what is considered true as opposed to what is considered not true, regarding day-to-day practical matters, are included in this zone. It is in this arena that human suffering is most evident.

Zone E is the fully developed imaginary world of the mind, where a higher state of consciousness applies, involving analytical processes such as the sciences related to reality and processes involving integral belief related to imagination, such as philosophy and religion etc. The relation between these objective and subjective arenas is shown in Figure 4.8.1.

The increasing space-time from conception to death represents the decrease in order of the physical unfolding reality of the individual in four dimensions, in accordance the laws of thermodynamics. After death that individual is dispersed physically in accordance the applicable chemistry, as science and the belief in one life maintains.

However according to the Reference Model the growth of the belief and imaginary worlds alters this description, as these involve Implicate space where ultimately the four-dimensional world becomes increasingly irrelevant and our notion of the boundary between life and death (i.e. between the individual and the whole thing) disperses.

Accordingly positive wave motion in Implicate space in the direction of increasing order expands the life domain in that space at the expense of death in four dimensional space-time, all related to the present state and the growth of wholeness, ultimately the Holomovement, as shown in Figures 4.7.1 and 4.7.2. As the Ordering Centre moves towards that state the wave motion will centre round the alignment of higher mean levels as described in the conclusions given in Section 5.0, from which a more whole conscious entity will be unfolded in the same sense as human beings are unfolded at our mean level, as implied by the above figures.

Alternatively wave motion close to the negative present state should likewise retain less whole conscious entities, but in an increasingly implicit disordered state ultimately leading to the decay of consciousness as that disorder takes over.

It is clear that the Model is replacing the limited three world view of Popper and Eccles associated with human consciousness referred to in Section 4.8 with an unlimited potentially infinite number of worlds associated with the concept of a universal consciousness.

Comprehension of the above processes offers to vastly improve on the confused existing concepts of the spatiotemporal variety such as humanism and non-spatiotemporal notions such as the afterlife and reincarnation.

These have evidently arisen from the failure of any of the religions, like science, to form a valid explanation for consciousness and awareness in terms of processes.

All this then explains the basis for religious belief which can be construed from the mean of the various religions described in Section 7.5.4.

The range of human personality and emotions has been discussed earlier in Section 4.8, as illustrated in Figures 4.8.12 and 4.8.13.

There is nothing particularly controversial about either of these diagrams, since their meaning is clear to all. They can be easily related to similar but more comprehensive diagrams by Eysenck (Ref 4.8.1) and Plutchik (Ref 4.8.2) well known to psychologists.

All of the above can be related primarily to the belief world with extensions in to the imaginary world, as will be apparent from the earlier figures in Section 4.8.

7.6.8.6. *Account for the political spectrum.*

Account for the economic spectrum.

The political spectrum has already been correlated with order and variance in Section 4.8 as illustrated in Fig 4.8.8.

Stable democracy based on free elections involving moderate constructive conservatism and socialism is compared with its opposite unstable autocracy involving fascist, communist and indeed religious totalitarianism.

The abiding signature of such totalitarian regimes is their propensity to force all of the citizens under their control to obey their dictates without question, or face long periods of incarceration or even execution if they object. Free elections and any form of free thinking are of course forbidden.

Order is equated with the honest, virtuous, responsible citizen and the democratic state and disorder with the corrupt, criminal, irresponsible individual and the correspondingly corrupt state.

Differential variance is equated with free enterprise and ego-centric individualism, a form of laissez faire which may be beneficial to society as a whole, but converts in the disordering direction to capitalist oligarchy which is not so.

Variance in the direction of unfolding similarity is equated with state control forms of radical socialism, which likewise converts to state control oligarchy which is again undesirable.

Economics as might be expected shares a considerable degree of commonality with politics, see Section 4.8, Fig 4.8.9.

In the ordered direction there are the stable mixed economies and in the opposite sense the unstable corrupt economies.

Differential variance unfolds to form private capital orientated economies with a varying degree of laissez faire. Unfolding similarity is equated with those economies more centrally controlled. Too much laissez faire or too much central control leads generally to the benefit of the few at the expense of the many.

Of course politics and economics are complex subjects but this simple analysis seems obvious to the author and again aligns directly with the basic principle of this book.

7.6.8.7. *What is meant by the common sense?*

What is its relation to the modern concept of political correctness?

What is generally termed the common sense can be defined quite simply using the Reference Model.

It is the view based on the average opinion in respect of a given topic or set of topics corresponding to the intersection of the present state and variance axes, but which looks outwards taking into account the spread away from that opinion in each direction along those axes.

Whereas the common sense looks outward in the sense described above, the politically correct view is to look inward so that all views are to be concentrated towards the mean with respect to both order and variance.

The common sense is a healthy attitude but can lead to offence for those distant from the mean. Political correctness avoids offence but leads to unhealthy absurdities, variety being abandoned in favour of similarity.

7.6.8.8. The background to the debate between the preservation of life and that of euthanasia.

The ethical stance taken by the human race is in general to preserve life. This is true not only where a generally peaceful situation applies but to a considerable extent even when war between nation states or other conflicts, political, religious or otherwise are in progress. This is because most human beings have a built-in aversion to killing another human being irrespective of any punishment arising from the framework of law.

There are of course, in times of war when the dehumanising effects of the conflict take over, when large numbers of individuals and groups of individuals succumb to the desire to wipe out the opposition by mass killing. When this is adopted as a matter of practice, as in the case of the Holocaust and the Second World War for example, outraged democracies pursue what is regarded quite rightly as a just war which involves killing those guilty of mass murder. In principle the just war involves defeating the enemy, but as far as possible preserving life to the extent that is possible and consistent with that goal.

In terms of the Reference Model all this is carried out in accordance the ordering principle and the relation between the mean level of order which defines the division between ordering stuff and disordering stuff, the former moving up the ordering structure and the latter the opposite. Ordering wave motion endeavours to preserve life, disordering wave motion tends not to.

What the Model allows for is the real state of an individual in the explicit world and the corresponding belief and imaginary states in the form of wave motion in the Implicate space of those worlds. Arising from the absence in the Model of any absolute present time order enfolded along the present state in the ordering direction is always available to be unfolded in explicit reality; just as in the opposite sense disorder is always available to be likewise unfolded in explicit reality. Religion allows for these other worldly states, but has failed completely to understand how this might be. Science meanwhile simply does not understand why this should be and decides this is not how it is, thereby reducing everything to the mean level. The degree of incomprehension by both those parties is removed by the Model and clarity substituted.

The dissipation of the birth and death life cycle boundaries as the Ordering Centre merges with the Whole Ordering Centre implies killing a person in the explicit sense associated with locality, does not remove their implicit contribution to non-local wave motion wave motion at higher or lower levels of order closer to the present state.

Religion is confused in thinking heaven or hell, to use religious terms, is entered after explicit death, for example; these states apply at all times as the present state contracts. Science simply regards heaven and hell as nonsense, as it makes no allowance for such contraction.

Much better therefore to abandon these simplified concepts and replace them with the admittedly more difficult and subtle description the Reference Model supplies.

This analysis can be extended to questions not involved with the kind of conflicts referred to above, i.e. those concerning suicide by individuals carried out independent of others and those suicides aided and abetted by others. These circumstances are of course played out against the aforementioned adherence in normal circumstances to preserving life. When individuals commit suicide they are likely to be acting on the assumption that in religious terms there is no afterlife or before life for that matter and any aiders or abetters likewise. There may be exceptions, where the thought is the afterlife might be better than the earthly one, extreme religious ideologists for example, but we must not confuse the latter with the typical suicide as was explained in answer to the previous question.

Suicides generally involve people in a state of despair due to circumstances in their lives they feel intolerable such as oppression, abuse, extreme pain or illness as applicable.

People without religious belief for example may be prone to thinking that suicide is their right where life is seen as a torment to themselves and others who out of love or duty nurse them, where that is applicable.

The Reference Model shows however that the real world is essentially the inanimate and biological world which provides the root system for life in general, but that implicate wave motion dominates in the belief and imaginary worlds where beginnings and ends are not as well defined.

In fact it is the ordered aspect of these worlds which is responsible for the instinct to preserve life as opposed to the disordered aspect which is increasingly less concerned.

Nature, red in tooth and claw, which seems ambivalent in this respect is essentially a more restricted form of ordering related primarily to the mean level, with limited extension in to the more complex worlds involved with the question being discussed here.

So the Model provides the basis for the belief in the preservation of life and also explains what is not commonly discussed, namely the reason for the limited human lifetime as measured explicitly. It is the generally held view that we must leave it to the tug of war between the second law of thermodynamics responsible for the aging process and the ordering process trying to preserve life, to decide on the moment of death. Where accidents involving death occur, for reasons other than aging (irrespective of our efforts to preserve life), then this is a matter for fate to decide (in effect the powers that be), or in the context of the Model the ordering/disordering process.

The question arises, notwithstanding this general view, whether there are circumstances when the individual is in say extreme pain or distress due to illness, that steps should be taken to curtail life as seen from the perspective of the earthly world.

We would after all love to depart this world like the wonderful 'one hoss shay' which having been designed and constructed to last an exact period of time and having served its owner faithfully without fault, keeled over and collapsed to dust at the moment that period expired. Reality is of course different, it is more complex.

It appears that for the unfortunate individual in extreme distress the only solution is the provision of relief in the form of painkillers to the point where death ensues because of their application. This is the cusp where the effort to relieve pain while preserving life is in balance with the inevitability of death. The failure to relieve pain and allow suffering is in principle to be avoided as is the deliberate act of causing death to avoid the suffering, these being on either side of the cusp. This is in effect a tuning process applied in an effort to achieve the best possible outcome in critical circumstances.

The Model implies that the principle of preserving human life should be safeguarded where possible and the undesirable consequences which would follow easing of that principle are all too clear. The many want the principle, the potential suicide does not. The concept of the tuning process is in line with the limited importance of the unfolded explicit world compared to that of the belief and imaginary worlds.

The same principle applies to the amount of effort that can and should be put in to maintain human lifespans generally. There is a limit to the natural length of a human life say of the order of 10^2 years and there is clearly a limit to the maintenance of a health service to support them. If the average human lifespan exceeds the resources available from that health service a dilemma will arise similar to that of the individual in distress as described above. So again a tuning process is necessary to maintain a balance.

The same principle again applies on an even larger scale to the human race. The human population has grown to the point where some regard it as a plague and the human race itself has a limited technological lifetime of the order of 10^3 years according to the answer given in Sections 7.2.6 and 7.2.8.6.

With regard to the size of the human population and its effect on the environment and incidentally vice versa, the environment on it, it will be necessary to limit that size and those effects by tuning in to the likely technological lifetime of the human race in terms of the explicit world.

REFERENCES

2.5.1   D. Bohm, *Wholeness and the Implicate Order*, Routledge and Kegan Paul, London,1980.

2.7.1   J.E. Kasper and S.A. Feller, *The Complete Book of Holograms: How they work and How to make them*, Dover Publications, 2001.

2.7.2   M. Schroeder, *Fractals, Chaos, Power Laws*, W.H. Freeman and Co, New York,1991.

2.7.3   B.B. Mandelbrot, *Fractals and the rebirth of iteration theory. In The Beauty of Fractals: Images of Complex Dynamical Systems*, H.O. Peitgen and P.H. Richter, Springer-Verlag, Berlin, 1986.

2.7.4   K.R. Popper and J. C. Eccles, *The Self and it's Brain*, Routledge and Kegan Paul, London,1983.

2.7.5   M.S. Gazzaniga, The Split Brain revisited, *Scientific American*, July 1998.

2.7.6   K. Pribam, Brain and Perception: Holonomy and Structure in Figural Processing, Psychology Press, London, 1991.

2.8.1   A. Koestler, *The Ghost in the Machine*, Hutchinson, London, 1967.

3.1.1   J.S. Bell, *Speakable and Unspeakable in Quantum Mechanics*, Cambridge University Press, Cambridge,1987.

3.2.1   N. Smart, *The World's Religions*, Cambridge University Press, Cambridge, 1998.

4.2.1   W.C. Elmore and M.A. Heald, *Physics of Waves*, Dover Publications Inc, New York, 1985.

4.5.1   B. d'Espagnat, The Quantum Theory and Reality, *Scientific American*, November 1979.

4.5.2   A. Aspect, P. Grangier and G. Roger, Phys. Rev. Lett. 47 (1981). A. Aspect, J. Dalibard and G. Roger, Phys. Rev. Lett. 49 (1982).

4.5.3   H. Everett, *The Theory of the Universal Wave Function, in The Many Worlds* Interpretation of Quantum Mechanics, ed B.S.

DeWitt and N. Graham, Princeton University Press, Princeton, New Jersey,1973,64.

4.5.4   B.S. De Witt and N. Graham, *The Many Worlds Interpretation of Quantum Mechanics*, Princeton University Press, Princeton, New Jersey, 1973, 155-165.

4.5.5   D. Bohm and B.J. Hiley, *The Undivided Universe*, Routledge, London,1993.

4.5.6   D. Bohm, *Quantum Theory*, Dover Publications Inc, New York, 1951.

4.6.1   B. Ernst, *The Magic Mirror of M.C. Escher*, Ballantine Books, New York, 1976.

4.8.1   H.J. Eysenck, *Fact and Fiction in Psychology*, Pelican Books, London, 1965.

4.8.2   R. Plutchik, *The Circumplex as a General Model of the Structure of Emotions and Personality*, American Psychological Association, 1997.

4.8.3   H. Price, *Time's Arrow and Archimedes Point*, Oxford University Press, Oxford, 1996.

4.8.4   J. Lovelock, Gaia: *A New Look at Life on Earth*, Oxford University Press, Oxford, 2000.

4.8.5   J. Lovelock, *The Vanishing Face of Gaia: A Final Warning*, Basic Books, New York, 2009

7.1.1   A.R. Lacey, *A Dictionary of Philosophy*, Routledge and Kegan Paul, London, 1976.

7.1.2   K. Popper, *Unended Quest*, Fontana/Collins, London, 1976.

7.2.1   C. Darwin, *The Origin of Species by means of Natural Selection*, Original publication 1876, Reprinted Cambridge University Press, Cambridge, 2009.

7.2.2   P. Teilhard de Chardin, *The Phenomenon of Man*, English Translation, Harper and Row, Colophon edition. New York, 1975.

7.2.3   E. Schödinger, *What is Life? with Mind and Matter and Autobiographical Sketches*, Cambridge University Press, Cambridge, 1967.

7.2.4   J.D. Barrow and F.J. Tipler, *The Anthropic Cosmological Principle*, Oxford University Press, Oxford, 1986.

7.2.5   A.G. Cairns-Smith, Seven Clues to the Origin of Life, Cambridge University Press, Cambridge, 1985.

7.2.6   R. Penrose, *Shadows of the Mind*, Vintage Books, London, 1995.

7.2.7   M. Minsky, *Matter, Mind and Models, In Semantic Information Processing*, (Ed M. Minsky), MIT Press, Cambridge, Massachusetts, 1968.

7.2.8   J. McCarthy, *Ascribing Mental Qualities to Machines, In Philosophical Perspectives in Artificial Intelligence,* (ed M. Ringle), Humanities Press, New York, 1979.

7.2.9   D.R. Hofstadter, *Godel, Escher, Bach, An Eternal Golden Braid*, Penguin Books, London, 1980.

7.2.10   D. R. Hofstadter and D.C. Dennett (ed), *The Mind's I*, Basic Books, Penguin, London,1981.

7.2.11   J. Searle, *Minds, Brains and Programs, in the Behavioural and Brain Sciences*, Vol 3, Cambridge University Press, Cambridge, 1980.

7.2.12   J. Searle, *Minds and Brains without Programs, In Mindwaves*, (ed C. Blakemore and S. Greenfield), Basil Blackwell, Oxford, 1987.

7.2.13   F. Drake, *Intelligent Life in Space*, Macmillan, New York, 1962.

7.2.14   B. Carter, Phil. Trans. R. Soc A370, 347 (1983); also in *The constants of nature*, ed W. H. McCrea and M. J. Rees (Royal Society London, 1983.

7.2.15   B. Carter, *The Anthropic Principle and its Implications for Biological Evolution*, Phil. Trans. R. Soc A310 (1512), 1983.

7.2.16   I. Crawford, Searching for extra-terrestrials, Where are they? *Scientific American*, July 2000.

7.2.17   A. J. LePage, Searching for extra-terrestrials, Where they could hide, *Scientific American*, July 2000.

7.2.18   G. W. Stenson, Searching for extra-terrestrials, Intergalactically speaking, *Scientific American*, July 2000.

7.3.1   R. Penrose, *The Emperor's New Mind*, Vintage Books, London, 1990.

7.3.2   R. Penrose, *The Road to Reality*, Vintage Books, London, 2005.

7.3.3   M. Tegmark, Parallel Universes, *Scientific American*, May 2003.

7.3.4   A.M. Turing, *On computable numbers with an application to the Entscheidungsproblem*, Proc. Lond. Math. Soc (Ser 2), 42, 1937.

7.3.5   K. Godel, *Uber formal unentscheidbare Satze der Principia Mathematica und verwandter Systeme 1*, Monatshefte fur Mathematik und Physik,38, 1931.

7.3.6   A. Church, *The calculi of lambda-conversion, Annals of Mathematics studies*, no 6, Princeton University Press, 1941.

7.3.7   S.C. Kleene, *Mathematical Logic*, John Wiley and Sons Inc, New York, 1967.

7.4.1   A. Guth, *The Inflationary Universe*, Jonathan Cape, London, 1997.

7.4.2   A. Linde, The Self Reproducing Inflationary Universe, *Scientific American*, November 1994.

7.4.3   R. Penrose, *Cycles of Time: An Extraordinary New View of the Universe*, Vintage Books, London, 2011.

7.4.4   P. Steinhardt, The Inflation Debate, *Scientific American*, April 2011.

7.4.5   D. Bohm, Phys. Rev, vol 85, p 166-193, 1952.

7.4.6   D. Bohm, Phys. Rev, vol 89, p 458-466, 1953.

7.4.7   D. Bohm, *A suggested interpretation of the quantum theory in terms of 'hidden variables, Quantum Theory and Measurement*, ed J. A. Wheeler and W. H. Zurek, Princeton University Press, 1983.

7.4.8   J. Al-Khalili et al, *Quantum, A Guide for the Perplexed*, Weidenfeld and Nicolson, London, 2003.

7.4.9   G.C. Ghirardi, A. Rimini, and T. Weber, *A General Argument against Superluminal Transmission through the Quantum Mechanical Measurement Process*, Lett Nuovo   Chim, 27, 1980.

7.4.10   R Bousso, and J. Polchinski, The String Theory Landscape, The Frontiers of Physics, *Scientific American*, Vol 15, No 3, 2006.

7.4.11   L. Smolin, The Atoms of Space and Time, The Frontiers of Physics, *Scientific American*, Vol 15, No 3, 2006.

7.4.12 S.W. Hawking, and R. Penrose, *The Singularities of Gravitational Collapse and Cosmology*, Proc. Roy. Soc, London, A314, 1970.

7.4.13 J.D. Bekenstein, *Black Holes and Entropy*, Phys. Review, D7, 1972.

7.4.14 S.W. Hawking, *Black Hole Explosions*, Nature 248, 1974.

7.4.15 S.W. Hawking, *Particle Creation by Black Holes, Communications in Mathematical Physics*, 43, 1975.

7.4.16 R. Stephens, G. t' Hooft, B. F. Whiting, *Black Hole Evaporation without Information Loss, Classical and Quantum Gravity*, 11 (3), 1994.

7.4.17 L. Susskind, *The World as a Hologram, Journal of Mathematical Physics*, 36 (11), 1995.

7.4.18 B. Greene, *The Elegant Universe*, Jonathan Cape, London, 1999.

7.4.19 B. Greene, *The Fabric of the Cosmos*, Allen Lane, London, 2004.

7.4.20 J. Maldacena, The Illusion of Gravity, *Scientific American*, Nov 2005.

7.4.21 J. D. Bekenstein, Information in the Holographic Universe, *Scientific American*, Aug 2003.

7.4.22 David Z. Albert, Bohm's Alternative to Quantum Mechanics, *Scientific American*, May 1994.

7.4.23 A. Whitaker, *Einstein, Bohr and the Quantum Dilemma*, Cambridge University Press, 1996.

7.4.24 P. R. Holland, *The Quantum Theory of Motion*, Cambridge University Press, 1993.

7.4.25 L. Smolin, *Einstein's Unfinished Revolution, The search for what lies beyond the quantum*, Allen Lane, Penguin Random House UK, 2019.

7.4.26 L. Deeke, B Grotzinger and H. H. Kornhuber, *Voluntary finger movements* in man: *cerebral potentials and theory*, Biol. Cybernetics,23, 99, Springer, 1976.

7.4.27 B. Libet, E. W. Wright Jr, B. Feinstein and D.K. Pearl, *Subjective referral of the timing for a conscious sensory experience*, Brain,102, 193, 1979.

7.5.1 G. Oppy, *Ontological Arguments, Feb 1996. Substantive revision 15 July 2011. Stanford Encyclopedia of Philosophy. Gödel's Ontological Argument.* Retrieved 2011-12-09.

7.5.2 R. Dawkins, *The Blind Watchmaker*, Penguin Books, London, 1988.

7.5.3 R. Dawkins, *The God Delusion*, Bantam Books, London, Oct 2006.

7.5.4 G.H. Smith, Atheism: *The Case Against God*, Prometheus Books, New York, 1989.

7.6.1 S. Greenfield, *Brain Story*, BBC Worldwide Ltd, London, 2000.

7.6.2 H.J. Eysenck, *Uses and Abuses of Psychology*, Penguin Books, London, 1953.

7.6.3 H.J. Eysenck versus L. Kamin, *Intelligence: The Battle for the Mind*, Pan Books, 1981.

7.6.4 Linda. S. Gottfredson, The General Intelligence Factor, *Scientific American*, Exploring Intelligence, Quarterly, Vol 9, No 4, 1998.

7.6.5 H. Gardner, A Multiplicity of Intelligences, *Scientific American*, Exploring Intelligence, Quarterly, Vol 9, No 4, 1998.

7.6.6 J.R. Flynn, IQ gains over time, R.J. Sternberg (Ed) *Encyclopaedia of Human Intelligence*, Macmillan, New York, 1994.

7.6.7 U. Neisser (Ed), *The Rising Curve, Long term gains in IQ and related measures*, APA Science Volume Series, American Philosophical Association, Washington DC, 1998.

7.6.8 M. Holloway, Flynn's Effect, *Scientific American*, Jan 1999.

7.6.9 Gary P. Ferraro and Elizabeth K. Briody, *The Cultural Dimension of Global Business*, Routledge, London and New York, 2016.

7.6.10 H. Cooper, *How to read Music*, Omnibus Press, London, 1982.

7.6.11 J. Horgan, *The Psychology of Terrorism*, Routledge, London, 2014.

7.6.12 R. Borum, *Psychology of Terrorism*, University of South Florida, Tampa FL, (monograph), 2004.

LIST OF FIGURES

INDEX

Flew, A., 341
Flynn effect, 382
forebrain, 373
formless void, 42
four-dimensional space-time, 32,
    116–17, 211, 330
Fourier series, 200–1, 272
Fourier transforms, 201, 275
fractals, 8, **28**, 28–9, 34, 69, 116,
    201–2
  holographic, 9, 85–6
  quantum state, 318
  time, 333
  unified Ordering Centre, 56–7
freewill, 103
freewill vs. determinism, 141–2, 159
Friedmann cosmological models,
    253
Friedmann equations, 255–6
Friedmann-Lemaitre-Robertson-
    Walker (FLRW) models,
    253–7, 269
frog's eye view, 21, **52**, 68, **71**, 76
frontal lobe, 373

Gabor, Dennis, 377
Gaia hypothesis, 102–3, 114, 119
galaxy cluster formation, 294
Galilean transformations, 236
Gardner, H., 381, 398
Gaunilo of Marmoutiers, 338
Gaussian distribution, 3, **23**, 23
Gazzaniga, M. S., 29, 376
Gell-Mann, M., 282
general relativity, 242–52
  black holes, 299
  Standard Reference Model, 317
genes, 163, 180
geodesics, 243
The Ghost in the Machine (Koestler),
    31
gluons, 290
God, probability of, 341, 356

Gödel, K., 90, 225, 229–31, 338
goodness vs. evil, 346, 354–5
Gottfredson, L. S., 377–8
Graham, N., 69
Grand Unification epoch, 293
Grand Unification Theory, 295,
    329–30
Grassman algebra, 221–2
gravity, 245, 329
gravity, quantum, 295–8
Greene, B., 301
Greenfield, S., 370
gauge theories, 284–5
Guth, A., 268–9, 297

hadron epoch, 293
hadrons, 288
hallucinogenic drugs, 157
Hameroff, S., 162
  microtubule structures, 162, 310,
    325
Hamilton, W. R., 206
Hamiltonian function, 206
Hamiltonian mechanics, 18
hatred/love, 368–9
Hawking, S., 257
  black holes, 299–300, 301
Hawking radiation, 300
Heald, M. A., 46
heaven and hell, 37–8, 346–7, 408
  Buddhism, 351–2
  Christianity, 348–9
  Hinduism, 351
  Islam, 350
Heisenberg, W., 272
Heisenberg uncertainty principle, 18,
    272, 275, 292, 305
helix, quantum theory, 275
hemispheres, brain, 29, 100, 374,
    376–7, 396
hereditary influence, intelligence,
    379
Hermitian forms, 213–14, 220

www.ingramcontent.com/pod-product-compliance
Lightning Source LLC
Chambersburg PA
CBHW062151270326
41930CB00009B/1495